Mineral Scale Formation and Inhibition

About the Editor

Zahid Amjad received his M.Sc. in Chemistry (1967) from Punjab University, Lahore, Pakistan, and a Ph.D. in Physical Chemistry from Glasgow University, Glasgow, Scotland. Dr. Amjad was Lecturer at the Insitute of Chemistry of Punjab University, and was Assistant Research Professor at the State University of New York at Buffalo (1977–1979) when he joined the Calgon Corporation. He is currently Research and Development Fellow in the Advanced Technology Group of The B.F. Goodrich Company in Brecksville, Ohio, where he has served since 1982. His areas of research include the development and applications of water soluble/swellable polymers, the adsorption of polyelectrolytes at a solid–liquid interface, and the control and removal of foulants from water purification apparatus—particularly from membrane-based processes. Dr. Amjad has authored or co-authored over 80 technical publications and is holder of 26 patents. He recently edited the book *Reverse Osmosis: Membrane Technology, Water Chemistry, and Industrial Applications*. He has been inducted into the National Hall of Corporate Inventors and is listed in *American Men and Women of Sciences, Who's Who in Technology*, and *Who's Who of American Inventors*. He is a member of several professional organizations.

Mineral Scale Formation and Inhibition

Edited by

Zahid Amjad
The B.F. Goodrich Company
Brecksville, Ohio

Plenum Press • New York and London

Library of Congress Cataloging-in-Publication Data

Mineral scale formation and inhibition / edited by Zahid Amjad.
      p.   cm.
    "Proceedings of an American Chemical Society symposium on mineral
scale formation and inhibition, held August 21-26, 1994, in
Washington, D.C."--T.p. verso.
    Includes bibliographical references and index.
    ISBN 0-306-45195-6
    1. Crystallization--Congresses. 2. Incrustations--Congresses.
3. Descaling--Congresses. I. Amjad, Zahid.
QD548.M56 1996
548'.5--dc20   95-45221
                                                                   CIP

Proceedings of an American Chemical Society symposium on Mineral Scale Formation and Inhibition, held August 21–26, 1994, in Washington, D.C.

ISBN 0-306-45195-6

© 1995 Plenum Press, New York
A Division of Plenum Publishing Corporation
233 Spring Street, New York, N. Y. 10013

10 9 8 7 6 5 4 3 2 1

Printed in the United States of America

To George H. Nancollas, for his many contributions to the field of crystal growth, dissolution, and inhibition, and for introducing this challenging field to so many of us

and also

to Rukhsana and Naureen, my devoted family, with much love.

PREFACE

This book documents the proceedings of the symposium, "Mineral Scale Formation and Inhibition," held at the American Chemical Society Annual Meeting August 21 to 26, 1994, in Washington, D.C. The symposium, sponsored by the Division of Colloid and Surface Chemistry, was held in honor of Professor George H. Nancollas for his pioneering work in the field of crystal growth from solution. A total of 30 papers were presented by a wide spectrum of scientists. This book also includes papers that were not presented but were in the symposium program.

The separation of a solid by crystallization is one of the oldest and perhaps the most frequently used operations in chemistry. Because of its widespread applicability, in recent years there has been considerable interest exhibited by academic and industrial scientists in understanding the mechanisms of crystallization of sparingly soluble salts. The salt systems of great interest in industrial water treatment area (i.e., cooling and boiler) include carbonates, sulfates, phosphates, and phosphonates of alkaline earth metals. Although not as common as calcium carbonate and calcium sulfate, barium and strontium sulfates have long plagued oil field and gas production operations. The build-up of these sparingly soluble salts on equipment surfaces results in lower heat transfer efficiency, increased corrosion rates, increased pumping costs, etc. In the laundry application, insoluble calcium carbonate tends to accumulate on washed fabrics and washing equipment parts, resulting in undesirable fabric-encrustation or scaling.

In reverse osmosis (RO), the precipitation of the scale forming salts as the feed is desalted is one of the major causes of RO membrane fouling. Scaling of the membranes by calcium carbonate, calcium sulfate, or barium sulfate can lead to loss in production, poor product water quality, premature membrane replacement, and unscheduled downtime.

The crystallization of sparingly soluble salts is also of primary importance in biological systems. Tartar, or dental calculus, primarily consists of salts of calcium, phosphate, and carbonate. Calcium oxalates are the main components of pathological deposits in the urinary tract. The medical community is seeing renewed interest in the regulation of these species.

The precipitation of scale forming salts is commonly controlled by the addition of an inhibitor to the feed water. Recently, inhibitor selection has become problematic because of the large number of inhibitors available. Proper selection of the inhibitor may be accomplished more easily if the effects of the scalant crystal-inhibitor interactions are understood for a given water chemistry. Another challenge facing scientists, especially from the industrial application perspective, is the development of new inhibitors that offer superior performance, especially under extremely stressed conditions (i.e., high pH, high temperature, high hardness, etc.). One key issue in operating systems for optimum performance is the appropriate predictive model to project the required inhibitor dosage. In many cases underdosing or overdosing of the inhibitor can lead to unexpected scaling, resulting in higher operating costs. In view of these challenges, understanding the mechanisms of scale formation and inhibition is of paramount importance.

   This book provides an introduction to the type and severity of scaling problems in both industrial and biological systems. The target audience academic and industrial scientists and engineers who may encounter the deposition of mineral scales on heat exchanger or RO membranes, and as tartar build-up on teeth or stones in biological systems, or who may apply chemicals to inhibit scaling. A wide range of expertise and experiences has been brought together to yield the first book on mineral scales in such a variety of applications. The first thirteen chapters of this book concentrate on scaling problems encountered in industrial systems particularly water purification, cooling water, and boiler water treatment systems. The next ten chapters examine the relationship of certain scales in biological systems. The final chapters address the solution chemistries of the interaction between scale forming salts and inhibitors in complex systems and predictive modeling of those systems. This book provides the reader with the latest developments in the area of crystal growth, dissolution, and inhibition. It is hoped that this book will stimulate interest in the development of new inhibitors to control scaling problems in such fields as industrial water treatment (i.e., cooling or boiler water), water purification (i.e., membrane separations or thermal distillation), medical and dental applications (i.e., bone growth or dissolution of renal stones or dental calculus), or the specialized analytical techniques needed for the progress of these disciplines.

ACKNOWLEDGMENTS

I express my appreciation to those who participated in the symposium in Washington, D.C. I thank all the authors for their hard work in writing their respective chapters. Financial support of the national and international scientists is gratefully acknowledged. I extend thanks to the American Chemical Society Division of Colloid and Surface Chemistry and corporate contributions from Alco Chemicals, Betz Laboratories, BFGoodrich Company, Proctor & Gamble, Unilever- USA, and WR Grace. Their assistance contributed substantially to the success of the meeting.

I am thankful to Dr. Victoria F. Haynes for her encouragement and continued support. I also wish to thank the management of The BFGoodrich Company for permitting me to organize this symposium and to edit this volume. I want to give special thanks to Jeff Pugh and Penny Frantz for their efficient and organized handling of the considerable correspondence associated with both the symposium and this book. Thanks are also extended to Pat Vann of Plenum Publishing Corporation for her continued interest in this project. Finally, a special thanks to my wife who has contributed in more ways than I can mention to the completion of this book.

CONTENTS

CONSTANT COMPOSITION KINETICS STUDIES OF THE SIMULTANEOUS CRYSTAL GROWTH OF SOME ALKALINE EARTH CARBONATES AND PHOSPHATES

George H. Nancollas and Anita Zieba

Chemistry Department
Natural Science Complex, State University of New York at Buffalo
Buffalo, New York 14260

ABSTRACT

Precipitation processes in aqueous systems frequently involve concomitant crystal growth, dissolution, and phase transformations. The elucidation of the overall mechanisms must therefore take into account more than a single reaction step. In order to limit complications due to phase transformations driven by changes in supersaturation, it is desirable to investigate these processes at constant thermodynamic driving force. This is the basis of the constant composition method that employs multiple ion specific electrodes in order to control the addition of titrants containing the precipitating ions to compensate for the changes induced by crystal growth. In this paper, the dual constant composition method is discussed by the presentation of two examples, the phase transformation of calcium phosphates and the simultaneous crystal growth of strontium and calcium carbonates from their mixed supersaturated solutions. In the former system, ion specific hydrogen and calcium electrodes were used to control titrants for the growth of octacalcium phosphate and the simultaneous dissolution of dicalcium phosphate dihydrate, respectively. For the carbonate studies, hydrogen ion and calcium ion electrodes were used to control titrants for strontium carbonate and calcium carbonate crystal growth, respectively. In this case, the crystal growth of both minerals is surface controlled, following rate laws parabolic relative supersaturations. It is interesting, however, that the addition of strontium ion during calcium carbonate crystallization induces the formation of a less thermodynamically stable polymorph, aragonite, with additional lattice incorporation of the foreign ion.

INTRODUCTION

The alkaline earth phosphates and carbonates are of importance in biological and industrial applications ranging from water treatment to studies of hard tissues such as bones and teeth.[1,2,3] It is now generally agreed that a number of calcium phosphate phases may be involved in these reactions such as dicalcium phosphate dihydrate (DCPD, $CaHPO_4 \bullet 2H_2O$), octacalcium phosphate (OCP, $Ca_4H(PO_4)_3 \bullet 2.5H_2O$) and hydroxyapatite (HAP, $Ca_5(OH)(PO_4)_3$).[4] In neutral and acidic solutions, DCPD has been found to be the first crystalline phase to precipitate *in vitro* but *in vivo*, there is some disagreement as to whether this phase is involved in the formation of early hard tissues such as embryonic chick and bovine bones and developing dentin.[5,6] In terms of a simple Ostwald Rule of Stages[7] precipitation of calcium phosphate would be expected to be sequential, in the order DCPD>OCP>HAP. It has been shown that this is the order of the rate constants for crystal growth but since DCPD grows so rapidly as compared with the thermodynamically most stable HAP, it may be difficult to detect HAP by analyzing the products of precipitation reactions in aqueous solution. However, since it is clear that these more acidic phases may be integrally involved in the precipitation reactions, there is considerable interest in investigating the development of more basic calcium phosphate phases in suspensions of DCPD crystallites. The present paper will discuss the application of the dual constant composition method to investigate the simultaneous dissolution or growth of DCPD and growth of OCP in solutions seeded with these phases.

The crystallization and dissolution of alkaline earth carbonates are important not only in fields such as oceanography and sedimentology but also, more recently, through the use of coral and modified coral surfaces as bone substitute materials. The system that has been most studied, the precipitation of calcium carbonate polymorphs, also has wide applications in reactions that take place in natural water systems in which less stable polymorphs of calcium carbonate may persist for long periods. When other alkaline earths are present in the calcium carbonate solutions, the formation of mixtures of precipitates occurs. Unfortunately, however, studies in mixed alkaline earth carbonate solutions have been limited to free-drift measurements in which the concentrations of the precipitating ions decrease during the reaction. Since the influence of the second alkaline earth ion may be strongly dependent upon the thermodynamic driving force, these studies are not able to elucidate the mechanisms of possible concomitant precipitations involving more than one sparingly soluble salt. Thus a study of calcium carbonate precipitation in the presence of strontium ion at pH values sufficiently high to favor co-precipitation, resulted in the formation of aragonite.[8] Moreover, it appeared that the incorporation of small amounts of strontium ion into a limited number of non-lattice sites in the calcium carbonate surface was influenced by the absolute strontium ion concentration in the solution as well as the presence of a another alkaline earth, barium ion, in the solution phase.[9] It was clear that answers to the questions posed by the results of these studies could only be obtained by measuring the simultaneous crystal growth of calcium carbonate and other alkaline earth salts in these mixed solution phases. In the present paper, the application of the dual constant composition, DCC, method to the simultaneous growth of mixtures of calcium carbonate and strontium carbonate is discussed. The results of crystal growth experiments at sustained supersaturations with respect to both salts point to the ability of both calcium carbonate and strontium carbonate phases to nucleate the strontium and calcium salts, respectively, in solutions supersaturated with respect to both phases.

THE SIMULTANEOUS GROWTH OF CALCIUM AND STRONTIUM CARBONATES - A DUAL CONSTANT COMPOSITION STUDY

Titrants Composition for DCC Experiments

The DCC method can readily be used to study the simultaneous growth of two crystals without a common ion, by adopting two electrodes, each sensitive to only one of the reactions. In such a relatively simple case two independent sets of titrant solutions can be used. The addition of one set of titrants would be controlled by an A ion specific electrode, yielding rates of reactions, in which A ions participate. Similarly, the use of a second set of titrants, controlled by B ion specific electrode would provide rate data for reactions involving B ions. In order to keep the composition of the reaction supersaturated solution constant, titrants designated for the growing crystals must contain lattice ions of each crystalline material together with background electrolytes so as to compensate for the dilution and ionic strength effects. The method could be extended to more complex systems, in which common ions are involved in both reactions, provided that the stoichiometry of the precipitating salts is well established. The simultaneous growth of calcium and strontium carbonates from solutions supersaturated with respect to both salts, presented in this paper, is a typical example of the latter system.

In the DCC experiments, titrants, designed for strontium carbonate crystal growth, were controlled by means of a glass electrode, while calcium carbonate growth titrants were controlled using a calcium electrode.[10] In the DCC method, both electrodes, coupled with a common reference electrode, were maintained at constant electromotive force (EMF) by means of computerized stepper motor-controlled titrant systems. For the growth of strontium carbonate, two titrant solutions were prepared: one containing both alkaline earth chlorides together with sodium chloride, and the other, sodium hydroxide and sodium bicarbonate. The compositions of titrant solutions were calculated so as to compensate for both dilution effects and the consumption of lattice ions due to crystal growth from equations 1-5:

$$(M'Cl_2)_t = 2(M''Cl_2)_{ws} + C_{eff}, \tag{1}$$

$$(NaCl)_t = 2(NaCl)_{ws} - 2\ C_{eff}, \tag{2}$$

$$(M''Cl_2)_t = 2(M''Cl_2)_{ws} \tag{3}$$

$$(NaHCO_3)_t = 2(NaHCO_3)_{ws} + C_{eff}, \tag{4}$$

$$(NaOH)_t = 2(NaOH)_{ws} + C_{eff} \tag{5}$$

In these equations, $M' = Sr$, $M'' = Ca$. Subscripts t and ws represent titrant and reaction working solutions, respectively. C_{eff} is the effective concentration of growing phase represented by the added titrant. The calcium electrode monitored calcium carbonate crystal growth and controlled the second set of titrants with $M' = Ca$, and $M'' = Sr$. Sodium chloride was added to the working solution so as to maintain constant ionic strength (0.10 mol l^{-1}), while sodium hydroxide was used to bring the pH to the required value, 8.50 ± 0.005.

Following the introduction of the seed crystals to the supersaturated solution, calcium, strontium and carbonate ions are consumed due to the growth of both carbonate phases. The changes in concentration of CO_3^{-2} ions result from the growth of both salts. The induced pH change, which is sensed by a glass electrode, is compensated by the addition of two titrant solutions (calculated from Eqs. (1)-(5) with $M' = Sr$, $M'' = Ca$). Following this addition, the working solution will contain the preset concentration of carbonate ions, together with an excess, $+\varepsilon$, of strontium ions (these titrants are added as if only $SrCO_3$ was precipitating) and a depletion, $-\varepsilon$, of calcium ions. Changes in calcium ion concentrations are caused by calcium carbonate growth. When the absolute value of the Ca electrode potential exceeds the preset value, the two titrant solutions (calculated from Eqs. (1)-(5) with $M' = Ca$, $M'' = Sr$) controlled by this electrode are added to working solution to restore the concentration of calcium ions to the preset value. This addition also momentarily raises carbonate ion concentration to a $+\varepsilon$ excess. The concentration of strontium ions approaches the initial value at a rate dependent upon the relative rates of strontium and calcium carbonate crystal growth. It follows that the carbonate ion depletion due to calcium and strontium carbonate crystals growth is compensated for twice: first by the addition of titrants controlled by the glass electrode, and second, by the addition of the titrants controlled by the calcium electrode. Thus, the subsequent addition of the strontium carbonate titrants will be delayed, thus compensating for the former double addition. This cycle is repeated throughout the experiments.

The exclusive growth of strontium or calcium carbonates, with or without the second cation, was studied using the CC method, which employed one set of two titrant solutions described by equations 1-5 (M' = Sr and M" = Ca for $SrCO_3$ crystal growth, and M' = Ca, and M" = Sr for $CaCO_3$ crystal growth). These were controlled by a pH glass electrode.

Determination of the Growth Rate Using DCC Kinetic Data

The rate of the crystal growth, J, is usually expressed in terms of overgrowth of the crystal, as a function of time

$$J = d(n-n_o)/dt, \tag{6}$$

where n_o is the number of moles of seed used to initiate the growth reaction ($n_o = 0$ for spontaneous precipitation) and n is the number of moles of crystal at time t. In CC and DCC experiments, an overgrowth of the crystals can be readily calculated from the titrants addition data using equation 7

$$(n-n_o) = C_{eff}V, \tag{7}$$

where V is volume of the titrant added at time t. Thus, the overall reaction rate is given by

$$J = (dV / dt) C_{eff}, \tag{8}$$

or by

$$J = (dV / dt) C_{eff} A^{-1}. \tag{9}$$

In equations 8 and 9 dV/dt is the slope of the titrant volume-time curve recorded during the reaction and the rate of the crystal growth is normalized to the total surface area of the growing crystal. This is usually estimated using equation 10

$$A = A_o (m/m_o)^{2/3} \tag{10}$$

where m_o and m are the masses of crystal present initially and at time t, respectively. A_o is the initial surface area of the crystals, calculated from the specific surface area, SSA, using equation 11

$$A_o = SSA \, m_o. \tag{11}$$

The mechanism of crystal growth is usually explored by using experimental rate data at different thermodynamic driving forces to confront the theoretical models. As a first step, the rate of crystallization of many sparingly soluble salts may be interpreted using an empirical kinetic equation of the form of equation 12:

$$J = ks \, K_{so}^{1/2} \sigma^n, \tag{12}$$

in which, for the alkaline earth carbonates,

$$1 + \sigma = \{[M^{2+}] [CO_3^{2-}]\}^{1/2}/K_{so}^{1/2}. \tag{13}$$

Here, K_{so} is the conditional solubility product under the experimental ionic strength conditions, k is the effective rate constant, s a function of the number of active growth sites on the crystal surfaces, and n is the effective order of the crystal growth reaction.

Alkaline Earth Carbonates Kinetics

It is interesting to compare various alkaline earth carbonates, since they are of fundamental importance in diverse areas such as limnology, sedimentology and oceanography.[8] The kinetics of calcium carbonate crystal growth and dissolution, especially its most stable polymorph, calcite, is well documented in the literature.[11,12,13,14] In contrast, other alkaline earth carbonates, including strontium carbonate, have been almost totally ignored.

For the Constant Composition, CC, growth of strontium carbonate crystals typical titrants addition plots as a function of time are shown in Figure 1. They indicate a greater rate of addition immediately following the introduction of seed crystals. This frequently observed phenomenon, also documented in crystal growth studies of other mineral phases, could be attributed to a reduction in active growth sites on the seed crystals due to a rapid high energy site mineralization.[4] In support of this hypothesis, it was found that the extent of the initial surge increased with seed concentration.[10] The use of seed, harvested at the end of a growth experiment (pregrown seed), as new initiating surfaces for crystal growth, markedly reduced the initial surge (Figure 1), again suggesting that changes in surface characteristics are important. It is worth noting (Figure 2) that calcite growth takes place without the initial rapid stage observed in the case of strontium carbonate. The influence of calcium ions on the rates of strontium carbonate crystal growth, and of strontium ions on calcium carbonate growth, also studied by the use of CC method, is shown in Figure 2. In these experiments, the concentrations of the additive ions were selected so as to achieve solutions undersaturated with respect to the additive carbonate salt. It can be seen that calcium ions markedly reduce the rate of strontium carbonate crystal growth as well as the initial surge. It is significant that, in contrast, the rate of calcite growth remains almost unchanged in the presence of strontium ions. The former results suggest the adsorption of calcium ions at high energy sites on the strontium carbonate seed crystals. Although undetectable analytically, evidence for calcium ion adsorption by strontium carbonate was

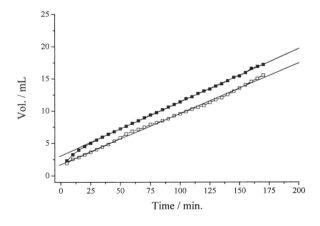

Figure 1. Crystallization of strontium carbonate at $\sigma = 0.25$, (pH = 8.50, temp. = 25 °C); supersaturated solutions seeded with regular (closed squares) and pregrown (open squares) strontium carbonate crystals. Plots of volume of titrant added, as a function of time. Lines refer to the linear regression of these plots after the initial surges.

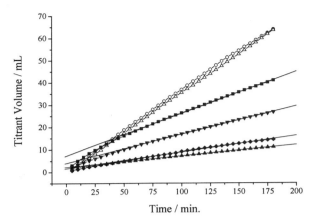

Figure 2. Crystallization of strontium carbonate at = 0.75 (pH = 8.50, temp. = 25°C) in the absence (closed squares) and presence of different amounts of calcium ions ((closed triangles) 6.5×10^{-4} mol l^{-1}, (closed diamonds) 3.5×10^{-4} mol l^{-1}, (closed inverted triangles) 0.5×10^{-4} mol l^{-1}) and crystallization of calcium carbonate at $\sigma = 1.20$ (pH = 8.50, temp. = 25°C) in the absence (open triangle) and presence (open circle) of 0.5×10^{-4} mol l^{-1} strontiom ions. Plots of volume of added titrant, as a function of time. Lines refer to the linear regression of these plots after the initial surges.

obtained from measurements of zeta potential. The values were markedly more positive in the presence of calcium ions than in the pure suspensions of $SrCO_3$, clearly indicating uptake of calcium ions.

A logarithmic rate plot of equation 13, for strontium carbonate crystal growth at supersaturations, σ, ranging from 0.25 to 1.70 is presented in Figure 3. The slope of this linear plot (with correlation coefficient, R = 0.99) corresponds to a value of n = 2.0 ± 0.3 suggesting a surface controlled spiral growth mechanism. It is interesting to note that the results of both CC[12] and free drift[13] kinetic studies of calcite crystal growth, have also shown a rate law second order in relative supersaturation.

DCC experiments of the simultaneous crystal growth of both alkaline earth carbonates confirmed the abilities of strontium and calcium carbonates to nucleate calcium and strontium carbonates, respectively. It can be seen from plots of titrant volume as a function of time in Figure 4, that following the addition of $SrCO_3$ seed crystals to solutions supersaturated with respect to both strontium and calcium carbonates, the nucleation of calcium carbonate is initiated immediately and both alkaline earth carbonates undergo crystal growth. Strontium carbonate can be also nucleated at calcite surfaces immediately upon adding these seed crystals to the mixed supersaturated solutions (Figure 4). The kinetics of the simultaneous strontium and calcium carbonate mineralization obtained from all DCC experiments are in agreement with the CC data. The rate of strontium carbonate crystal growth observed in DCC experiments is considerably less than that calculated for the same supersaturation in pure strontium carbonate solutions (Figure 3). This again reflects the

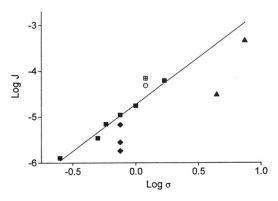

Figure 3. Logarithmic plot of constant composition strontium carbonate crystal growth rates (pH = 8.50, temp. = 25°C) against relative supersaturation σ, from pure supersaturated solutions (closed squares) and in the presence of added calcium ions at undersaturation (closed diamonds) or supersaturation (closed triangles) levels with respect to calcite. The plot also presents data for calcium carbonate growth at $\sigma = 1.20$ (pH = 8.50, temp. = 25°C), from pure supersaturated solutions (four-point star) and with added strontium ions at undersaturation (open square) or supersaturation (open circle) levels with respect to $SrCO_3$.

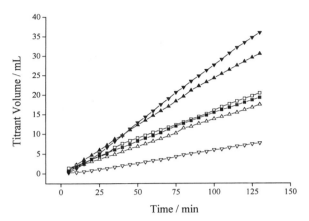

Figure 4. Crystallization of strontium/calcium carbonates (pH = 8.50, temp. = 25°C) in DCC experiments at σ_{SrCO_3} = 4.50 and σ_{CaCO_3} = 1.20; supersaturated solutions seeded with strontium carbonate ((open square) $SrCO_3$, (closed square) $CaCO_3$), calcium carbonate ((open inverted triangle) $SrCO_3$, (closed inverted triangle) $CaCO_3$), and a mixture of strontium and calcium carbonates ((open triangle) $SrCO_3$, (closed triangle) $CaCO_3$). Plots of volume of titrant added, as a function of time.

inhibitory effect of calcium ions on $SrCO_3$ growth. The rate of calcium carbonate crystal growth observed in DCC experiments is only little influenced by the presence of strontium ions (Figure 3).

Physico-chemical analysis of the solid phases grown in DCC experiments provided additional information about the simultaneous growth of strontium and calcium carbonates. Typical X-ray diffraction spectra for the solid phases grown in DCC experiments show peaks characteristic for strontium carbonate, calcite (calcite peaks are present only when calcite is introduced as seed) and aragonite. The observed small shift of all aragonite peaks towards lower 2θ angle, compared to that of pure aragonite peaks, suggests the incorporation of strontium ions into the growing aragonite. Thus, although strontium ion does not influence the growth rate of calcium carbonate, it favors the formation of the metastable polymorph, aragonite. Fourier transformation infrared spectroscopy and energy dispersive X-ray analysis confirmed the nucleation and subsequent growth of calcium and strontium carbonates in all DCC experiments, irrespective of the seed used.

THE APPLICATION OF THE DUAL CONSTANT COMPOSITION METHOD TO CALCIUM PHOSPHATES

Titrants Composition for DCC Experiments

For DCPD and OCP, the application of the DCC method was tested (a) in solutions supersaturated with respect to both phases to investigate the simultaneous crystal growth of mixtures of DCPD and OCP[15], and (b) in solutions undersaturated in DCPD and supersaturated in OCP in order to investigate the transformation reaction DCPD → OCP. In the latter case, changes in solution concentration during DCPD dissolution were compensated for by the addition of a single acidified titrant solution, controlled by a ion specific calcium electrode. This titrant solution contained hydrochloric acid and sodium chloride, with concentrations calculated from equations 14 and 15

$$[HCl]_t = 2[CaCl_2]_{ws} - [KH_2PO_4]_{ws} - [KOH]_{ws} \tag{14}$$

$$[NaCl]_t = [KH_2PO_4]_{ws} + [NaCl]_{ws} + [KOH]_{ws} \tag{15}$$

In equations 14 and 15, $[CaCl_2]_{ws}$ and $[KH_2PO_4]_{ws}$ are the total concentrations of calcium and phosphate ions in the working solution. $[KOH]_{ws}$ and $[NaCl]_{ws}$ are the concentrations of potassium hydroxide and sodium chloride added to the working solution to restore the initial pH and ionic strengths, respectively, to their original values. Changes in lattice ion concentrations accompanying the nucleation and growth of OCP during the reaction were compensated by the addition of two titrant solutions, controlled by a glass electrode. These titrant solutions contained potassium dihydrogen phosphate with potassium hydroxide (equations 16 and 17), and calcium chloride with sodium chloride (equations 18 and 19).

$$[KH_2PO_4]_t = 2[KH_2PO_4]_{ws} + 3\,C_{eff} \tag{16}$$

$$[KOH]_t = 2[KOH]_{ws} + 5\,C_{eff} \tag{17}$$

$$[CaCl_2]_t = 2[CaCl_2]_{ws} + 4C_{eff} \tag{18}$$

$$[NaCl]_t = 2[NaCl]_{ws} - 8\,C_{eff} \tag{19}$$

C_{eff} is the effective concentration of the titrants for OCP growth while subscripts ws and t, represent working and titrant solutions, respectively. As for the alkaline earth carbonates growth kinetic studies, sodium chloride was introduced so as to maintain constant ionic strength (0.15 mol l^{-1}) while potassium hydroxide was used to keep the pH at the desired value, 7.40 ± 0.005.

Calcium Phosphates Kinetics

DCC experiments of the simultaneous growth of OCP and dissolution of DCPD were made at physiological pH of 7.40, 37°C and ionic strength 0.15 mol l^{-1} in NaCl in solutions undersaturated with respect to DCPD ($\sigma_{DCPD} = -0.12$) but supersaturated in OCP ($\sigma_{OCP} = 2.35$). Typical experimental results are shown in Figure 5 as plots of DCPD dissolution percent (calculated from equation 20) and OCP growth (calculated from equation 8), as functions of time.

$$\text{Dissolution \%} = (n_o - n) / n_o = C_{eff}V / n_o \tag{20}$$

It can be seen that DCPD dissolution commenced immediately upon the introduction of DCPD seed crystals while nucleation of OCP occurred after an induction period of about

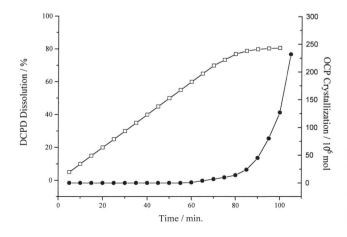

Figure 5. Extent of OCP (closed circle) precipitation and DCPD (open square) dissolution in DCC experiment at $\sigma_{OCP} = 2.35$ and $\sigma_{DCPD} = -0.12$ (pH = 7.40, temp. = 37°C).

55 minutes. Following this period, the growth of OCP predominated, calling only for the addition of the appropriate titrant solutions (equations 16-19).

Scanning electron micrographs showed that the new OCP crystallites formed near the edges of the DCPD crystals[16] and subsequently completely covered them. As anticipated, the induction time for OCP nucleation decreased with increasing supersaturation and by increasing the amount of initiating DCPD surface. The latter observation clearly points to the specific nucleation of OCP at the dissolving DCPD surfaces. Analysis of the kinetics data showed that the DCPD dissolution was probably surface controlled by the desorption of lattice ions rather than by simple diffusion of these ions away from the dissolving surface. A more extensive constant composition study of the dissolution of DCPD[17] has shown that although volume diffusion provides a substantial resistance to dissolution at higher under-saturations, surface processes become rate determining at lower driving forces as reflected by the relatively high activation energy and the insensitivity of the rates of dissolution to changes of stirring speed. These changes with thermodynamic driving force emphasize the need to investigate the growth and dissolution reactions at constant thermodynamic driving force in order to obtain mechanistic information.

REFERENCES

1. Moreno E.C., Brown W.E., and Osborn G., 1960, Solubility of dicalcium phosphate dihydrate in aqueous solution., Soil Sci. Soc. Am. Proc. 24 : 94-98.
2. Meyer J.L., and Eanes E.D., 1978, A thermodynamic analysis of the secondary transition in the spontaneous precipitation of calcium phosphate., *Calcif. Tissue Res.*, 25 : 209-216.
3. Brown W.E., Eidelman N., and Tomazic B., 1987, Octacalcium phosphate as a precursor in biomineral formation., *Adv. Dent. Res,*, 1 : 306-313.
4. Nancollas G.H., 1989, In vitro studies of calcium phosphate crystallization in Biomineralization, Eds. Mann S., Webb J., and Williams J.P., VCH Verlagsgesellschaft, Weinhein,.
5. Glimcher M. J., Bonar L.C., Grynpas M.D., Landis W.J., and Roufosse A.H., 1981, Recent studies of bone mineral : is the amorphous calcium theory valid ?, *J. Cryst. Growth*,53 : 100-119.
6. Johnsson M.J., and Nancollas G.H., 1992, The role of brushite and octacalcium phosphate in apatite formation., CRC Crit. Rev. Oral Biol. Med. 3 : 61-81.
7. Ostwald W., 1897, Studien uber die bildung und umwandlung fester korper., *Z. Phys. Chem.*, 22 : 289-324.
8. Cowan J.C., and D.J. Weintritt , 1976, Water - Formed Scale Deposits, Gulf Publ. Co., Houston, TX.
9. Pingitore N.E., Jr., and Eastman M.P., 1976, The coprecipitation of Sr^{+2} with calcite at 25 °C and 1 atm., *Geochim. Cosmochim. Acta.*, 50 : 2195-2202.
10. Zieba A. and Nancollas G.H., 1994, Constant Composition Kinetics Studies of the Simultaneous Crstal Growth of Alkaline Eart Carbonates. The Calcium - Strontium System, *J. Crystal Growth*, 144: 311-319.
11. Chiang P.P., and Donohue M.D., 1988, A kinetic approach to crystallization from ionic solution. I. Crystal growth., *J. Colloid Interf. Sci.*, 122 : 230-250.
12. Kazmierczak T.F., Tomson M.B., and Nancollas G.H., 1982, Crystal growth of calcium carbonate. A controlled composition kinetic study., *J. Phys. Chem.*, 86 : 103-107.
13. Christoffersen J., and Christoffersen M.R., 1990, Kinetics of spiral growth of calcite crystals and determination of the absolute rate constant., *J. Cryst. Growth*, 100 : 203-211 .
14. Nancollas G.H., Kazmierczak T.F., and Schruttringer E., 1981, A controlled composition study of calcium carbonate crystal growth : the influence of scale inhibitors., National Association of Corosion Science, 37 : 76-81.
15. Ebrahimpour A., Jingwu Zhang, and Nancollas G.H., 1991, Dual constant composition method and its application to studies of phase transformation and crystallization of mixed phases.,*J. Cryst. Growth*, 113 : 83-91.
16. Jingwu Zhang, Ebrahimpour A., and Nancollas G.H., 1992, Dual constant composition studies of phase transformation of dicalcium phosphate dihydrate into octacalcium phosphate., *J. Colloid Interf. Sci.*, 152 : 132-139.
17. Jingwu Zhang, and Nancollas G.H., 1992, Interpretation of dissolution kinetics of dicalcium phosphate dihydrate., *J. Cryst. Growth*, 125 : 251-269.

2

CALCITE GROWTH INHIBITION BY FERROUS AND FERRIC IONS

Joseph L. Katz and Katrin I. Parsiegla

Department of Chemical Engineering
The Johns Hopkins University
Baltimore, Maryland 21218

ABSTRACT

The inhibiting effects of ferric and ferrous ions on calcite growth were investigated using a free drift, seeded growth technique (at 25°C, pH 7.0 and 8.0, 0.15 M alkalinity and an initial supersaturation of 8). For the first 60 minutes after growth was initiated, the inhibiting effectiveness of ferric ions at pH 7.0 was essentially the same as at pH 8.0. Differences were observed at longer times. Ferrous ions were much less effective inhibitors than ferric at pH 7.0; however, their inhibiting effectiveness was dramatically improved by the addition of oxygen. At pH 8.0, essentially the same inhibiting effectiveness was observed with ferrous ions as with ferric, probably because they immediately oxidize to ferric ions even without added oxygen. With decreasing alkalinity the inhibiting effectiveness of ferric ions decreased. Initial studies on other metal ions showed that, at low alkalinities, zinc and copper are effective calcite growth inhibitors.

INTRODUCTION

A serious problem encountered in many industrial processes is the precipitation of minerals onto the walls of water handling equipment from process water that has become supersaturated with respect to these minerals. These deposits, especially on heat transfer surfaces in cooling water systems, lead to a loss of efficiency and to partial or even total blockage of water flow. In most cases, these scales are composed of calcium carbonates, phosphates, and sulfate hydrates[1,2]; calcite (one of the crystalline forms of calcium carbonate) is the most common scale forming mineral. Several techniques are used to prevent or control scale deposition[3,4], e.g., lowering the pH of the water by adding acid and thus increasing the solubility of calcium carbonate[5], pretreatment of the water with ion exchange resins which replaces calcium ions by sodium ions, adding complex-forming agents such as EDTA and thus sequestering the calcium from the solution, or adding certain chemical substances which inhibit the growth of scale forming minerals at very low concentrations. These substances are believed to inhibit growth by adsorbing onto the mineral surface and blocking sites on the surface that were otherwise used in the growth process. Typical scale inhibitors are organic polyphosphonates[6,7], inorganic polyphosphates[1,8], polyacrylic acids, polycarboxylic acids, polyacrylamids[9] and naturally occurring polyamino acids.[10] In previous publications[11,12] we showed that iron ions are very effective calcite growth inhibitors. At a concentration as small as 5 μM total iron, calcite growth is almost completely inhibited for more than 8 hours. Thus their addition to water may prove to be a useful procedure for inhibiting scale formation and growth. This paper presents some results of an ongoing effort to investigate calcite growth inhibition by metal ions, with particular emphasis on iron ions.

EXPERIMENTAL

The apparatus, materials, and experimental method used in this work are identical to those described in a previous paper.[11] However, a brief description here may be helpful. 100 ml of 0.15 M sodium bicarbonate buffer solution is equilibrated in a thermostated, 250 ml Teflon container by bubbling through it a CO_2/N_2 gas mixture. The desired pH is obtained by choosing the appropriate partial pressure of CO_2 (0.77 atm for pH 7.0 and 0.075 atm for pH 8.0). This solution is made *saturated* with respect to calcite by adding the appropriate volume of 0.1 M or 0.01 M $CaCl_2$ solution. Calcite seed crystals are then added, and allowed to equilibrate for 5 to 10 minutes before the solution is made *supersaturated* by adding more $CaCl_2$ solution, thus initiating growth. No further addition of calcium chloride or any other substance is made, so the calcium concentration decreases with time; this is called the free drift seeded growth method. The pH remains very constant during each experimental run because the solution (deliberately) is very strongly buffered by sodium bicarbonate. Calcium concentration and pH are monitored throughout the entire experimental run. Calcite growth in the presence of iron is studied by adding a small volume of a freshly prepared 10^{-3} M $FeCl_3$ (or $FeCl_2$) solution *before* adding the seeds. The seeds are then added and allowed to equilibrate between 5 to 30 minutes in this solution, which then is made supersaturated by adding more $CaCl_2$ solution.

Unless otherwise stated, all measurements were at 25°C, an initial supersaturation (S_i) of 8, a seed concentration of 5 g/l, an alkalinity (Alk) of 0.15 M, an ionic strength (I) of 0.15 M and at pH 7.0 or 8.0. The calcite seeds used were Mallinckrodt reagent grade calcite. Their specific surface area (measured by BET single point nitrogen adsorption analysis) is 0.32 m^2/g; the average edge length (measured in an SEM) is 10 μm. In all experiments, dry seeds were added to the system. All iron concentrations reported in this paper are the total iron concentrations ([FeT]) that were actually added; they are not the concentrations of the free or dissolved iron ions that are actually present in a solution in equilibrium with iron hydroxides.

RESULTS AND DISCUSSION

Figures 1a and 1b show calcite growth in the presence of various ferric ion concentrations at pH 7.0 and 8.0, respectively, at otherwise identical solution conditions. As is apparent from Figs. 1a and 1b, growth inhibition increases with increasing iron concentration. At pH 7.0, the inhibiting effect of ferric ions increases dramatically over the small concentration range, 2 μM [FeT] to 5 μM [FeT]. At pH 8.0, the inhibiting effect of ferric ions increases more gradually; the range of concentration required to go from weak inhibition to almost total inhibition is 1 μM [FeT] to 7 μM [FeT]. Note that at both pH's the order of magnitude of ferric ion concentration required to inhibit calcite growth is the same. This is

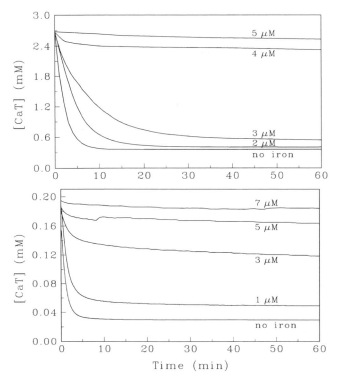

Figure 1. (a) Effect of various ferric ion concentrations on calcite growth at pH 7.0. (b) Effect of various ferric ion concentrations on calcite growth at pH 8.0.

not what one would expect since the solubility of ferric ions is about ten times smaller at pH 8.0 than at pH 7.0 (due to precipitation as ferric hydroxide). Figures 2a and 2b show calcite growth over more extended lengths of time, in the presence of 5 and 7 μM ferric ion concentration at pH 7.0 and 8.0, respectively. One now sees quite different characteristics. At pH 7.0, growth is almost completely inhibited for several hours; the calcium concentration then decreases rapidly but levels out again at a concentration which is significantly higher than the concentration that would be in equilibrium with calcite in the absence of iron. Increasing the ferric ion concentration from 5 to 7 μM nearly doubles the length of time of almost complete inhibition. At pH 8.0, the calcium concentration decreases continuously with time. The growth rate is slower both at the beginning and at the end, and is slightly faster in between, but the S-shape of these growth curves is not nearly as pronounced as it is at pH 7.0. Increasing the ferric ion concentration does not significantly increase growth inhibition.

     Another observation may be worth while noting. At pH 7.0, the seeds had to be equilibrated with ferric ions for 30 minutes prior initiation of growth; 0, 5 or 10 minutes pre-equilibration resulted in less effective inhibition and 60 minutes gave the same effectiveness as 30 minutes. However, at pH 8.0, no significant difference in inhibiting effectiveness could be observed when the seeds were pre-equilibrated for 0, 1, 3, or 10 minutes; 30 minutes of pre-equilibration caused *decreased* inhibiting effectiveness, i.e., a slightly faster growth rate was observed. Thus, a pre-equilibration time of 30 minutes was used for all the

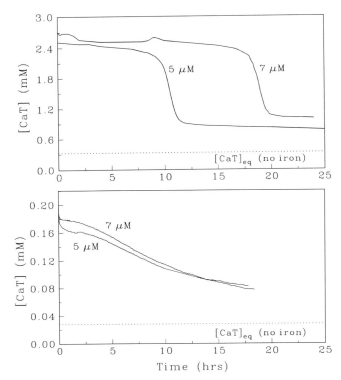

Figure 2. (a) Effect of ferric ions on calcite growth over an extended length of time at pH 7.0. (b) Effect of ferric ions on calcite growth over an extended length of time at pH 8.0.

growth curves shown in Figs. 1a and 2a while those shown in Figures 1b and 2b were obtained using a pre-equilibration time of 5 to 10 minutes.

Figure 3 shows the effect of various concentrations of ferrous ions on calcite growth at pH 7.0. Comparing Figure 3 with Figure 1a shows that ferrous ions are much less effective

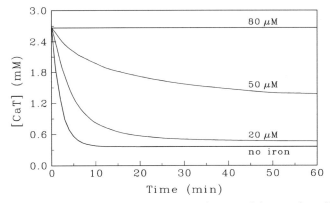

Figure 3. Effect of various ferrous ion concentrations on calcite growth at pH 7.0.

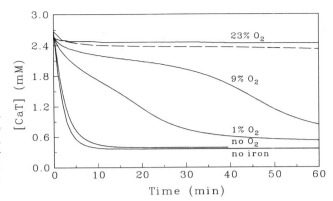

Figure 4. Effect of various oxygen partial pressures on calcite growth inhibition by 4 μM ferrous ions at pH 7.0. Also shown for comparison is the effect of 4 μM ferric ion concentration (the dashed line).

growth inhibitors than ferric ions; more than a tenfold higher ferrous ion concentration is required to achieve the same inhibiting effectiveness. However, adding a small concentration of oxygen to the gas stream results in a dramatic increase in inhibiting effectiveness of ferrous ions, as can be seen in Figure 4. Figure 4 shows that a 4 μM total ferrous ion concentration is sufficient to cause significant growth inhibition when as little as 1% oxygen is added to the gas mixture and that adding 23% oxygen results in almost complete growth inhibition. Figure 4 also shows that a 4 μM ferrous ion concentration *in the presence* of 23% oxygen causes essentially the same total growth inhibition as 4 μM ferric ions. Figure 5 shows that, at pH 8.0, a 3 μM ferrous ion concentration is slightly more effective than 3 μM ferric ion concentration, and that at 4 μM concentration, ferrous and ferric ions are equally effective. However, ferrous ions probably are not the active agent in the inhibition process. At pH 8.0, ferrous ions are rapidly oxidized to ferric ions.[13] Even though the oxygen concentration was minimized by adding hydrogen to the feed gas and passing it through a bed of hot copper turnings, there probably was enough O_2 left in the system to oxidize ferrous ions[11], so the growth inhibition observed with ferrous ions at pH 8.0 may be actually due to ferric ions that were produced by in situ oxidation.

At this point, a brief discussion on two ways of quantifying inhibition effectiveness may be useful. Calcite growth curves in the presence of iron, as shown in Figures 1a and 1b, typically level off at an almost constant calcium concentration. Thus, growth inhibition can be defined as the difference in final calcium concentration in the presence of iron, C_f, to that in the absence of iron, C_{eq}, divided by the difference in initial calcium concentration C_i to the equilibrium concentration, i.e., $I\% \equiv (C_f - C_{eq})/(C_i - C_{eq}) * 100\%$. (Since the final cal-

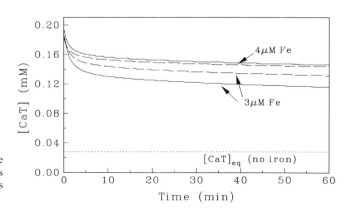

Figure 5. Comparison of calcite growth inhibition by ferrous ions (dashed lines) and ferric ions (solid lines) at pH 8.0.

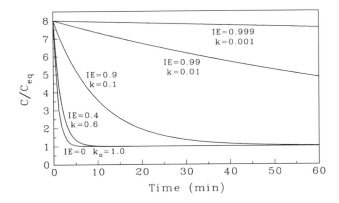

Figure 6. Simulated growth curves for 1st order growth. k is the rate constant and IE is inhibiting effectiveness.

cium concentrations in the presence of iron slightly decrease with time, they are compared at a fixed length of time after growth was initiated, typically 60 minutes.) However, this simple method of quantifying growth inhibition is useful only in those cases where a relatively constant supersaturation persists for a sufficient length of time.

A second way of quantifying growth inhibition is to compare growth rates measured in the presence of inhibitors to those measured in their absence.[14,15,16] The inhibiting effectiveness, IE, is then defined as $IE \equiv (k_o - k)/k_o$, where k_o and k are the rate constants in the absence and presence of inhibitor, respectively. However, this definition expresses growth inhibition on a linear scale. The following example calculation suggests that a logarithmic scale may be more appropriate. The decrease in supersaturation (C/C_{eq}) with time which would occur at different inhibiting effectivenesses, IE, for a reaction which *exactly* followed first order rate law, $C/C_{eq} = 1 + (C_i/C_{eq} - 1)e^{-kt}$, where $C_i / C_{eq} \equiv 8$ and $k_o \equiv$ 1.0 min^{-1}, was calculated and is shown in Figure 6. (The corresponding growth rate constants, k, are also shown.) As can be seen, even at an inhibiting effectiveness of 0.99, a number which would be considered by most to imply *almost complete* growth inhibition, the calcium concentration still decreases quite strongly with time. Furthermore, at an inhibiting effectiveness of 0.4, a not very small number, almost *no change* in growth curve can be observed. One possible improvement would be to use a logarithmic definition, $IE_L \equiv -log(k/k_o)$. This would result in a linear scale which, in this sample calculation, goes from 0.0 for no inhibition to 3.0 for essentially complete inhibition.

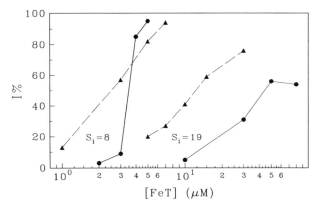

Figure 7. Comparison of calcite growth inhibition by ferric ions at pH 7.0 (solid lines) and 8.0 (dashed lines), and initial supersaturations, S_i, of 8 and 19.

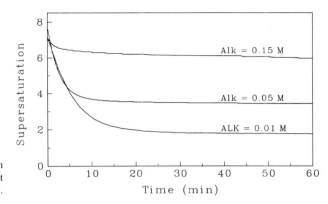

Figure 8. Effect of 5 μM ferric ion concentration on calcite growth at three different alkalinities (at pH 8.0).

Figure 7 shows a comparison of the I% caused by ferric ions at pH's 7.0 and 8.0 and at initial supersaturations (S_i) of 8 and 19. Inhibition is quantified by comparing final calcium concentrations in the presence and absence of iron using the concentrations measured 60 minutes after growth was initiated. Four observations can be made: First, under these conditions, ferric ions are more effective inhibitors at pH 8.0 than at pH 7.0, except for the region of almost complete growth inhibition at $S_i = 8$. (Note that in an earlier paper[11] we reported that ferric ions are more effective at pH 7.0 than at pH 8.0. This was based on results obtained at $S_i = 8$ and [FeT] > 4 μM and is consistent with these results.) Second, at pH 7.0 and $S_i = 8$, the inhibiting effectiveness of ferric ions increases strongly over a small concentration range of ferric ions. Third, higher supersaturations required higher iron concentrations for the same I%. Fourth, increasing ferric ion concentration results in increased growth inhibition. However, at pH 7.0 and $S_i = 19$, growth inhibition increases with increasing total ferric ion concentration only up to 50 μM [FeT]. This may be due to precipitation of ferric hydroxide at this high ferric ion concentration, thus removing ferric ions from solution.

The results presented thus far were all at a relatively high alkalinity of 0.15 M; a value chosen so that the pH would remain constant despite the growth of calcite. However, industrial water systems usually operate at a much lower alkalinity, typically less than 0.005M. Figure 8 shows calcite growth at alkalinities of 0.15 M, 0.05 M and 0.01 M in the presence of 5 μM ferric ions at pH 8.0. (The ionic strength was held constant at 0.15 M by adding the appropriate amount of NaCl.) At 0.05 M alkalinity, no effort was made to hold pH constant during growth, but it changed by only 0.05 pH units. At 0.01 M alkalinity, pH was kept constant by continuously titrating 0.1 M sodium hydroxide into the system using a Metrohm 665 Dosimat. As can be seen in Figure 8, ferric ions are less effective inhibitors at lower alkalinities.

A possible cause for the decrease in growth inhibition with decreasing alkalinity is that changing the solution alkalinity from 0.15 to 0.05 or to 0.01 M at pH 8.0 causes both the Ca^{2+}/CO_3^{2-} and the Ca^{2+}/HCO_3^- ratios to increase approximately 220 fold. Thus the calcite surface becomes more positively charged at lower alkalinities, and positively charged ionic ferric species are less strongly adsorbed. Another possible cause is that ferric ions can form complexes with carbonate and maybe with bicarbonate ions. (There is evidence in the literature for the existence of ferric carbonate complexes.[17,18]) Thus, the total dissolved iron concentration in equilibrium with colloidal ferric hydroxide would be higher than in the absence of any carbonate or bicarbonate ions, and fewer Fe^{3+} ions precipitate as ferric hydroxide. With increasing alkalinity, both the carbonate and the bicarbonate concentrations increase. Thus more ferric ions remain in solution and available for adsorption onto the

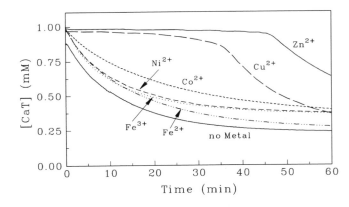

Figure 9. Effect of various metal ions on calcite growth; [MeT] = 10 μM, pH = 8.0, S_i = 7, Alk = 0.01M, I = 0.01M.

calcite surface. Note that both explanations assume that calcite growth is inhibited by ionic ferric species and that the two explanations are compatible with each other. Complex formation with bicarbonate ions also may explain the surprisingly weak dependence of growth inhibition by ferric ions on pH (see Figures 1a and 1b).

Figure 9 shows some initial results on the effect of other transition metal ions on calcite growth, at pH 8.0, Alk = 0.01M, I = 0.01M, S_i = 7 and a total metal ion concentration of 10 μM. During the runs shown in Figures 9 and 10, pH was not kept constant. The largest change in pH occurred in the very beginning of each run due to the addition of $CaCl_2$ solution *and*, when the inhibitor is not effective, due to the growth of calcite. As a result, in the absence of inhibitor, the pH decreased from about 8.00 to about 7.74 within the first 6 minutes after growth was initiated and then slowly increased to about 7.90 over the next 54 minutes. When there was almost complete initial inhibition (e.g., with zinc or copper ions) the initial drop in pH was only about 0.05 pH units. A pH change from 8.0 to 7.9 changes the equilibrium calcium concentration and thus the supersaturation by about 20%. Therefore, the results shown in Figures 9 and 10 should be viewed only as preliminary results showing the growth inhibiting potential of some transition metal ions. Figure 9 shows that, under these conditions, ferric and ferrous ions are not very effective inhibitors but zinc and copper ions are quite effective. The effect of doubling the concentration of the zinc or the copper is shown in Figure 10. One sees that at 20 μM zinc is a very effective inhibitor, and that copper is quite good. Note that these curves show S-shapes similar to those obtained in the presence of ferric ions at pH 7.0 and Alk = 0.15 M (Figure 2a).

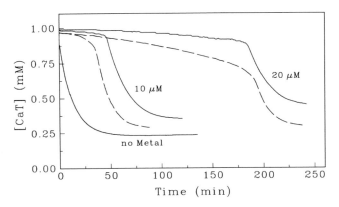

Figure 10. Effect of various concentrations of zinc ions (solid lines) or copper ions (dashed lines) on calcite growth; pH = 8.0, S_i = 7, Alk = 0.01M, I = 0.01M.

SUMMARY AND CONCLUSION

As little as 5 μM total ferric ion concentration was sufficient to inhibit calcite growth almost completely at pH 7.0, 0.15 M alkalinity and an initial supersaturation of 8. A slightly higher ferric ion concentration (7 μM) was needed to obtain about the same *initial* inhibiting effectiveness at pH 8.0 as at pH 7.0. Growth curves in the presence of ferric ions had similar shapes at pH 7.0 and 8.0 (Alk = 0.15 M, S_i = 8) in the first 60 minutes after growth was initiated. However, when observed over several hours, growth curves at ferric ion concentrations high enough for almost complete growth inhibition were strongly S-shaped at pH 7.0 but decreased quite smoothly with time at pH 8.0. The inhibiting effectiveness of ferrous ions was much less than that of ferric, at pH 7.0 and 0.15 M alkalinity. However, with the addition of oxygen, ferrous ions became as effective as ferric ions. At pH 8.0, ferrous ions showed about the same inhibiting effectiveness as ferric, probably because inhibition is actually due to ferric ions produced in situ by oxidation. Ferric ions become less effective at lower alkalinities. Thus, under some conditions, adding ferric ions may be an insufficiently effective scale inhibition method. However, at low alkalinities, zinc and, to a lesser extend, copper ions are quite effective calcite growth inhibitors.

ACKNOWLEDGMENT

Support by National Science Foundation grant CTS - 9214829 is gratefully acknowledged.

REFERENCES

1. Elliot, M. N., 1969, Scale Control by Threshold Treatment, Process Technology Division, U.K.A.E.A. Research Group (Atomic Energy Research Establishment, Harwell, Berkshire, UK) AERE-R5696.
2. Nancollas, G. H., 1979, The growth of crystals in solution, *Adv. Colloid Interface Science* 10:215-252.
3. BETZ, 1980, Handbook of Industrial Water Conditioning (BETZ Laboratories Inc., Trevose, PA) p. 12.
4. Hartung, R., and Marturana, D., 1992, Water treatment: Invest in a quality program, *Chemical Engineering* 99:98-105.
5. Glater, J., York, J. L., and Campbell, K. S., 1980, in: Principles of Desalination Part B, 2nd Edition, Ed. K.S. Spiegler and A.D.K. Laird (Academic Press, New York) chp. 10. p. 627.
6. Reddy, M. M., and Nancollas, G. H., 1973, Calcite crystal growth inhibition by phosphonates, *Desalination* 12:61-73.
7. Nancollas, G. H., Kazmierczak, T. F., and Schuttringer, E., 1981, A controlled composition study of calcium carbonate crystal growth: The influence of scale inhibitors, *Corrosion-Nace* 37:76-81.
8. Reference [3] p. 197.
9. Glater, J., York, J. L., and Campbell, K. S., 1980, in: Principles of Desalination Part B, 2nd Edition, Ed. K.S. Spiegler and A.D.K. Laird (Academic Press, New York) chp. 10. p. 672.
10. Sikes, C. S., 1994, Mechanistic study of polyamino acids as antiscalants, *Corrosion 94*, paper no. 193.
11. Katz, J. L., et al., 1993, Calcite growth inhibition by iron, *Langmuir* 9:1423-1430.
12. Takasaki, S., et al., 1994, Calcite growth and the inhibiting effect of iron (III), *J. Crystal Growth* 143:261-268.
13. Sung, W., and Morgan, J. J., 1980, Kinetics and products of ferrous iron oxygenation in aqueous systems, *Environmental Science and Technology* 14:561-568.
14. Christoffersen, J., and Christoffersen, M. R., 1981, Kinetics of dissolution of calcium hydroxyapatite, *J. Crystal Growth* 53:42-54.
15. Giannimaras, E. K., and Koutsoukos, P. G., 1988, Precipitation of calcium carbonate in aqueous solutions in the presence of oxalate anions, *Langmuir* 4:855-861.
16. Reddy, M. M., and Wang, K., 1980, Crystallization of calcium carbonate in the presence of metal ions, *J. Crystal Growth* 50:470-480.

17. Zvyagintsev, O. E., and Lopatto, Y. S., 1962, Tetranuclear oxocarbonate-ferrate(III) complexes, *Russian Journal of Inorg. Chem.* 7:657-659.
18. Zaitsev, L. M., 1956, Carbonate compound of iron, *Zhur. Neorg. Khim. (Russ.J.Inorg.Chem.)* 1:2425-2428.

3

CARBONATE PRECIPITATION IN PYRAMID LAKE, NEVADA

Probable Control by Magnesium Ion

Michael M. Reddy

U.S. Geological Survey
3215 Marine Street, Boulder, Colorado, 80303

ABSTRACT

Magnesium ion inhibition of calcium carbonate (calcite) formation explains present-day controls on carbonate formation in Pyramid Lake. Concentrations of magnesium ion are sufficient to reduce calcium carbonate nucleation rates and calcite formation rates in present-day supersaturated lake water. Calcium carbonate nucleation and crystal growth measurements in the presence of magnesium ion are consistant with whole-lake whitings and carbonate mound formation in and around Pyramid Lake.

INTRODUCTION

Calcium carbonate mineralization occurs in hardwater lakes of the United States and elsewhere.[1-4] Pyramid Lake, Nevada, a present-day lake remnant of paleolake Lahontan, in contrast to other hardwater lakes, has unique physical and chemical characteristics which modify in-lake calcium carbonate mineralization (Figures 1 and 2).[5-8] Other high-salinity and high-alkalinity lakes in this region of the Great Basin are remnants of paleolakes Lahontan (Walker Lake) or Russell (Mono Lake) (Figure 1). Like Pyramid Lake, Walker Lake and Mono Lake exhibit unusual carbonate mineralization.

An ongoing research question involving carbonate mineralization reactions in surface waters, with significance in the areas of global carbon budgets and global warming, concerns the appearance of massive amounts of suspended calcium carbonate in surface waters (whitings).[9-13] Much concerning the nature of whitings remains uncertain. Pyramid Lake appears to be an appropriate location to investigate the mechanism of whiting formation and the water conditions that cause whitings to happen. Carbonate mineralization processes in specific lakes and lake basins depend strongly on a number of geochemical and biological factors. To investigate specific processes, such as calcium carbonate precipitation within a water column, it is useful to study lakes where these processes are well developed.

Mineral Scale Formation and Inhibition, Edited by Zahid Amjad
Plenum Press, New York, 1995

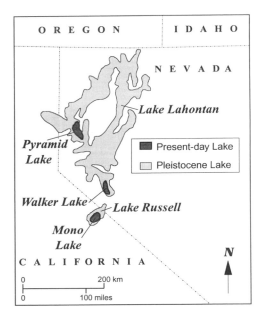

Figure 1. Remnant highly-saline, highly-alkaline lakes, Pyramid Lake, Walker Lake, and Mono Lake and the paleolakes Lake Lahontan and Lake Russell (Redrawn from Bishoff et al., 1993).

Figure 2. View of Pyramid Lake from the shore (Photograph by author, 1992). Characteristic large carbonate mounds are on the left and are well above the present lake level. A tree close to the present lake shore gives scale.

Figure 3. A cross-section of a carbonate mound along the shore of Pyramid Lake exhibiting the characteristic radial distribution of crystals found near the mound center. The key at the center of the photo is about one inch in diameter (photograph by author, 1992).

Calcium carbonate mineralization in Pyramid Lake differs from carbonate formation in less-saline, less-alkaline lakes in several respects, including:

- The occurrence of large carbonate mounds in the near-shore region (Figures 2 and 3).
- The formation of a less stable form of calcium carbonate (aragonite) rather than the thermodynamically stable polymorph calcite.[14]
- The occurence of massive calcium carbonate precipitation episodes, whitings, in the Pyramid Lake water column.[14]

Carbonate mineralization at Pyramid Lake, from analogy to laboratory studies of calcium carbonate formation, is regulated by specific chemical factors:

- Water column supersaturation,
- Carbonate mineral nucleation and crystal growth rates, and
- The presence of inhibitors such as magnesium and phosphate ions.[1,15,16,17,18,19,20,21]

The working hypothesis of this paper is that Pyramid Lake magnesium ion concentrations control calcium carbonate formation rates. Laboratory and in-lake calcium carbonate formation studies at Pyramid Lake are in progress to test this hypothesis. Field studies were initiated in the summer of 1992; laboratory studies of calcium carbonate formation are continuing.

PURPOSE

This report outlines calcium carbonate nucleation and crystal growth kinetics in the presence and absence of magnesium ion. From analogy with these laboratory measurements, it is reasonable to consider that carbonate nucleation (homogeneous nucleation from labile supersaturated lake water) and crystalization (crystal growth at a mineral surface from a metastable supersaturated solution)[22] are the basis for carbonate formation in the Pyramid Lake Basin. Nucleation leads to whitings; crystallization leads to carbonate mound formation.

SCOPE

Discussion in this report will be limited to laboratory studies dealing with the nucleation and crystal growth of calcium carbonate in the presence and absence of magnesium ion at 25°C. Summaries of Pyramid Lake water chemical analyses used here are from published data.[23,24] The average Pyramid Lake alkalinity calculated using the published data of Benson agrees well with the recently measured lake alkalinity.[25] A survey of carbonate deposits in the Pyramid Lake Basin is available.[25]

BACKGROUND

Calcium carbonate formation in a test tube or in a lake is initiated by nucleation, often involving formation of unstable polymorphs and hydrates.[22] Unstable calcium carbonate polymorphs will ultimately transform to calcite, the thermodynamically stable form of calcium carbonate at standard temperature and pressure. Unstable polymorphs may persist in solution for extended intervals in the presence of substances (termed "inhibitors") which inhibit the formation of a thermodynamically more-stable phase. Typical inhibitors include magnesium and phosphate ions, both of which reduce calcite formation rates.[22] Solution temperature influences formation of unstable calcium carbonate polymorphs and their subsequent transformation. Calcite forms at 25 °C in inhibitor-free solutions with moderate supersaturations (supersaturation values of 10 to 20). Calcite formation in the presence of magnesium ion enhances the formation of unstable calcium carbonate polymorphs. Natural waters which have unusual solution compositions (i.e., high concentrations of magnesium or phosphate ions) often form unstable calcium carbonate polymorphs. Calcim carbonate nucleation rates differ among polymorphs and hydrates; nucleation rates increase dramatically with increasing calcium carbonate solution supersaturation (Figure 4.).

CALCIUM CARBONATE NUCLEATION

A convenient measure of calcium carbonate nucleation is the time necessary to form solid carbonate from a supersaturated solution. This time, referred to as "induction time", is measured by changes in solution light scattering and composition. Nucleation is often difficult to characterize because of the presence of heterogeneous nucleation sites[22] and because growth often accompanies nucleation.

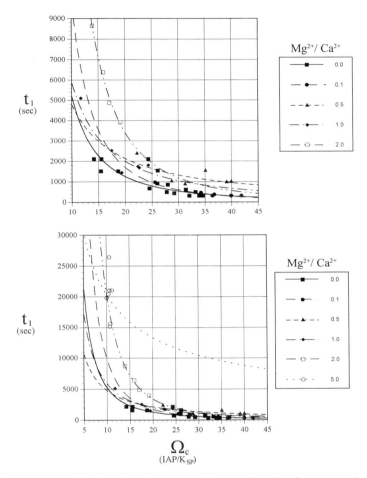

Figure 4. Calcium carbonate induction time, in seconds, plotted as a function of supersaturation for a solution containing 4 millimolar total carbonate ion concentration and a range of solution magnesium ion to calcium ion concentration ratios.[16] A. Induction times for solutions containing magnesium ion to calcium ion concentration ratios from 0 to 2. B. Induction times for solutions containing magnesium ion to calcium ion concentration ratios from 0 to 5.

Solution supersaturation, the solution variable directly controlling nucleation, is defined as the ratio of the calcium carbonate ion activity product to the calcium carbonate (calcite) solubility product (details of the calculation are discussed in reference 16). Nucleation rate is inversely proportional to the induction time; as induction time increases, nucleation rate decreases.

As supersaturation values decrease to about 10, the time for calcium carbonate nucleation becomes large, and the nucleation rates become very small (Figure 4). Solutions with supersaturations less than 10 are termed "metastable". This designation implies that carbonate minerals will not form in such solutions without the presence of a suitable growth surface or substrate. In the presence of a carbonate surface, calcium carbonate may precipitate from a metastable solution. Carbonate mineral formation will continue until the solution reaches equilibrium (i.e., calcite saturation). In Pyramid Lake, carbonate mineral growth from metastable supersaturated solution formed large carbonate mounds (Figures 2 and 3).

When Pyramid Lake water supersaturtion values are above the metastablity limit, nucleation of calcium carbonate is expected to occur rapidly in the water column (Figure 4).

Magnesium ion controls, in part, calcium carbonate induction times and nucleation. Induction times increase as the solution magnesium ion concentration and magnesium-to-calcium ion ratio increase (Figure 4). Average Pyramid Lake magnesium ion concentration (about 4×10^{-3} moles per liter, and magnesium-to-calcium ion ratio about 20) is sufficient to increase the calcium carbonate induction time and to reduce calcium carbonate nucleation (Figure 4). For supersaturation of 25, changing solution magnesium to calcium ratios from 0-1 roughly doubles the induction time. At lower supersaturations (i.e., about 20) or at higher magnesium-to-calcium ion ratios, induction times were about three times that of the magnesium free solution. Elevated lakewater magnesium concentration increases carbonate induction times, reducing nucleation rates. Removal of the magnesium ion from lake water due to calcium carbonate nucleation (as a mixed calcium-magnesium carbonate) may increase the nucleation rates.

Magnesium ion content and polymorphic composition of nucleated calcium carbonate has been reported elsewhere.[16] Briefly, the predominate crystalline components of the precipitate, over a range of magnesium ion concentrations and supersaturations, were vaterite and aragonite. Formation of calcite from the metastable solutions was retarded by the magnesium ion. Some aged solutions containing low concentrations of magnesium ion developed calcite as a final precipitate. In these experiments, vaterite (which was initially formed) was transformed to calcite during the course of the experiment. Magnesium ion content of the nucleated carbonate phase was a function of nucleation and ripening conditions.[16] Magnesium ion content was greatest for the initial precipitate. As the precipitate aged, its magnesium ion content decreased.[16] Nucleation sensativity to a solution calcium carbonate supersaturation has confounded past attempts to predict the onset and duration of whitings in Pyramid Lake.[14] Supersaturation is currently (1994) below the critical value for nucleation (10) except during summer (Figure 5). Supersaturations were calculated for a range of temperatures and calcium ion concentrations with WATEQ4F.[26] Temperatures were varied from 0 to 25°C for both calcite and aragonite supersaturation (Figure 5). Maximum supersaturation for calcite (at 25°C) is above 10 while that for aragonite is less than 10. In the absence of magnesium ion, calcite would nucleate and grow in present-day Pyramid Lake water. However, since calcite formation is inhibited, lake supesaturations can increase to about 10. Lake whitings would be expected to occur only when the supersaturation with respect to aragonite is above 10.

CALCIUM CARBONATE (CALCITE) CRYSTAL GROWTH

A seeded growth method has been used to characterize the rate of calcite formation from metastable supersaturated solutions. Crystal growth data has been interpreted using a parabolic rate equation which relates the crystal growth to the product of three terms: 1) the growth surface available for reaction; 2) the crystal growth rate constant; and 3) the square of the reaction driving force.[16,22] This rate law, which has a parabolic reaction driving force term indicating that the rate determining step is a surface reaction, allows a convenient representation of inhibitor ions influence on calcite formation rates.[16,22] During carbonate mound formation, the initial calcium carbonate polymorphs and their subsequent transformation are uncertain. Also, lake water and inflow spring water chemical compositions at the time of mound formation are not known. In this report, only the growth of calcite is considered. Additional studies of other polymorphs and hydrates are important for the application of lake deposits in paleoclimate reconstruction partly because isotope fractionation factors may be regulated by the polymorphic composition of the carbonate deposit.

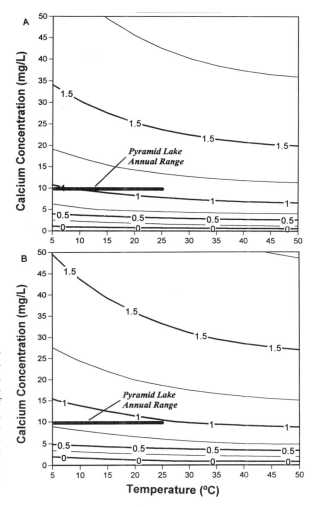

Figure 5. Saturation indices for Pyramid Lake, Log (Ion Activity Product/Solubility Product) calculated with WATEQ4F.[26] Supersaturation isopleths (in log units) are plotted as a function of lake water calcium concentration and temperature. The Pyramid Lake temperature range was 0 to 25°C. A) Saturation indices calculated with respect to calcite as the mineral phase. B) Saturation indices calculated with respect to aragonite as the mineral phase.

CALCIUM CARBONATE FORMATION INHIBITORS

At a fixed calcium carbonate supersaturation level and temperature, inhibitor ion concentration levels regulate carbonate mineral polymorphic compositions and formation rates (Figure 6).[15,27]

Following nucleation, or in the presence of a suitable growth surface, calcium carbonate minerals form in supersaturated solution. Unlike calcite crystal growth on pure calcite seed crystals in magnesium-free solutions, calcite growth in solutions containing 10^{-3} moles per liter magnesium is dramatically reduced (Figure 6). As shown in Figure 6, for the solution containing 1 millimolar magnesium ion, neither solution pH nor calcium ion concentration decrease during the experiment. In the absence of magnesium ion, the solution rapidly approaches calcite equilibrium. Crystal growth is inhibited more than crystal nucleation by millimolar magnesium ion concentrations (compare Figure 6 with Figure 4), presumably because nucleation involves several polymorphs, only some of which are sensitive to magnesium ion growth inhibition.

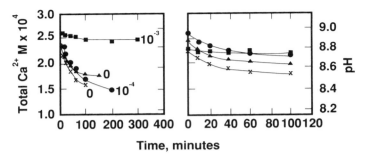

Figure 6. Total calcium concentration and pH of the supersaturated solution plotted as a function of time for seeded crystal growth in the presence and absence of added magnesium ion. Molar concentrations of added magnesium ions are indicated at the end of the curves.[18]

Inhibition of calcite growth by magnesium ion would be expected to occur in Pyramid Lake. The parabolic growth rate expression[16] describes the effect of magnesium ion concentration levels on calcite formation rates. Rate data for calcium carbonate seeded growth in the presence and absence of magnesium ion plotted as the integrated rate function versus time illustrates calcite growth rate reduction (Figure 7) by magnesium ion. Line slopes in Figure 7 are proportional to the calcite growth rate constant; the decrease in slope with increasing magnesium concentration demonstrates calcite growth rate reduction in solutions containing 10^{-3} moles per liter magnesium ion. The Pyramid Lake average magnesium ion concentration is four times the highest magnesium ion concentration shown in Figure 7, and would thus be expected to strongly inhibit calcite growth. The carbonate polymorphs found in Pyramid Lake surface sediments and water is aragonite, not calcite.[28]

Calcite crystal growth reduction caused by magnesium ion is shown for a range of magnesium concentrations by plotting the reduced rate constant (k/k_o) versus magnesium

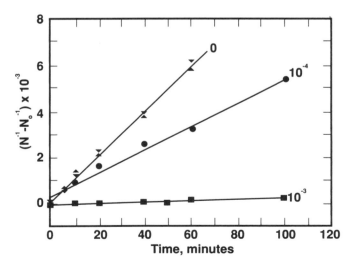

Figure 7. Plots of the integrated rate function versus time for growth of calcite seed crystals in the presence and absence of magnesium ions. Molar concentrations of added magnesium ions are indicated at the end of each line.[18]

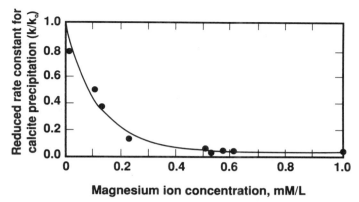

Figure 8. Calcite crystallization reduced-rate constant (k/k_o) as a function of the concentration of magnesium ions in solution.[16]

ion concentration (Figure 8). The calcite reduced growth rate constant (the ratio of the rate constant in the presence and absence of magnesium ion) decreases with increasing magnesium concentration to very low values at magnesium concentrations greater than one millimolar.

It is not clear from laboratory studies how locally elevated supersaturation, developed by high calcium concentration spring water inputs and their subsequent modification by mixing with lake water, modifies carbonate growth inhibition. At present (1994), one location at the north shore of Pyramid Lake exhibits carbonate formation.[29] Lake water magnesium ion concentrations stabilize elevated calcium concentrations and supersaturations allowing mineralization where spring inputs cause locally elevated supersaturation.

Most springs along the shore of Pyramid Lake are not at present associated with carbonate precipitation. Hot springs at the north end of Pyramid Lake have high calcium and low magnesium and alkalinity concentrations.[25] Calculations suggest that mixing of low-alkalinity, calcium rich spring water with Pyramid Lake water leads to non-linear variation in supersaturation as a function of the mixing ratios. Supersaturation of some mixtures is above that of lake water, suggesting that maximum supersaturation and mineralization occurs at the lake water / spring water mixing zone. Mixing zone precipitation may lead to detailed crystal morphology present in Pyramid Lake carbonate mounds (Figure 2).

MECHANISM OF MAGNESIUM ION INHIBITION OF CALCITE FORMATION

In studies of natural phenomena, such as calcium carbonate formation inhibition in lakes, it is useful to characterize the process mechanism. Magnesium ion inhibition of calcite crystal growth has been interpreted in terms of a Langmuir adsorption isotherm model (Figure 9). The Langmuir model assumes that inhibitor ions adsorb at crystal growth sites on the calcite suface.[15,16,27] This model may also apply to magnesium ion calcite crystal

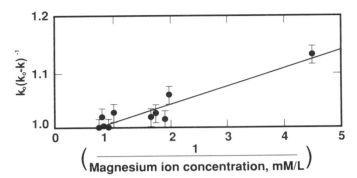

Figure 9. Langmuir isotherm plot of $k_o/(k_o-k)$ as a function of the reciprocal of the added magnesium ion concentration in solution for calcite seeded crystallization.[16]

growth inhibition in present-day Pryamid Lake. The Langmuir model for phosphate and glycerophosphate[15] indicated that the current Pyramid Lake phosphate concentration (about one micromolar at the sediment-water interface[25]) will reduce the calcite crystal growth rate constant only slightly. Although phosphate is an effective calcite growth inhibitor, its concentration in Pyramid Lake surface water is too low to impact calcite reaction rates. Pyramid Lake sediment phosphate concentrations are higher (up to 0.5 millimolar) and will inhibit calcite crystal growth.[25] Mono Lake carbonate mineralization is regulated by elevated water column phosphate concentrations.[24]

IN-LAKE CALCIUM CARBONATE CRYSTALLIZATION RATES

A test of calcium carbonate formation inhibition in Pyramid Lake, using an in-lake crystal growth experiment, was begun in the summer of 1992. This procedure measures calcite and aragonite crystal growth rates for single crystals suspended in Pyramid Lake.[30] Initial results from August and September 1992 (no growth in the supersaturated lake water) suggests the importance of crystal growth inhibitors, such as magnesium ion, in reducing calcium carbonate formation from Pyramid Lake onto a carbonate surface. Field observations demonstrated carbonate deposition on an artificial substrate at a site of spring water input (Micaela B. Reddy, personal communication, 1993). A can was observed to be fully encrusted with a mineral deposit. At this location carbonate mineralization occurred during the last few years. Additional laboratory and field experiments are in progress to evaluate the role of magnesium ion inhibition of carbonate formation in Pyramid Lake.

CONCLUSIONS

Magnesium ion inhibits calcium carbonate nucleation and calcite crystal growth. Calcium carbonate formation inhibition by magnesium ion concentrations in Pyramid Lake water (about 4×10^{-3} moles per liter) at the maximum summer lake supersaturation values (about supersaturations of 10) is consistent with laboratory data.

Calcium carbonate nucleation is reduced by millimolar magnesium ion concentrations. Induction times, compared to magnesium free solutions, increase in solutions containing millimolar magnesium ion. Whitings in Pyramid Lake result from calcium carbonate nucleation and are influenced by water column supersaturation and also perhaps by magnesium ion concentration levels. Massive lake whitings, observed during the 1980's, indicate that at that time supersaturations were well above 10. Elevated Pyramid Lake supersaturations may reflect high calcium ion loads transported to the lake during peak discharge. Carbonate nucleation studies of lake water may have application in characterizing and predicting Pyramid Lake whiting episodes. Continued research in this area is warrented.

Present-day Pyramid Lake maximum supersaturation occurs in late summer, corresponding to the maximum lake water temperature, and is below the value necessary for nucleation. An increase in maximum summer supersaturation appears to be necessary to cause whiting episodes now (1994). Whitings in other large lakes (i.e., the North American Great Lakes) occur at lower supersaturations, presumably because inhibitor concentrations in these lakes are lower than in Pyramid Lake.[3] Crystal growth kinetic characterization is the key to understanding carbonate formation in and about Pyramid Lake, and at other Great Basin Lakes. Continued research on the influence of inhibitors on carbonate mineralization reaction rates is needed.

ACKNOWLEDGMENT

The author acknowledges the help and encouragement of the Pyramid Lake Paiute Tribal Council and the staff of the Pyramid Lake Fisheries Resources Laboratory during the course of this work. Larry Benson, of the U.S. Geological Survey, Boulder, Colorado, first asked the author to study carbonate formation processes in the Great Basin Lakes. Field work was done with the assistance of L. Benson, S. Charlton, C. Reddy, J.A. Reddy, M.B. Reddy, and P. Schuster. Computer modelling was done by S. Charlton. Word processing was done by C. Gunther. Reviews by L. Benson and T. McConnaughey were most helpful in relating field studies to laboratory investigations.

REFERENCES

1. Reddy, M.M., 1975, "Kinetics of Calcium Carbonate Formation", Proceedings of the International Association of Theoretical and Applied Limnology, v 19, p. 429-438.

2. Johnson, D.L., Jiao, J., DosSantos, S.G., and Effler, S.W., 1991, "Individual particle analysis of suspended materials in Onondaga Lake, New York", Environmental Science and Technology, v 25, p. 736-743.

3. Strong, A.E. and Eadie, B.J., 1978, "Satellite observations of calcium carbonate precipitation in the Great Lakes", Limnology and Oceanography, v 23, p. 877-887.

4. Dean, W.E. and Fouch, T.D., 1983, Lacustrine Environment, in Scholle, P.A., Bebout, D.G., and Moore, C.H., eds., Carbonate depositional environments, Tulsa, The American Assoication of Petroleum Geologists, p. 98-121.

5. Galat, D.L., Lider, E.L., Vigg, S., and Robertson, S.R., 1981, "Limnology of a large, deep, North American terminal lake, Pyramid Lake, Nevada, U.S.A", *Hydrobiologia*, v 82, p. 281-317.

6. Hamilton-Galat, K. and Galat, D.L., 1983, "Seasonal variation of nutrients, organic carbon, ATP, and microbial standing crops in a vertical profile of Pyramid Lake, Nevada", *Hydrobiologia*, v 105, p. 27-43.

7. Benson, L.V., Currey, D.R., Dorn, R.I., Lajoie, K.R., Oviatt, C.G., Robinson, S.W., Smith, G.I., and Stine, S., 1990, "Chronology of expansion and contraction of four Great Basin lake systems during the past 35,000 years*"*, *Palaeogeography Palaecolimate Paleoecology*, v 78, p. 241-286.

8. Benson, L.V., 1993, "Factors affecting C-14 ages of lacustrine carbonates: Timing and duration of the Last Highstand Lake in the Lahontan Basin", *Quaternary Research*, v 39, p. 163-174.

9. Robbins, L.L. and Blackwelder, P.L., 1992, "Biochemical and ultrastructural evidence for the origin of whitings: A biologically induced calcium carbonate precipitation mechanism", *Geology*, v 20, p. 464-468.

10. Shinn, E.A., Steinen, R.P., Lidz, B.H., and Swart, P.K., 1989, "Perspectives: Whitings, a sedimentologic dilemma", *Journal of Sedimentary Petrology*, v 59, p. 147-161.

11. Milliman, J.D., Freile, D., Steinen, R.P., and Wilber, R.J., 1993, "Great Bahama Bank Aragonitic Muds - Mostly Inorganically Precipitated, Mostly Exported", *Journal of Sedimentary Petrology*, v 63, p. 589-595.

12. Boss, S.K. and Neumann, A.C., 1993, "Physical Versus Chemical Processes of 'Whiting' Formation in the Bahamas", *Carbonate Evaporite*, v 8, p. 135-148.

13. Friedman, G.M., Robbins, L.L., and Blackwelder, P.L., 1993, "Biochemical and ultrastructural evidence for the origin of whitings: A biologically induced calcium carbonate precipitation mechanism", Comment and Reply: *Geology*, v. 21, p. 287-288.

14. Galat, D.L. and Jacobsen, R.L., 1985, "Recurrent aragonite precipitation in saline-alkaline Pyramid Lake, Nevada", *Archives of Hydrobiology*, v 105, p. 137-159.

15. Reddy, M.M., 1978, "Kinetic inhibition of calcium carbonate formation by wastewater constituents", in Rubin, A.J., ed., The chemistry of wastewater technology, Ann Arbor, Michigan, Ann Arbor Science Publishers, Inc., p. 31-58.

16. Reddy, M.M., 1986, "Effect of magnesium ion on calcium carbonate nucleation and crystal growth in dilute aqueous solutions at 25° celsius, in Studies in Diageneses, Denver, Colorado, U.S. Geological Survey", Bulletin 1578, p. 169-182.

17. Reddy, M.M. and Nancollas, G.H., 1976, "The crystallization of calcium carbonate IV. The effect of magnesium, strontium and sulfate ions" *Journal of Crystal Growth*, v 35, p. 33-38.

18. Reddy, M.M. and Wang, K.K., 1980, "Crystallization of calcium carbonate in the presence of metal ions. I. Inhibition by magnesium ion at pH 8.8 and 25°C", *Journal of Crystal Growth*, v 50, p. 470-480.

19. Reddy, M.M. and Gaillard, W.D., 1981, "Kinetics of calcium carbonate (calcite)-seeded crystallization: The influences of solid-solution ratio on the reaction constant", *Journal of Colloid Interface Science*, v 80, p. 171-178.

20. Nancollas, G.H. and Reddy, M.M., 1971, "The crystallization of calcium carbonate II. Calcite growth mechanism", *Journal of Colloid Interface Science*, v 37, p. 824-830.

21. Reddy, M.M., Plummer, L.N., and Busenberg, E., 1981, "Crystal growth of calcite from calcium bicarbonate solutions at constant pCO_2 and 25 degrees C: A test of a calcite dissolution model", *Geochimica et Cosmochimica Acta*, v 45, p. 1281-1289.

22. Reddy, M.M., 1988, "Physical-chemical mechanisms that affect regulation of crystallization", in Sikes, C.S. and A.P. Wheeler, eds., Chemical aspects of regulation of mineralization, Proceedings of a symposium sponsored by the Division of Industrial and Engineering Chemistry of the American Chemical Society, Mobile, University of South Alabama Publications Services, p. 1-8.

23. Benson, L.V., 1984, "Hydrochemical data for the Truckee River drainage system, California and Nevada", U.S.Geological Survey, Open-File Report, v 84-440, p. 1-35.

24. Bischoff, J.L., Fitzpatrick, J.A., and Rosenbauer, R.J., 1993, "The solubility and stabilization of ikaite ($CaCO_3*6H_2O$) from 0° to 25°C: Environmental and paleoclimatic implications for Thinolite Tufa", *Jour. Geology*, v 101, p. 21-33.

25. Benson, L.V., 1994, "Carbonate Deposition, Pyramid Lake Subbasin, Nevada. 1. Sequence of Formation and Elevational Distribution of Carbonate Deposits (Tufas)", *Palaeogeography Palaecolimate Paleoecology*, v 109, p. 55-87.

26. Ball, J.W. and Nordstrom, D.K., 1991, "User's manual for WATEQ4F, with revised thermodynamic data base and test cases for calculating speciation of major, trace, and redox elements in natural waters", U.S.Geological Survey, Open-File Report, v 91-183, p. 1-189.

27. Reddy, M.M., 1977, "Crystallization of calcium carbonate in the presence of trace concentrations of phosphorus-containing anions I. Inhibition by phosphate and glycerophosphate ions at pH 8.8 and 25° C", *Journal of Crystal Growth*, v 41, p. 287-295.

28. Reddy, M.M., Unpublished.

29. Reddy, M.B., Personal Communication, 1994.

30. McConnaughey, T.A., LaBaugh, J.W., Rosenberry, D.O., Striegl, R.G., Reddy, M.M., Schuster, P.F., and Carter, V., 1994, "Carbon budget for a groundwater-fed lake: Calcification supports summer photosynthesis", *Limnology and Oceanography*, v 39, p. 1319-1332.

4

THE USE OF MODERN METHODS IN THE DEVELOPMENT OF CALCIUM CARBONATE INHIBITORS FOR COOLING WATER SYSTEMS

R. V. Davis, P. W. Carter, M. A. Kamrath, D. A. Johnson, and P. E. Reed

Nalco Chemical Company
One Nalco Center, Naperville, Illinois 60563

INTRODUCTION

Cooling water systems represent the single largest area of industrial water use. Most of these systems operate by evaporative cooling, allowing the system water to become supersaturated with respect to dissolved mineral salts. In such a cooling system, cool water is pumped across metal heat exchange tubes which are in contact with a hot process. The heated water then enters a cooling tower and, through evaporation, becomes cooled. The remaining cool water reenters the system to repeat the process. Since heated system water is lost through evaporation, the concentration of dissolved solids in the system water increases. As a result, the cooling water can become highly supersaturated with respect to mineral salts. Calcium carbonate, for example, can deposit on the heat transfer surfaces of the cooling system, thus lowering heat transfer efficiency, increasing pumping costs and requiring frequent system cleaning. Substoichiometric deposition control chemicals have long been the most cost effective solution to the problem of mineral salt deposition.[1] The addition of a small amount of deposit treatment chemical can have a dramatic effect on the formation of mineral scales. Continuous improvement of inhibitor chemistries is necessary due to constantly progressing technical, financial, and environmental treatment requirements. A key to the efficient development of acceptable calcium carbonate control chemistries is an understanding of the influence of inhibitor structural parameters on the fundamental processes involved in scale formation.[2]

This paper will examine some of the methods used to study calcium carbonate deposit formation and to determine the modes of action of the chemical inhibitors. The inhibitors which will be discussed are shown in Figure 1. The use of time lapse video microscopy to investigate the fundamental processes involved in calcium carbonate deposit formation will be discussed. Studies of overall calcium carbonate formation kinetics in the presence of inhibitor using pH static precipitation techniques will be presented. *In situ* crystal growth experiments using in situ atomic force microscopy and constant composition methods will

Mineral Scale Formation and Inhibition, Edited by Zahid Amjad
Plenum Press, New York, 1995

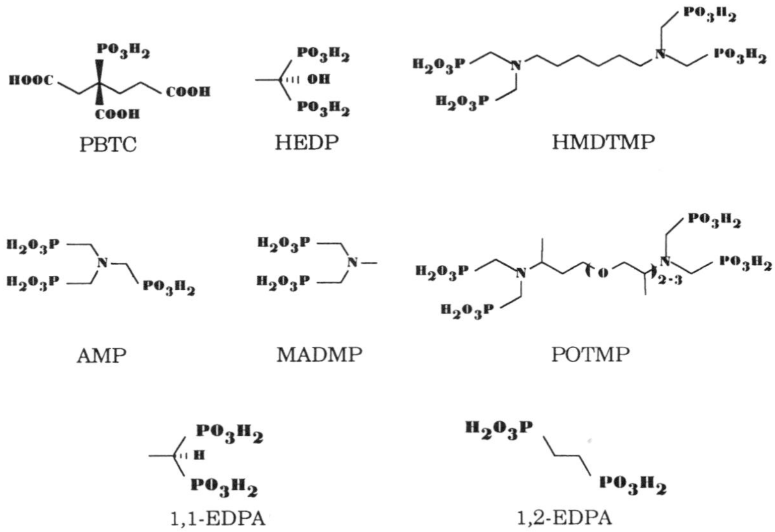

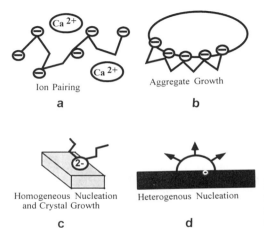

Figure 1. Molecular structures of representative phosphonate inhibitors: PBTC (2-phosphonobutane-1,2,4-tri-carboxylic acid), HEDP (1-hydroxyethylidene-1,1-diphosphonic acid), HMDTMP (hexamethylene-N, N, N′, N′-diamine tetramethylenephosphonic acid), AMP (aminotrimethylenephosphonic acid), MADMP (methylaminodimethylenephosphonic acid), POTMP (polyoxypropylenediamine-N, N, N′, N′-tetramethylenephosphonic acid), 1,1-EDPA (ethane-1,1-diphosphonic acid), and 1,2-EDPA (ethane-1,2-diphosphonic acid).

be presented. Emphasis will be placed on the use of these methods to develop an understanding of the influence of chemical structure on calcium carbonate control. The relationship between calcium carbonate control performance and molecular size, connectivity, and functionality will be discussed.

MECHANISM OF DEPOSIT FORMATION

The mechanism of scale formation is complex, with many parallel processes contributing to mineral deposition.[3] Some of the more important fundamental processes

Figure 2. Schematic representation of processes involved in initial stages of scale formation: (a) ion pairing, (b) prenucleation aggregate growth, (c) and (d) particle nucleation and growth.

involved in scaling are shown schematically in Figures 2 and 3. The kinetic pathway toward mineral scale deposition begins with electrostatic interactions between dissolved anions and cations resulting in ion pairing. Ion pairs are the building blocks of larger molecular assemblies or prenucleation aggregates. Molecules which can complex either constituent of the ion pair are termed chelating agents and can disrupt ion pair formation (Figure 2a). The ion pair concentration increases at higher supersaturations and can aggregate to form larger particles. The aggregates are in dynamic equilibrium with the solution, and polyelectrolytes can adsorb to prenucleation aggregates and influence the growth and dissolution kinetics (Figure 2b).

The free energy of a particle nucleation event is dependent upon aggregate size. The aggregate size determines the volume and surface free energy contributions to the overall free energy. When an aggregate grows to some critical size (driven by supersaturation) nucleation of a solid state particle occurs.[4] These larger, nucleated precipitates do not readily redissolve. "Homogeneous" nucleation refers to particles nucleated in the bulk solution (Figure 2c), while nuclei generated at a surface are formed by "heterogeneous" nucleation (Figure 2d). In actuality, both types of nucleation events probably occur at a heterogeneous interface, but isolating the nucleation site (dust, impurity, etc.) in bulk solution is quite difficult.

Once particles are nucleated, either at a macroscopic surface or in solution, there are several potential deposit growth mechanisms.[5] The most important physical processes which contribute to deposit formation are:

- Direct ion incorporation into the solid state lattice
- Nucleation at the surface of a crystallite
- Particle agglomeration
- Particle adsorption onto a heat transfer surface

In addition to these processes, solid state phase transitions which occur during precipitation will influence the kinetics of the processes described above (Figure 3c). At high supersaturation ratios (>5), the nucleated phases may not be thermodynamically favored. For example, calcium carbonate monohydrate is not the thermodynamically favored phase under most solution conditions. However, if the kinetics of ion dehydration are slow, a hydrated calcium carbonate phase may be nucleated. This is a simple illustration of a kinetically controlled nucleation event. Under most cooling water conditions, the calcium carbonate particles nucleated tend to be amorphous with some hydration initially. Phase transitions to more thermodynamically favored anhydrous phases such as calcite or aragonite occur over time.

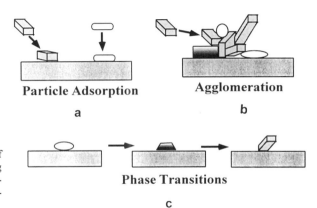

Particle Adsorption   **Agglomeration**

a   b

Figure 3. Schematic representation of post-nucleation processes contributing to scale deposition: (a) particle adsorption, (b) agglomeration, (c) phase transitions.

Phase Transitions

c

One would like to follow the complex crystallization kinetics by characterizing the individual processes. The methods described below represent current analytical techniques for evaluating calcium carbonate precipitation under conditions relevant to cooling water applications.

TIME LAPSE VIDEO MICROSCOPY

Time lapse video microscopy (TLVM) has been developed as an in situ characterization technique for calcium carbonate scaling phenomena in simulated cooling water systems. Nucleation, growth, agglomeration, and adsorption processes are observed and quantified directly using TLVM. Images of a glass heat exchanger in a simulated cooling system depicts the early stages of a deposition event (Figure 4a). In this experiment, the calcium carbonate supersaturation was increased by slow addition of hardness and alkalinity over time until a nucleation event occurred. Optical microscopy characterizes particle sizes, morphologies, and frequencies during scale deposition. The dynamics are captured on video as illustrated in the image obtained from the same location two hours later. Figure 4b depicts multiple crystal outgrowths from a central calcium carbonate agglomerate. In addition, smaller particles from solution have adsorbed/agglomerated around the edges of the central growth. The smaller particles have a lower attachment energy and easily desorb under the

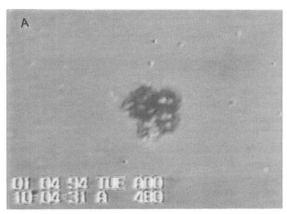

Figure 4. Time lapse video micrographs of calcium carbonate nucleation and growth at a glass heat transfer surface in the absence of inhibitor. The images depict two hours of growth at a central cluster followed by adsorption of smaller particles at the edges of the cluster. The full horizontal scale is 75 microns.

shear flow conditions of the experiment. The full field of horizontal view in these images is 75 microns.

Figures 5a and 5b display the results of a similar process simulation performed in the presence of 20 ppm of an organic phosphonate inhibitor. Nucleation occurs first at a preexisting calcium carbonate particle about 5 microns in diameter (center of Figure 5a). Within twenty minutes of the initial nucleation event, multiple nucleation events are observed uniformly across the 75 micron field of view. The glass surface becomes decorated with calcium carbonate particles 3-10 microns in diameter. The calcium carbonate nucleation frequency, growth rate, and particle morphology are dramatically altered in the presence of an organic inhibitor molecule.

Although the previous two experiments present phenomena occurring on a glass surface, the heat exchangers of cooling water systems are typically made of carbon steel, stainless steel, copper or copper alloy. Since the surface has an influence on heterogeneous nucleation events, a change in surface structure and composition could have a significant impact on the process of deposit formation. Time lapse video microscopy experiments performed using a copper heat exchange surface indicate that calcium carbonate deposition at the copper surface is mechanistically different from deposition on glass. One such experiment performed under the same conditions as the process simulation example of Figure 5 except on copper is summarized in Figure 6. Figure 6a shows the copper surface prior to the formation of any deposit. In addition to unique surface functionalities and interfacial chemistry, the copper surface possesses a much larger surface roughness than glass. When

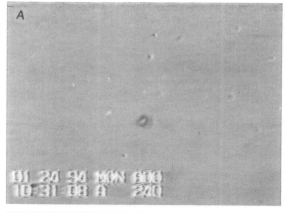

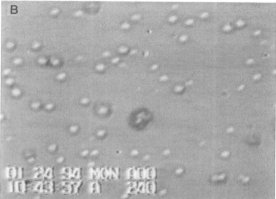

Figure 5. Time lapse video micrographs of calcium carbonate nucleation and growth at a glass heat transfer surface in the presence of inhibitor. Nucleation occurs at a central preexisting particle, followed by uniform decoration of the glass heat transfer surface. The images are separated by 14 minutes and the full horizontal scale is 75 microns.

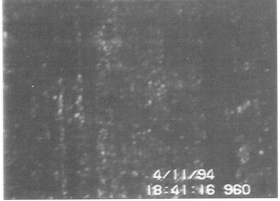

Figure 6. Time lapse video micrographs of calcium carbonate nucleation and growth at a copper heat transfer surface in the presence of inhibitor. Top, surface is rougher than that of the glass and clean of calcium carbonate. Bottom, within twenty minutes the surface has become completely covered by a uniform submicron particle calcium carbonate deposit. The full horizontal scale is 100 microns.

scale formation occurs in the presence of copper, shown in Figure 6b, the particle size is much smaller and the number of surface nucleation events is much higher. Within twenty minutes of the initial nucleation event, the entire surface has been uniformly covered with calcium carbonate. No signs of particle adsorption or agglomeration were observed. The unique nucleation sites and interfacial chemistry of a copper surface cause fundamental changes in the calcium carbonate scale deposition process.

pH STATIC PRECIPITATION

Static pH precipitation techniques are the most common experiments used to study calcium carbonate formation in the presence of inhibitors. Ion pair formation, aggregation, nucleation, crystal growth, agglomeration and phase transitions, all processes key to the deposition of calcium carbonate, occur during pH static calcium carbonate precipitation. By observing the changes in calcium and inhibitor concentration over time, such experiments provide a measure of the kinetics of overall calcium carbonate formation in the presence of an inhibitor.

The results of a pH static precipitation experiment performed using PBTC (2-phosphonobutane-1,2,4-tricarboxylic acid) are shown in Figure 7. Note that in the absence of inhibitor rapid precipitation of calcium carbonate takes place with the nearly complete loss

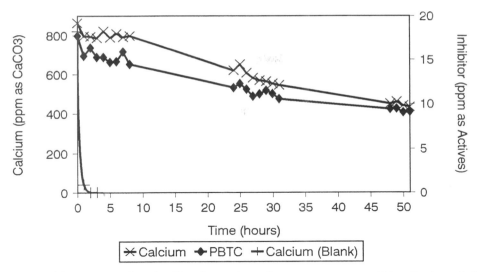

Figure 7. Calcium concentration of a pH static experiment decreases to near zero within a few hours in the absence of inhibitor. In the presence of PBTC, calcium and inhibitor concentrations decrease slowly over a 52 hour precipitation time.

of soluble calcium within a few hours. In the presence of PBTC, however, the rate of calcium carbonate formation is drastically lower and two distinct kinetic regions of calcium carbonate formation are evident. Filterable particles appear within the first hour, after which calcium carbonate precipitation and growth takes place very slowly over the next several hours. During this slow formation stage the rate of inhibitor consumption is greater than the rate of soluble calcium loss. After 8 to 22 hours the rate of calcium carbonate formation increases and remains relatively constant up to the end of the experiment. During this second stage of calcium carbonate formation the inhibitor depletion rate is somewhat lower than that of the first few hours. These two different kinetic regions reflect the differences in the fundamental processes which are dominant at different stages of calcium carbonate precipitation inhibited by PBTC.

The calcium depletion profiles for several other inhibitors are displayed in Figure 8. The data for PBTC is included for comparison. Although not shown, in each case the inhibitor depletion profile is well correlated to the loss of soluble calcium. Although the four selected inhibitors contain many similarities, there are significant differences between their action at both early and later stages of calcium carbonate formation. For example, PBTC and polyoxypropylenediamine-N, N, N', N'-tetramethylenephosphonic acid (POTMP)[6] exhibit a very slow formation region over the first eight hours of the experiment whereas hexamethylene-N, N, N', N'-diaminetetramethylenephosphonic acid, (HMDTMP) and the phosphonomethylated polymer (Phos polymer) display relatively constant calcium carbonate formation rates throughout the experiment. The most dramatic change in the rate of calcium carbonate formation occurs in the presence of POTMP. In the case of this inhibitor, a massive calcium carbonate formation (nucleation and/or growth) event occurred between 8 and 24 hours after initiation of the experiment. Approximately 75% of the soluble calcium has been removed after this change in precipitation kinetics. Clearly, the dominant process occurring at this second stage of calcium carbonate formation cannot be controlled by POTMP. Unfortunately, static pH precipitation studies are unable to reveal the nature of the funda-

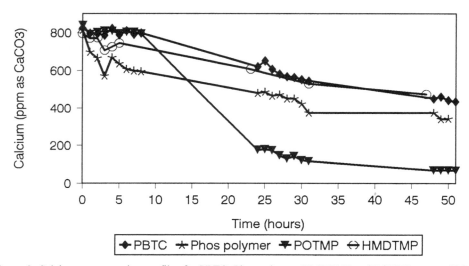

Figure 8. Calcium concentration profiles for PBTC, Phos polymer, HMDTMP and POTMP over a 52 hour precipitation time. The unique shape of the POTMP depletion profile compared to the other inhibitors demonstrates the time dependence of inhibition activity. The POTMP inhibitor has the lowest soluble calcium level at the end of the trial.

mental calcium carbonate precipitation process occurring at the two kinetic regions for these inhibitors.

CALCITE CRYSTAL GROWTH

While pH static methods allow the evaluation of overall calcium carbonate inhibitor performance, they reveal little mechanistic information about the processes occurring during separate kinetic regimes of precipitation. Techniques which isolate individual processes are required to separate the inhibitor's influence on each kinetic step of the deposition pathway.

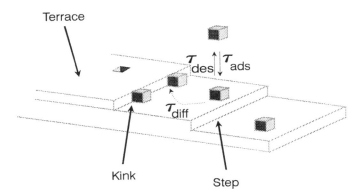

Figure 9. Schematic representation of ordered crystal growth depicting terrace, step, and kink surface sites. Adsorption, desorption, and surface diffusion rates influence the observed rate of crystal growth.

Atomic force microscopy and the constant composition method are examples of techniques which have been used for isolating crystal growth and crystal growth inhibition.

ATOMIC FORCE MICROSCOPY (AFM)

Calcite crystal growth, shown schematically in Figure 9, can be characterized *in situ* on submicron length scales using AFM. The crystal grows by incorporation of cations and anions (or ion pairs) to surface sites. Typically, adsorption of solution species onto the surface is followed by some combination of surface diffusion, dehydration and incorporation into the crystal lattice. Generally, surface sites such as step edges and kinks are the most favorable positions for the incorporation of adsorbed ions into the crystal lattice.[7]

Using atomic force microscopy it is possible to observe calcite crystal growth *in situ*. Figure 10a shows an AFM image of the (104) cleavage plane of a calcite crystal under a supersaturated calcium carbonate solution. The higher topographical regions are lighter in color. The image illustrates a series of terraces separated by step heights of 4-5 angstroms. Note that at early stages under solution the steps appear relatively jagged. Images obtained every 1.5 minutes during growth are shown in Figures 10a - 10d. As seen in the sequence of images, growth occurs at the step edges so that the steps appear to move perpendicular to the step axis. Furthermore, by the end of the series of images the steps have become relatively smooth and parallel to one another. Preferential incorporation at kink sites reduces the roughness along the step axis, and parallel step axes results from stabilization of a low energy crystallographic plane of calcite. In summary, high energy surface sites fill in rapidly and the remaining lower energy planes are manifested in the resultant surface topography.[8]

The addition of inhibitor species disrupts the crystal growth process. Studies calcite growth in the presence of orthophosphate[9] and in the presence of 1-hydroxyethylidene-1,1-diphosphonic acid (HEDP)[10] indicate that growth inhibition by phosphorus containing inhibitors occurs through the preferential adsorption at growth sites such as step edges and kinks by an inhibitor species. Growth may then continue only through a less favored, slower pathway (i.e. nucleation at a terrace site). Other phosphonate inhibitors such as PBTC likely

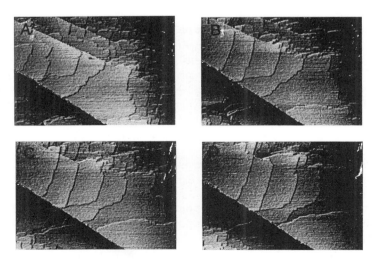

Figure 10. Atomic force micrographs illustrating step advancement on the (104) crystallographic face of calcite. The images obtained in situ demonstrate the dynamics of surface structure evolution on the submicron scale. The images are 2800 x 2800 nanometers with a vertical scale of 10 nanometers.

Figure 11. Atomic force micrographs illustrating complete crystal growth inhibition by PBTC on the (104) crystallographic face of calcite. The images obtained in situ are 2800 x 2800 nanometers with a vertical scale of 10 nanometers.

operate through a similar inhibition mechanism. Atomic force micrographs of calcite crystal growth in the presence of PBTC are presented in Figure 11. These images are a continuation of the AFM images shown in Figures 10a -10d. After normal calcite growth, a solution containing PBTC was injected into the cell. As displayed in the images, no change in the crystal surface due to growth is detected by the AFM over a period of 12 minutes. After flushing the PBTC containing solution from the cell with the previous supersaturated growth solution, normal crystal growth was reinitiated.

CONSTANT COMPOSITION

Although atomic force microscopy can aid in the understanding of the mechanism of crystal growth and crystal growth inhibition, it is not as useful in understanding the chemistry taking place at the crystal surface. In order to better understand the interaction of inhibitors with the crystal surface the constant composition method, developed by Professor George Nancollas,[11] may be used. This method is a technique for quantifying the kinetics of crystal growth and has been used to investigate the influence of inhibitor structure on calcite growth.[12,13]

In a typical experiment, calcite crystals are introduced to a solution with a calcite supersaturation of ca. 2. As crystal growth occurs the solution pH drops. An autotitration system maintains a constant test solution pH through the addition of a carbonate/bicarbonate buffer. A calcium solution is also simultaneously added. The concentrations of the buffer and the calcium solution are chosen such that a constant test solution composition is maintained throughout the experiment. Normal crystal growth is then allowed to proceed until 5 milliliters of titrant have been added, after which crystal growth inhibitor is introduced. The change in the rate of addition of titrant to the test solution, directly correlated with the rate of crystal growth, provides information about the effectiveness of the potential growth inhibitor.

By performing a series of crystal growth experiments in the presence of different concentrations of inhibitor it is possible to describe the concentration/activity relationship for a given inhibitor as shown in Figure 12. As seen in the figure, the addition of a relatively small amount of PBTC has a large influence on the kinetics of crystal growth. In the absence of added inhibitor, crystal growth continues unchanged after 5 milliliters of titrant addition. In general, the relationship between growth control and inhibitor concentration is linear. The addition of PBTC produces a period of time over which crystal growth is impeded. Eventually, however, the ability of the inhibitor to control the growth of the crystal diminishes and rapid normal crystal growth is reinitiated. The amount of time over which growth is controlled is dependent on the concentration of inhibitor. Increasing the inhibitor

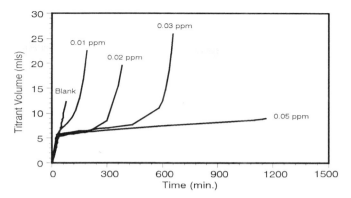

Figure 12. Calcite growth inhibition by PBTC at varying concentrations. Growth control is linearly dependent on the amount of PBTC added after 5 mls titrant addition. In the absence of PBTC the growth rate is unchanged after 5 mls titrant addition.

concentration increases the length of the inhibition period. Note that although the rate of calcite growth is greatly reduced in the presence of PBTC, crystal growth and titrant consumption continues throughout the experiment. One interpretation of these results is that surface bound PBTC is slowly incorporated into the crystal through secondary crystal growth processes. Dissolved PBTC replaces the PBTC consumed at the crystal face until the surface coverage necessary for crystal growth control can no longer be maintained. Rapid growth characteristic of uninhibited calcite is then observed. Since maintaining the necessary surface coverage of PBTC is critical to growth control, the presence of excess inhibitor lengthens the period of growth control. The amount of the excess inhibitor determines the amount of time over which crystal growth is controlled.

By comparing the crystal growth inhibitive properties of different inhibitors it is possible to probe the influence of chemistry on growth control. An examination of different derivatized acrylic acid polymers of similar molecular weight, shown in Figure 13, demonstrates the lower activity of polymers relative to small molecule phosphonates such as PBTC.

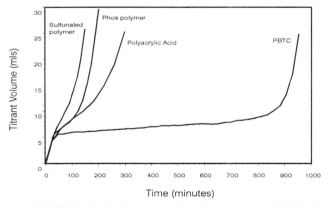

Figure 13. Calcite growth inhibition by derivatized acrylic acid polymers and PBTC. The polymers display lower growth inhibition activity than PBTC. Derivitization of polyacrylic acid with phosphonate or sulfonate lowers performance.

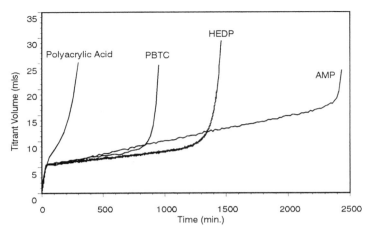

Figure 14. Calcite growth inhibition by polyacrylic acid, PBTC, HEDP and AMP. Increasing phosphorus content appears to improve performance.

Whereas introduction of phosphonate groups can improve the performance of a small molecule inhibitor, the addition of pendant phosphonate groups to an acrylic acid polymer lowers the performance. In this case, optimizing the structure and charge density on the polymer backbone appears to outweigh any phosphonate group advantage. Even lower performance is observed for sulfonate derivatives.

A comparison of calcite growth inhibition provided by PBTC, HEDP, aminotrimethylenephosphonic acid (AMP) and an acrylic acid polymer is shown in Figure 14. Clearly, the best growth control is provided by AMP and the least is provided by the polymer at equal concentrations. Although at first glance the data also seems to suggest that the number of phosphonate functional groups determines an inhibitors calcite growth activity, this is not entirely the case. Whereas the presence of phosphonate moieties has long been known to enhance calcite growth inhibition, molecular structure plays an equal, if not superior, role in calcite growth inhibition. This is most clearly demonstrated in Figure 15, a comparison of the calcite growth inhibition provided by 1,1 and 1,2 diphosphonates. As shown in the figure, HEDP and ethane 1,1 diphosphonic acid (1,1-EDPA) display high levels of inhibition. Ethane 1,2 diphosphonic acid (1,2-EDPA), in contrast, is a poor calcite growth inhibitor even at 2.5 times the concentration of HEDP. The large difference in growth inhibition activity between the 1,1 and 1,2 diphosphonates suggests that molecular structure, rather than phosphonate content, is the critical parameter determining calcite growth control activity. This is further supported by the observation that, even though 1,2-EDPA has a much higher phosphonate content, PBTC displays substantially higher calcite crystal growth inhibition characteristics than 1,2-EDPA.

Molecular structure also plays a strong role in the calcite growth inhibition performance of aminomethylene phosphonates. As seen in Figure 16, drastically different calcite growth kinetics are obtained in the presence of a phosphonomethylated polymer, a phosphonomethylated monoamine and a pair of phosphonomethylated diamines. The polymer and monoamine derivatives display relatively poor performance as growth inhibitors in comparison with PBTC. Both compounds display almost no deflection of the calcite growth curve after addition of inhibitor. Other phosphonomethylated polymers and monoamines have been found to similarly display poor calcite growth inhibition. In contrast, phosphonomethylated diamines typically yield better crystal growth inhibition and both

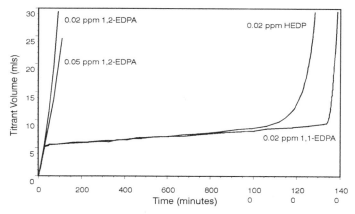

Figure 15. Calcite growth inhibition by diphosphonic acids. 1,1-EDPA and HEDP show high activity whereas 1,2-EDPA displays little activity at 0.02 ppm. At 2.5 times this concentration 1,2-EDPA shows little improvement.

HMDTMP and POTMP display better calcite growth control than methylaminodimethyle-nephosphonic acid (MADMP). The two diamines, however, produce vastly differing levels of performance. While HMDTMP yields growth control approaching that of PBTC, POTMP demonstrates very poor calcite growth inhibition characteristics and only slightly alters the rate of crystal growth. In this case, the structure of the diamine backbone rather than phosphonate placement appears to determine inhibitor performance. Thus, a seemingly small change in the diamine structure has a profound influence on calcite growth inhibition performance.

These results offer an explanation for the differences in calcium carbonate inhibition observed between HMDTMP and POTMP in pH static precipitation experiments (Figure 8). The change to higher precipitation and growth kinetics detected after the initial stages the experiment performed in the presence of POTMP may have been the result of a calcium carbonate phase transition. The relatively amorphous calcium carbonate initially formed may

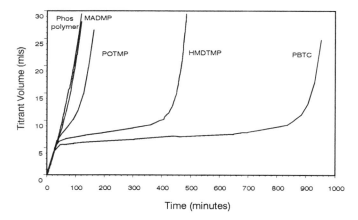

Figure 16. Calcite growth inhibition by phosphonomethylated acrylic acid polymer (Phos polymer), MADMP, POTMP, HMDTMP and PBTC. POTMP, though similar in structure to HMDTMP, shows poor calcite growth inhibition.

have been readily controlled by POTMP. Growth of a more crystalline calcium carbonate, similar to calcite, which would be favored in subsequent stages could not be adequately inhibited. Since HMDTMP appears to perform as both an amorphous and crystalline calcium carbonate inhibitor, its overall performance surpasses that of POTMP.

SUMMARY

The processes involved in calcium carbonate scale deposition are often quite complex. Analytical techniques have been presented which examine *in situ* dynamics or isolate calcite growth characteristics. Time lapse video microscopy and atomic force microscopy have captured dynamics of scale deposition and crystal growth which has not been readily observable previously. Precipitation experiments and constant composition experiments provide additional kinetic data on the calcium carbonate precipitation process complementary to other characterization techniques.

REFERENCES

1. Burger, R., 1990, Cooling Water Technology, Fairmont Press, Lilburn.
2. Kamrath, M. A., R. V. Davis, D. A. Johnson, "A Novel Approach to the Evaluation of Calcium Carbonate Scale Inhibitors", Corrosion/94, Paper No. 198, National Association of Corrosion Engineers
3. Khamskii, E. V., 1969, Crystallization From Solutions, Consultants Bureau, New York.
4. Chalmers,B., 1964, Principles of Solidification, Robert E. Krieger Co., Huntington.
5. Brice, J.C., 1986, Crystal Growth Processes, John Wiley and Sons, New York.
6. Chen, Thomas, and Shih-Ruey (Calgon Corporation), European Patent Publication Number 516382, "Polyether Polyamino Methylene Phosphonates for High pH Scale Control", December 2, 1992.
7. Carter, P. W., A.C. Hillier, M.D. Ward, 1994, "Nanoscale Surface Topography and Growth of Molecular Crystals: The Role of Anisotropic Intermolecular Bonding", *J. Am. Chem. Soc.*, 116, 944.
8. Carter, P. W., A.C. Hillier, M.D. Ward, 1994, "Molecular Single Crystal Interfaces: Topographical Structure and Crystal Growth", *Mol. Cryst. Liq. Cryst.* 242, 53.
9. Dove, P. M. and M.F. Hochella, 1993, "Calcite Precipitation Mechanisms and Inhibition by Orthophosphate: In Situ Observations by Scanning Force Microscopy", *Geochim. Cosmochim. Acta*, 57, 705.
10. Gratz, A. J. and P.E. Hillner, 1993, "Poisoning of Calcite Growth Viewed in the Atomic Force Microscope (AFM)", *J. Crystal Growth*, 129, 789.
11. Tomson, M. B. and G.H. Nancollas, 1978, "Minerization Kinetics: A Constant Composition Approach", *Science*, 200, 1059.
12. Nancollas, G. H., T.F. Kazmierczak, and E. Schuttringer, 1981, "A Constant Composition Study of $CaCO_3$ Crystal Growth: The Influence of Scale Inhibitors", Corrosion, 226, 18.
13. Amjad, Z. 1987, "Kinetic Study of the Seeded Growth of $CaCO_3$ in the Presence of Benzenetricarboxylic Acid", *Langmuir*, 3, 224.

5

SCALE CONTROL BY USING A NEW NON-PHOSPHORUS, ENVIRONMENTALLY FRIENDLY SCALE INHIBITOR

L. A. Perez[1] and D. F. Zidovec[2]

[1] Betz Water Management Group
   1 Quality Way, Trevose, Pennsylvania 19053
[2] Betz PaperChem
   7510 Baymeadows Way, Jacksonville, Florida 32256

ABSTRACT

A new non-phosphorus scale inhibitor that prevents different types of scales under industrial process conditions has been developed. The new compound can be applied to control calcium carbonate scale in the cooling industry, both recirculating cooling towers and once through systems. The inhibitor has shown excellent performance in preventing calcium carbonate, barium sulfate, and calcium oxalate precipitation under conditions typical to the paper industry. Applications of this new non-phosphorus anti-scaling agent (NPA) can be extended to the oil industry since as it has shown excellent scale inhibition for calcium carbonate and barium sulfate and acceptable inhibition for calcium sulfate under typical oil industry conditions. The new inhibitor is stable in the presence of chlorine and it possesses tolerance to relatively high calcium, iron, and aluminum concentrations.

INTRODUCTION

Scale formation is a serious problem in many areas such as cooling water technology, desalination, paper industry, and oil production. The calcium carbonate polymorphs, calcium sulfates, barium sulfate, calcium phosphates and in some cases, like in the paper and food industry, calcium oxalates are all involved in scale formation.[1,2] A scale problem causes production losses by reducing the heat transfer capacity of heat exchangers, emergency shutdown, clogging of pipelines, etc. The formation of these scales is affected by several parameters such as temperature, rate of heat transfer, pH of the water, and the character and concentration of ions present in the water.

In the past three decades, vast quantities of water have been used for cooling due to the large industrial growth and development. In an effort to conserve water, it has been necessary to minimize the blowdown which has resulted in an increase in the cycles of

Mineral Scale Formation and Inhibition, Edited by Zahid Amjad
Plenum Press, New York, 1995

concentration in the tower. This, in turn, increases the chances of deposition of sparingly soluble salts on the heat exchangers.

In the paper industry, scales such as calcium carbonate and barium sulfate are very common. This scale formation may cause capacity problems, increased steam consumption, reduced liquor flow, poor washing, increased salt cake loss, and loss or reduced quality of production. Also in this industry, calcium oxalate deposition frequently appears in sulfite and NSSC (neutral sulfite semi-chemical) evaporators. Lower steam economy, heat transfer coefficients and production rates result from these deposits.[3]

In the oil industry, barium, strontium, and calcium sulfates, and calcium carbonate are the predominant components of the scale layers. Barium and strontium sulfates being in the primary scale layers and the calcium sulfates and carbonates in the secondary layers. The development of these scale layers results in production losses.[4]

In the past, phosphate containing compounds, such as sodium tripolyphosphate and sodium hexametaphosphate have been used to prevent such scaling. However, due to environmental problems (eutrophication of sea areas, rivers and lakes) and the tendency of these compounds to form calcium phosphate scales, their use has been reduced in favor of other scale inhibitors, such as phosphonic acids and polyacrylates. In the case of phosphonates, it is well known that even at moderate conditions of calcium hardness, pH, and temperature, they can react with calcium ion and precipitate as a calcium phosphonate scale former salt, which can induce the precipitation of other scale formers. In addition, phosphonates can also produce eutrophication. Also, it has been established that at high phosphonate concentrations, the protective corrosion film formation can be reduced and an increase in corrosion, especially pitting, is observed.

Polyacrylic and polymaleic acids have a limited effectiveness, especially at high cycles of concentrations and pH. Moreover, in waters containing high concentrations of silica, their effectiveness is greatly reduced. As in the case of the phosphonates, these also have tendency to form insoluble calcium salts.

In the present study, we present results on scale inhibition testing conducted with a new scale inhibitor based upon polyepoxysuccinic acid that has been developed and patented by Betz Laboratories.[5,6] This new non-phosphorus inhibitor, in addition of being "environmentally friendly", provides better scale control than other treatments typically used.

EXPERIMENTAL

Calcium Carbonate Testing

Static beaker and dynamic tests were conducted by using double distilled water and reagent grade chemicals. The static beaker test involved the adding of the treatment to a solution containing calcium and carbonate ions at the pH studied. The beakers were incubated in a water bath for 18 hours at 70 °C. After cooling, a measured portion was filtered and the calcium concentration measured by using the standard ethylenediamine tetraacetic acid titration method. Murexide was used as the indicator. Percent inhibition was calculated from the titration of the treated, stock and control solutions.

Two types of dynamic tests were performed. The first one was conducted in a Bench Top Unit (BTU) in order to evaluate the efficacy of the treatment as scale control agent for cooling water systems. These recirculator units are designed to provide a realistic measure of the ability of a treatment to prevent corrosion and scale formation under heat transfer conditions. The treated water is circulated through a corrosion coupon by-pass rack and a heat exchanger tube contained in a Plexiglas™ block. The heat exchanger is fitted with an electrical heater that allows to control of the heat load in the 0 to 16,000 BTU/ft²/hr. range.

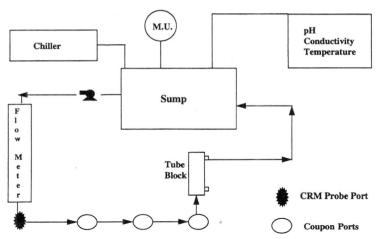

Figure 1. Schematic representation of a Bench Top Unit.

The water velocity at the surface of the heat exchanger can be varied between 0 and 4.5 ft./sec. The pH and the temperature of the circulating water are automatically controlled. The treated water is prepared by the addition of the component ions to deionized water. Provisions for continous makeup and blowdown are made by pumping freshly treated water from supply tanks to the sump of the unit, with overflow from the sump serving as blowdown. The efficacy of the treatment is assessed by the amount of deposit that is formed on the heat exchanger surface and other surfaces present in the unit. The absence of bulk precipitation is also taken into account. Figure 1 shows a schematic representation of the BTU.

The second dynamic method used has proved to be effective to simulate in the laboratory scaling conditions existing in paper mills and in the oil industry. In the test apparatus, four tube lines are combined to form a supersaturated solution by mixing four solutions that contain the ions of the scale former salt. One of these lines will usually contain the scale control product being tested. The supersaturated solution is pumped at constant flow rate through a capillary stainless steel tube, which is installed in a thermostatic bath. As scaling occurs, the pressure needed to keep the flow through the capillary increases. The change in pressure with time is monitored continuously. Effective scale inhibitors are recognized by a reduced or completely halted increase in pressure. Figure 2 shows a schematic representation of this type of unit.

Barium Sulfate Testing

Static barium sulfate beaker tests were conducted by adding the treatment to a solution containing barium and sulfate ions at the required temperature and pH. The beakers were incubated in a water bath at 60°C for 1 hour. The hot solutions were then filtered through 0.22 μm filter paper and analyzed for barium by inductively coupled plasma atomic emission spectroscopy (ICP). Percent inhibition was calculated from the barium concentration of the treated, stock and control solutions.

In the case of the paper industry, the efficacy of most barium sulfate inhibitors is affected by the presence of aluminum ions. Therefore, tests were also conducted in the presence of aluminum ions.

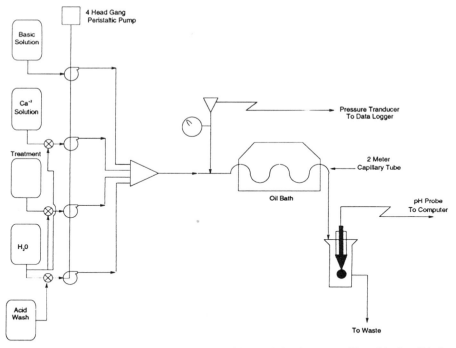

Figure 2. Flow diagram of unit used to simulate conditions existing in paper mills and in the oil industry.

Calcium Oxalate Testing

Calcium oxalate tests were conducted by using the static beaker test. In this test, the treatment was added to a solution containing calcium and oxalate ions at the required temperature and pH. The beakers were incubated in a water bath at 60°C. After 1.5 hours of incubation, the beakers were removed from the water bath and a measured portion of the solution was filtered and the calcium concentration measured by ICP. Percent inhibition was calculated from the analyses of the treated, stock and control solutions.

Calcium Sulfate Testing

Static beaker tests were used to evaluate the efficacy of inhibitors for calcium sulfate scale control. As in the other beaker tests, the treatment was added to the solution containing the ions of the scale former salt (calcium and sulfate). The beakers were incubated at 50°C for 24 hours. After the incubation time and filtration, calcium was analyzed by EDTA titration with the murexide indicator. The percent inhibition was calculated from the measured calcium concentration of the treated, stock and control solutions.

Chlorine Tolerance Testing

Chlorine tolerance was tested by means of the static beaker test used for calcium carbonate inhibition. The effect of different levels of chlorine on the percent inhibition was measured.

Calcium Tolerance Testing

Calcium tolerance beaker studies were conducted by adding the treatment to a solution containing calcium ion at the required level. Ten millimoles of sodium borate were used as a buffer to maintain the pH at the value studied. The beakers were incubated in a water bath at 70°C for 18 hours. Turbidity measurements were performed to the solution in the absence and presence of the treatment after the incubation period.

RESULTS AND DISCUSSION

Calcium Carbonate Scale Inhibition

Static beaker test results showed that the new non-phosphorus anti-scaling agent (NPA) outperformed other typical scale inhibitors. Figure 3 shows a bar graph comparing the percent inhibition obtained at the specified test conditions (supersaturation ratio [SSR] equal to 12.2 - defined as the ratio of the product of the ionic activity of the calcium and carbonate ions at the test conditions to the solubility product). As can be seen, NPA provided better scale control than hydroxyethylidenediphosphonic acid (HEDP) and two polyacrylic acids (PAA) of different molecular weights, when tested at the same treatment level (0.5 ppm active). Another static beaker test was conducted at a higher supersaturation ratio (157). At these more stressful conditions, NPA also showed to be more effective than HEDP (used as the only treatment) and than Polymaleic acid (PMA) and PAA. The conditions and results for this second test are illustrated in Figure 4.

Tests simulating typical cooling tower water conditions were conducted with a BTU unit using the scale inhibitor alone and in blends with other typical components of cooling water treatment programs. As shown in Table 1, under the conditions specified (SSR = 70; Langelier Saturation Index = 2.2), NPA provided excellent scale prevention on the admiralty

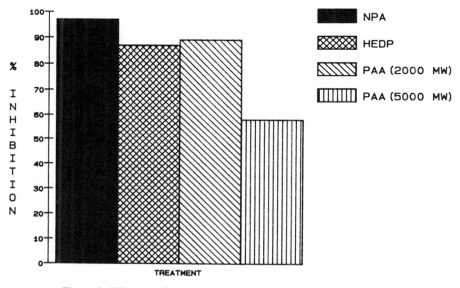

Figure 3. NPA outperformed other known scale inhibitors at SSR = 12.2.

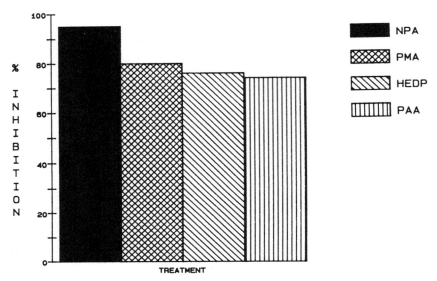

Figure 4. CaCO₃ inhibition by NPA and other scale inhibitors at SSR 157.

heat transfer surface, while other known non-phosphorus scale inhibitors, PMA and PAA, failed to keep the heat exchanger surface free of deposit.

Table 1 also shows that NPA gave excellent performance when tested in programs that use zinc ion as the low carbon steel (LCS) corrosion inhibitor. In this case, a second polymer that is commonly used to keep zinc ion in solution and to disperse clay was also

Table 1. Prevention of calcium carbonate scale formation under typical cooling conditions

B. DYNAMIC TESTING
CONDITIONS: 600 ppm Ca as CaCO₃, 200 ppm Mg as CaCO₃, 438 ppm CO₃ as CaCO₃,
250 ppm Alkalinity as CaCO₃,
234 ppm SO₄, 139 ppm Na, 426 ppm Cl, 50 ppm SiO₂, 3 ppm TTA, pH = 8.50 at 120 Fahrenheit
(8.8 at room temperature), experimental temperature = 48.89 Celsius, (120 Fahrenheit), Flow
Velocity = 2.7 ft/sec, Heat Input = 550 Watts, Retention Time = 1.4 days, System Volume = 12 L,
Test Duration = 7 days, SSR = 70 (LSI 2.2)

Treatment - ppm	Heat Exchanger		
	Metal	Deposition	Corrosion
NPA - 15	Admiralty	NONE	NONE
PMA - 15	Admiralty	SLIGHT	NONE
PAA(MW ~ 2000) -15	Admiralty	HEAVY	NONE
Polymer 1 (Typically used for Zn(OH)2 inhibition) - 20	Admiralty	HEAVY	NONE
NPA -15 Polymer 1 - 5 Zinc - 2	Mild Steel	NONE	Almost NONE
NPA - 5 Polymer for Silt Dispersion - 5 Mild Steel Corrosion Inhibitor - 1	Mild Steel	NONE	Almost NONE
NPA - 15 Iron - 10 Iron and Zinc Dispersant - 10 Zinc - 2	Mild Steel	NONE	Almost NONE

Table 2. Prevention of calcium carbonate formation by NPA in comparison to PAA

B. DYNAMIC TESTING

CONDITIONS: 450 ppm Ca as $CaCO_3$, 200 ppm Mg as $CaCO_3$, 400 ppm Alkalinity as $CaCO_3$, 300 ppm SO_4, 241 ppm Na, 320 ppm Cl, 50 ppm SiO_2, 3 ppm TTA, pH = 8.80 at 120 Fahrenheit (9.1 at room temperature), experimental temperature = 48.89 Celsius, (120 Fahrenheit), Flow Velocity = 2.7 ft/sec, Heat Input = 308 Watts, Retention Time = 1.4 days, System Volume = 12 L, Test Duration = 7 days, SSR = 128 (LSI 2.6)

Treatment - ppm	Heat Exchanger		
	Metal	Deposition	Corrosion
NPA -10 Dispersant for Suspended solids - 5 Mild Steel Corrosion Inhibitor -1	Admiralty or Mild Steel	NONE	NONE for Admiralty Almost NONE when Mild Steel was used
PAA -15 Dispersant for Suspended Solids -5 Mild Steel Corrosion Inhibitor -1	Mild Steel	Slight	Almost NONE

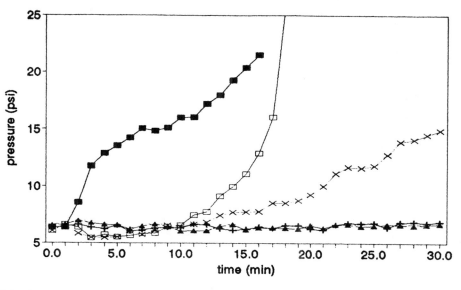

Figure 5. Calcium carbonate inhibition by NPA under typical paper mills conditions.

present. Iron did not affect the performance of NPA as a calcium carbonate inhibitor as illustrated in Table 1.

Under more stressful conditions, as illustrated in Table 2, (SSR = 128), NPA was able to control scale formation when blended with other typical components of cooling water treatment programs. Polyacrylic acid (PAA), however, failed when it was used in the program as the NPA replacement.

All tests conducted under typical cooling tower water conditions indicate that NPA is able to control calcium carbonate scale formation with superior performance than other known calcium carbonate scale inhibitors.

Dynamic testing conducted to simulate typical scale in paper mills showed that NPA was also effective in preventing deposition under the conditions studied (150 ppm Ca^{2+} as $CaCO_3$, 3700 ppm Alkalinity as $CaCO_3$; pH = 12.5; T = 101°C). As shown in Figure 5, twenty five (25) ppm active of NPA was sufficient to control deposition on the surface of the capillary tube. NPA was shown to be as effective as the same level of ethylenediamine tetrametyle-nephosphonic acid (EDTMP), and superior to a maleic acid copolymer (tested up to 60 ppm active treatment level) and aminotrimethylenephosphonic acid (AMP) (up to 40 ppm active).

Barium Sulfate Scale Inhibition

NPA has shown to be a superior barium sulfate scale inhibitor compared to other materials typically used in the paper and oil industry. As it can be seen in Figure 6, one (1) ppm active NPA was able to provide complete scale control under the beaker test conditions (2 ppm Ba^{2+}; 1000 ppm SO_4^{2-}; 60°C; 1 hour test duration). The most remarkable fact from

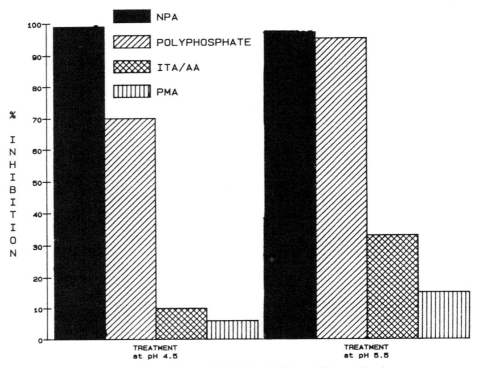

Figure 6. Barium sulfate scale inhibition by NPA at different pH values.

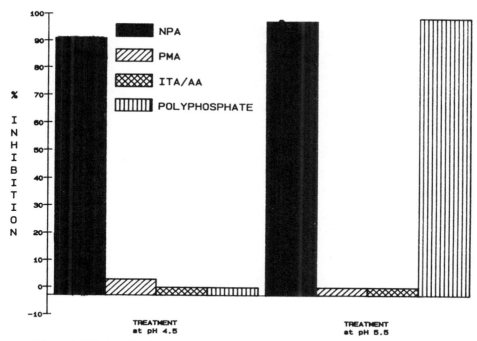

Figure 7. Effect of aluminum ion on the performance of NPA and other known inhibitors.

the test is that NPA was not affected by pH changes. When the test was performed at pH 4.5, NPA markedly outperformed the other inhibitors: polyphosphate, a copolymer of itaconic acid and acrylic acid, and polymaleic acid. At a higher pH (5.5), NPA showed the same efficacy as at pH 4.5. The other inhibitors had a moderate increase in their performance but still were inferior to NPA.

In paper machines operating at acid pH, aluminum ions are usually present, causing a decrease of the efficacy of most barium sulfate inhibitors. NPA showed to be affected to a lesser degree in typical tests conducted to simulate the presence of 5 ppm Al^{3+}. Figure 7 shows the conditions and test results from which it can be observed that NPA outperformed the other scale inhibitors studied. As in the previous study, NPA was not affected by changes in pH.

The stability of efficacy with respect to pH renders NPA a tremendous improvement in the prevention of barium sulfate scale in the paper and oil industries, at typical ranges.

Calcium Oxalate Scale Inhibition

In the paper industry, calcium oxalate deposition has been observed in sulfite mills from the weak liquor storage tank throughout the evaporators and to the liquor nozzles before final combustion. In the hypochlorite bleaching stage, with a pH of 7 to 9, and a strong oxidizing environment which promotes oxalate formation from wood derivatives, conditions are ideal for calcium oxalate formation.[6] The same is true for the chlorine dioxide stage. Tests were conducted to study the performance of NPA as a calcium oxalate inhibitor under conditions typical to the paper industry. In studies conducted at relatively low pH (4.0), NPA outperformed other typical scale inhibitors (PAA and a terpolymer of maleic acid/ethyl

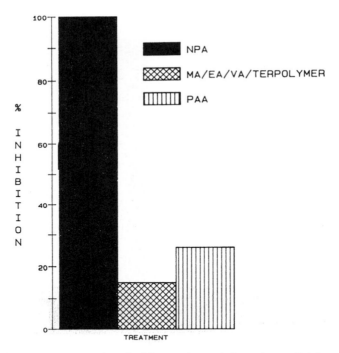

Figure 8. Prevention of calcium oxalate scale formation at pH 4.0.

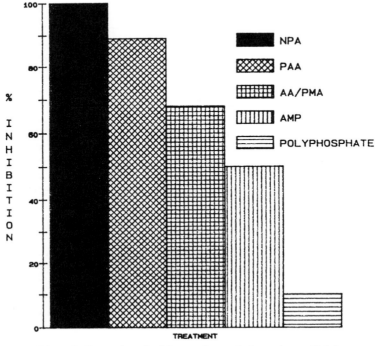

Figure 9. Prevention of calcium oxalate scale formation at pH 9.5.

acrylate/ vinyl acetate [MA/EA/VA]). As in the case of the barium sulfate scale inhibition, NPA was not affected by the changes in pH. Figures 8 and 9 illustrate the test conditions and results of the studies performed at different pHs.

Calcium Sulfate Scale Inhibition

NPA showed moderate efficacy in preventing calcium sulfate scale formation. As it can be observed in Table 3, higher active treatment levels of NPA were required to achieve the same efficacy of other commonly used calcium sulfate scale inhibitors (AMP, PAA, TENTMP). Test conditions and results are also given in Table 3.

Chlorine Tolerance Test

Chlorine is the most widely used biocide in several industries. Its powerful oxidizing properties affect the performance of typical scale inhibitors. As illustrated in Figure 10, NPA was not affected by the presence up to 10 ppm free chlorine in the water. The figure also illustrates how NPA compares with respect to other scale inhibitors.

Calcium Tolerance Test

As mentioned in the introduction, most inhibitors have the tendency to precipitate as their calcium salt at relatively high calcium concentrations and pH. This becomes a limitation in applications where high level of calcium are present. In the case of the cooling water technology, for example, this limitation reduces the number of cycles possible. NPA showed excellent calcium tolerance. Under typical cooling water calcium concentrations (~500 ppm

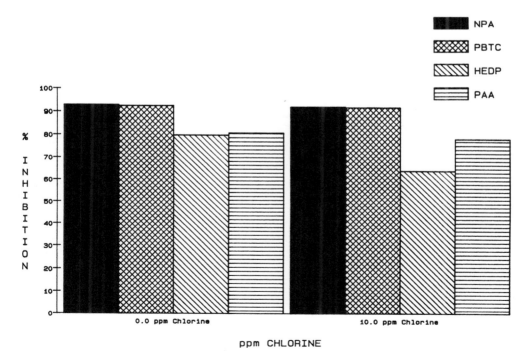

Figure 10. Chlorine tolerance of NPA and other known $CaCO_3$ inhibitors.

Table 3. Prevention of calcium sulfate scale formation by NPA

STATIC BEAKER TEST CONDITIONS:
5000 ppm Ca as CaCO₃   pH = 7.0
4800 ppm SO₄   Temperature = 120 Fahrenheit
2300 ppm Na   Test Duration = 24 hours

TREATMENT	% INHIBITION		
	1.0 (ppm active)	3.0 (ppm active)	5.0 (ppm active)
NPA	14.6	48.9	92.7
AMP	84.3	97.4	97.8
PAA (2000 MW)	95.9	98.4	98.6
PAA (5000 MW)	82.5	98.5	99.0
TENTMP	98.2	97.9	98.4

Ca²⁺ as CaCO₃) NPA outperformed typical phosphorus and non-phosphorus containing scale inhibitors. This is illustrated in Figures 11 and 12, where the test conditions and results are presented. The same results were observed under conditions in which the calcium level was increased to 1000 ppm as CaCO₃, as shown in Figures 13 and 14.

The relatively high calcium tolerance showed by NPA makes it suitable for treatment programs to control scale formation in cooling towers run at high number of cycles and consequently at relatively high pHs, resulting in not only economical but also environmental benefits, by way of water conservation.

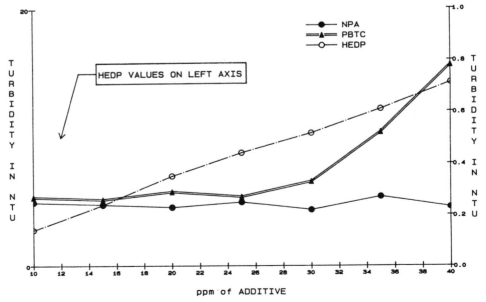

Figure 11. Comparison of NPA calcium tolerance with other typical phosphonate inhibitors at low ([Ca] = 500 ppm as CaCO₃) calcium levels.

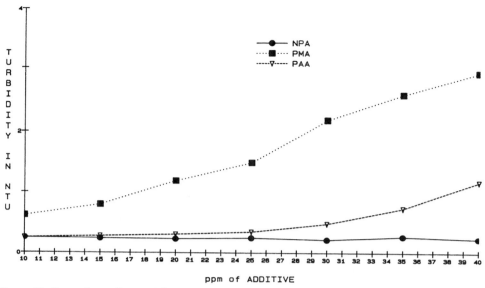

Figure 12. Comparison of NPA calcium tolerance with other non-phosphorus scale inhibitors at low ([Ca] = 500 ppm as $CaCO_3$) calcium levels.

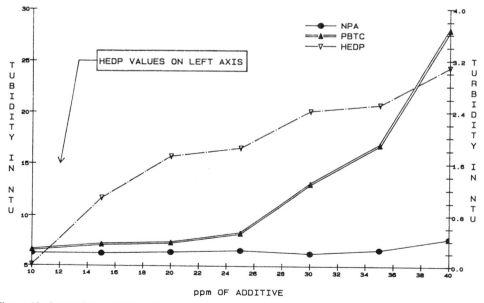

Figure 13. Comparison of NPA calcium tolerance with other typical phosphonate inhibitors at high ([Ca] = 1000 ppm as $CaCO_3$) calcium levels.

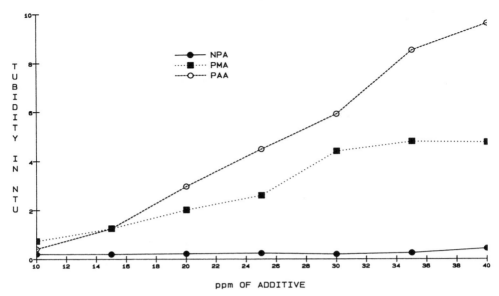

Figure 14. Comparison of NPA calcium tolerance with other non-phosphorus scale inhibitors at high ([Ca] = 1000 ppm as $CaCO_3$) calcium levels.

Aquatic Toxicology of NPA

One of the most important aspects of the development of an "environmentally friendly" program is the effect of the treatment on natural systems. As mentioned before, phosphate and phosphonate based programs represent an environmental concern due to their eutrophication potential. NPA is of very low toxicity with respect to aquatic life, as shown in Table 4.

Table 4. Aquatic toxicology of NPA

AQUATIC TOXICOLOGY OF NPA

Fathead Minnow
   96 Hours Static Acute Bioassay
   LC50: 1680 mg/L
   No effect Level: 1350 mg/L

Daphnia magna
   48 Hours Static Acute Bioassay
   LC50: 1635 mg/L
   No effect Level: 870 mg/L

Selenastrum Capricornutum (Green Algae)
   96 Hours Growth Inhibition Bioassay
   IC-10:   833 mg/L
   IC-50: >3333 mg/L

CONCLUSIONS

NPA has showed to be a very effective scale inhibitor for the most typically found scale former salts (calcium carbonate, calcium oxalate, barium sulfate, and, to some extent, calcium sulfate). The performance of this new inhibitor is not affected by the pH of the industrial process under consideration. This was demonstrated in the cases of barium sulfate and calcium oxalate scale prevention. The inhibitor is very calcium tolerant and its performance is not affected by the addition of chlorine to the water. Aquatic toxicology studies showed that NPA is practically harmless to the environment. The results indicate that NPA based programs to prevent scale formation in different industrial application will provide better performance than typically used scale inhibitors.

ACKNOWLEDGMENTS

The authors wish to thank Profesor G. H. Nancollas, who introduced them to the field of scale and scale prevention during the years when they were his Ph.D students. In addition, we wish to acknowledge the invaluable cooperation of Betz Laboratories in the publication of this article.

REFERENCES

1. Cowan J.C., and Weintritt D.J., 1976, Water formed scale deposits, Gulf Publishing Co., Houston, 29-30.
2. Nancollas G.H., White W., Tsai F., and Maslow L., (1979), The kinetics and mechanisms of formation of calcium scale minerals - The influence of inhibitors, Corrosion, 35:304-308.
3. Betz Handbook of Industrial Water Conditioning, (1980), Pulping process, 8th edition: 282 - 292.
4. Perez L., and Nancollas G.H., (1988), The gowth of calcium and strontium sulfates on barium sulfate surfaces, Scanning Microscopy, 2:1437-1443.
5. Brown J.M., McDowell J.F, Chang K.T., (1991), Methods of controlling scale formation in aqueous systems, U.S. Patent 5,062,962.
6. Perez L. A., Freese D.T., Rockett J.B., and Carey W.S., (1994), Methods of controlling scale formation in aqueous systems, U.S. Patent 5,326,478.

6

POLYMER MEDIATED CRYSTAL HABIT MODIFICATION

Allen M. Carrier and Michael L. Standish

Alco Chemical
Division of National Starch and Chemical
P. O. Box 5401
Chattanooga, Tennessee 37406

INTRODUCTION

Prevention of crystal growth and modification of crystal morphology in aqueous systems has received considerable attention recently due to changes in the water treatment industry instigated by economic and environmental pressures. Water costs, especially in the western United States, are forcing a shift in the manner that the water is managed. Historically, a concentration of minerals to the point of precipitation was reduced by discharging the process water and making up with fresh water or by altering precipitation kinetics or macroscopic crystal structure of mineral scale via water soluble polymers. The economic and environmental concerns require the water manager to maximize the number of cycles the process water must experience before being discharged to the environment. This increases the driving force on precipitation in the system and a new generation of polymers must be created that can be effective under these stressed conditions. The use of synthetic polymers has been widespread since the early 1960's[1] and has been our primary focus for several years. We have found that polymers containing substantial amounts of maleic acid are effective calcium carbonate scale inhibitors under these stressed conditions but historically suffer from poor cost effectiveness due to non-aqueous polymerization routes. We have developed polymers containing substantial(>50 mole%) maleic acid prepared via an aqueous polymerization route. The first commercial product from this technology is Aquatreat® AR-980.

EXPERIMENTAL

The preparation and exact chemical composition of Aquatreat AR-980 are proprietary.[2] All molecular weights are determined via gel permeation chromatograph by the published method[3], available from Alco Chemical.

Mineral Scale Formation and Inhibition, Edited by Zahid Amjad
Plenum Press, New York, 1995

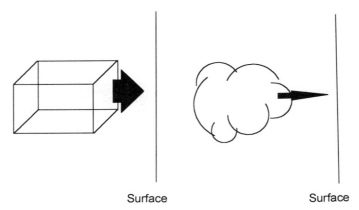

Surface   Surface

Figure 1. Crystal modification of calcite by polymer.

Stressed Calcium Carbonate Scale Inhibition Test

This test is designed to evaluate polymers for calcium carbonate stabilization in high pH and high electrolyte conditions and compare them to a known reference. The test can quantitatively compare polymer effects on calcium carbonate threshold inhibition, dispersancy, and surface adhesion. The aqueous test solutions needed are: 0.2 M sodium carbonate; a 160 ppm concentration of calcium prepared from calcium chloride dihydrate; and a solution 4.0 M in sodium chloride and 0.5 M in sodium hydroxide. Polymer test solutions should be prepared as desired (NOTE: Ensure the polymers are neutralized to the same degree; this permits accurate testing of active polymer for comparison purposes). To reduce experimental error, the test solutions should be added in the following manner:

To 50 ml of 0.2 M sodium carbonate in a 250 ml Erlenmeyer flask is added the test polymer solution (typically in the 0-100 ppm polymer range). To this mixture add, simultaneously, 50 ml of the sodium chloride/sodium hydroxide solution and 100 ml of the calcium chloride solution. Randomly test the pH of various solutions. The pH should be 12.3 to 12.7. Place all test solutions, including blanks prepared without polymer, into a 50 degree Celsius water bath for 24 hours. Remove, let cool to room temperature and gently invert 3 times. Filter the resulting slurry through a 0.2 micron membrane filter. The calcium carbonate remaining in the flask and collected on the membrane filter are separately dissolved in 14% nitric acid. The filtrate, dissolved scale from the membrane and the dissolved scale from the flask are analyzed and quantified for calcium using Atomic Absorption spectrophotometry. A Perkin-Elmer 5100 EC Atomic Absorption Spectrophotometer is recommended.

RESULTS AND DISCUSSION

Requirements for New Polymers

For any new polymer to enjoy commercial success in today's marketplace it must have an overall cost performance exceeding other available technologies. In general, where calcium carbonate scale is the major concern, poly(acrylic acid) has been the standard, due to it's relative low cost. In recent years, the market has become more demanding. These demands are due to several factors: increased water reuse which causes a buildup of scaling minerals beyond that previously seen; poorer quality of the make-up water as cleaner waters are more in demand for potable consumption; and operation of systems under more scaling conditions by continuing to push the systems beyond the initial design limits. Where few systems existed in the recent past requiring polymers to work under these conditions, more and more now require high performance polymers.

Modern water treatment requires the polymer be stable in a formulation with various other compounds. These additional materials function to control corrosion, adjust pH, react with hardness ions and scavenge oxygen. All the necessary components must be stable (no phase separation) over several months at extremes of ambient temperature (from 0 to 45 degrees Celsius). Any new polymer must formulate or its applicability will be greatly limited.

Functionality of Polymers

Polymers offer the water treatment market several important functionalities:

- Threshold Inhibition
- Dispersion
- Crystal Modification
- Adhesion/Cohesion Modification

Threshold inhibition is defined as the ability of a substance to maintain the solubility of an otherwise insoluble material beyond its normal limits at sub-stoichiometric levels. In this way polymers differ from traditional chelants such as ethylenediamine tetraacetic acid (EDTA) or nitrilo triacetic acid (NTA) since these work only in a stoichiometric manner. A dispersion is a stable mixture of insoluble materials distributed more or less uniformly throughout a liquid. By functioning as effective dispersing agents, polymers can keep insoluble material from depositing in unwanted areas. Crystal modification is best described in Figure 1.

The untreated calcium carbonate (calcite) is roughly cubic in its morphology which lends to significant adhesion to a surface. The surface may be a heat transfer area, a place where the fluid velocity is low or it may be another growing calcite crystal. By treating a solution which will precipitate calcium carbonate with polymer one can substantially alter the shape of any growing crystal. The precipitated material cannot adhere as easily to surfaces and can more easily be removed from the system.

Design of New Polymers

It was our goal to design a new polymer that would have the necessary functionality under stressed conditions. Stressed conditions are seen in systems that have any or all of the following properties:

- High pH
- High Ionic Strength
- High Carbonate Alkalinity
- High Hardness

Traditional polymers used in water treatment do not function under these conditions, often due to insolubility of the polymer in the complex matrix. Polymer compositions consisting exclusively or primarily of maleic acid do show a degree of functionality in these stressed systems. These materials are not widely used due to their cost, brought about by their method of manufacture. Maleic acid containing polymers are typically made in organic solvent and are then placed in aqueous solution via a solvent transfer process.

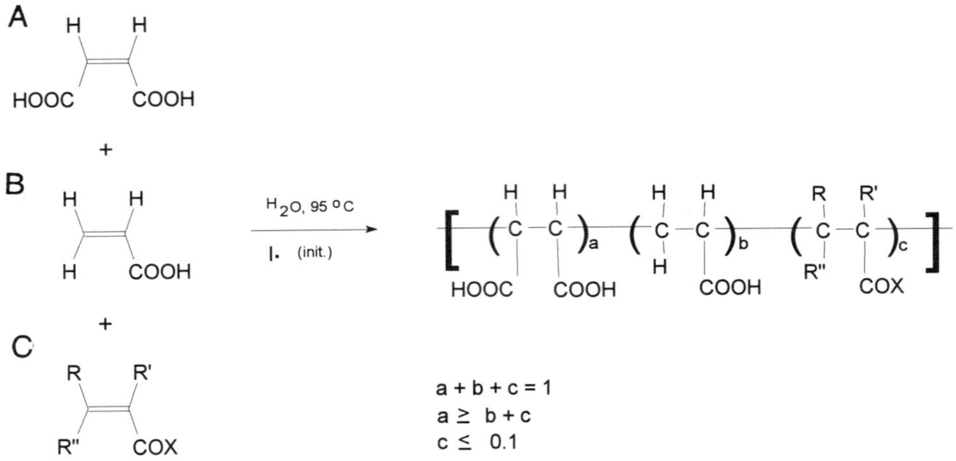

Figure 2. Generalized reaction scheme for preparing high maleic acid containing polymers.

To prepare a more cost viable polymer, it would be necessary to develop a means to effectively create high maleic acid containing polymers via an aqueous polymerization process. The generic family of products that can be produced by our process are depicted in Figure 2.

The process involves creating the polymer using aqueous free radical polymerization techniques. The polymer compositions are all greater than 50 mole % maleic acid with the remainder consisting of acrylic acid and a third monomer. The third monomer comprises 10 mole % or less. Through judicious use of reaction conditions it is possible to create polymers with up to 65 mole % maleic acid by this process. The reactions are carried out under ambient atmosphere in conventional batch reactors. Though idealized in Figure 2 as a regular block terpolymer, there is no indication the polymers are anything other than totally random. Molecular weights of the polymers are typical of those used for water treatment and are less than 10,000 daltons. Polymers prepared by this technique are obtained as clear, amber colored solutions of low viscosity at approximately 40% solids.

The polymerization process must also produce a compound with a high degree of monomer incorporation. State and Federal regulations can classify certain materials as hazardous if the levels of residual monomers are too high. Also in most formulations, unreacted maleic acid would precipitate from solution if more than about one weight percent remained unpolymerized. Our process provides for a typical total residual monomer level of less than 0.1 weight %.

Evaluation of Polymers under Stressed Conditions

The applicability of laboratory testing to field conditions is a subject of great controversy. This controversy can be avoided, or at least greatly reduced, if the conditions under which the test are run are clearly given and the performance of a reference material known to function in field applications is provided. In developing a static test for evaluating polymers under stressed conditions we needed to operate at conditions of electrolyte, pH and polymer dosage which are unreasonable in commercial water treatment. These conditions, however, were the only ones we explored that gave qualitative results consistent with field applications, i.e. poly(acrylic acid) did not perform and poly(maleic acid) did perform. In developing this test we also decided to extend the realm of static testing somewhat. Previous tests[4] provided quantitative results on the ability of a polymer to function as a threshold inhibitor but did nothing to quantify the total mineral level to ensure a proper mass balance was maintained. Our test provides a quantitative mass balance of the calcium introduced into each experiment to determine not only the level of threshold inhibition but also a means to quantify the dispersancy and the adhesive/cohesive modifying ability of the polymer. Results from our evaluation are shown in Figure 3.

To understand the significance of the data presented in Figure 3 it will be necessary to provide some operational definitions:

- Filtrate Sample: Filtered sample containing calcium inhibited by the material. This is representative of the *threshold inhibition* afforded by the material.
- Membrane Sample: Solid material collected on filter membrane. This is representative of the *dispersancy* effect of the polymer.
- Adherent Sample: Material collected as attached to the surface of the sample flask. This is representative of the polymer effect on the *adhesive/cohesive* properties of the formed crystals.

The poly(acrylic acid) shown has a weight average molecular weight of about 2500 and is typically used to prevent calcium carbonate scale in non-stressed conditions. The poly(maleic acid) shown has a weight average molecular weight of about 800 and is the commercially available product typically used. No commercially available samples of a 2500 molecular weight poly(maleic acid) are available. Attempts in our lab to prepare such a material were not successful. The performance of Aquatreat AR-980, with a weight average molecular weight of about 2500, is given in the middle of Figure 3. The data show that under the conditions of the test the poly(acrylic acid) has virtually no threshold inhibiting properties. Most of the calcium has precipitated as calcium carbonate and is dispersed throughout the system and caught on the membrane filter. About 8% of the calcium carbonate was attached to the flask. It is interesting to note the lack of change in the performance of poly(acrylic acid) over a fairly broad treatment

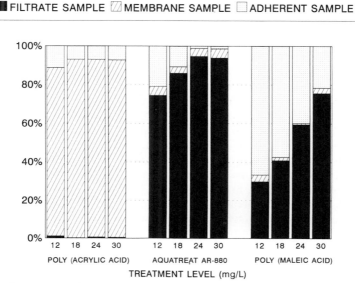

Figure 3. Calcium carbonate scale evaluation under stressed conditions.

Figure 4. SEM of calcium carbonate grown with various polymers.

range. The poly(maleic acid) shows significantly improved threshold inhibiting properties which correlate to field data. The threshold inhibition increases with increasing dosage as expected. There is little dispersing of calcium carbonate scale which is also consistent with field observations and the effects remain essentially constant over the dosage range. There is significant scale adhesion to the surface of the test vessel which decreases as threshold inhibition increases. We also examined a wide range of alternate polymer technologies to determine their efficacy in our test. Of all the compositions examined, the poly(maleic acid) was the best of the commercially available products at threshold inhibition. The AR-980 shows substantially improved threshold inhibition over the entire treatment range. At the higher dosages, the majority of the precipitating scale is caught in the membrane filter indicating the material is not adhering to the test vessel.

During our initial testing of AR-980, it was this significant change in the adhesion phenomenon that was the most interesting to us. In virtually all polymer treatment systems examined the scale produced was quite tenacious to the surface of the test flask. It was clear even on a macroscopic level that some polymers altered the crystal morphology by the physical form of the precipitate. In each experiment with AR-980, even when we altered the test conditions to produce more scale, the crystals would not adhere to the flask. Our exploration into this phenomenon led us to some startling results.

Polymer Modified Crystal Morphology

We grew and harvested crystals for study by optical and Scanning Electron Microscopy using a wide variety of experimental and commercially available polymers. The optical

Figure 5. SEM of calcium carbonate grown with various polymers.

microscopy confirmed what we observed visually: the crystals of calcium carbonate produced while AR-980 was in solution looked significantly different from those produced with any other type of polymer. The various crystals harvested with different polymers in solution were examined by SEM. The results of the various polymers are shown in Figures 4 and 5.

Figure 4 (A) shows the cubic calcite which grows when there is no polymer present. The crystals are reasonably large, about 10 microns along each edge and offer substantial surface area to adhere to a surface or to bind cohesively to each other. Figure 4 (B) and (C) show crystals grown in the presence of commercially available poly(acrylic acid/maleic acid) typically used in the detergent industry. These polymers have a weight average molecular weight of about 20,000 and an acrylic acid:maleic acid molar ratio of about 3.5:1. The crystals show only minor changes from the typical calcite structure even maintaining approximately the same size. Figure 4 (D) shows a commercially available poly(acrylic acid) with a molecular weight of about 5000. The crystals are substantially altered showing the trend toward a spherical distortion (Figure 1) with about a 10 micron diameter.

Figure 5 (A) shows calcium carbonate precipitated in the presence of the poly(acrylic acid) used in our stress test experiments. The precipitate is substantially distorted showing a spherical shape with a diameter of roughly 1 micron. The calcium carbonate tends to form aggregates of these small spheres which are about 8 microns in diameter. Figure 5 (B) is a commercially available poly(acrylic acid/maleic acid) with a weight average molecular weight of about 5,000 and an acrylic acid:maleic acid molar ratio of about 2.2:1. The crystals are distorted into platelets which grow together to form a "cauliflower" type of morphology. These aggregates are 8-12 microns in size. Figure 5 (C) is an experimental poly(acrylic acid/maleic acid) with a weight average molecular weight of about 2500 and an acrylic acid:maleic acid

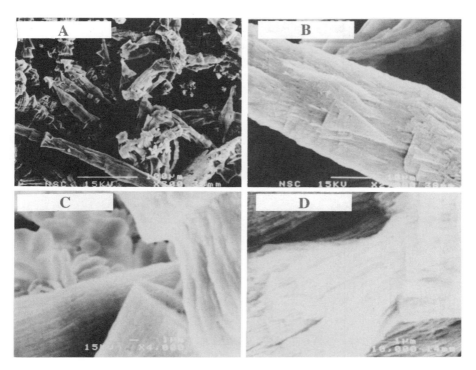

Figure 6. Calcium carbonate grown in the presence of Aquatreat AR-980.

molar ratio of 1:1. The precipitated calcium carbonate is distorted in a manner similar to 5 (B). In Figure 5 (D) are calcium carbonate crystals grown with Aquatreat AR-980 present. These crystals are , to the author's knowledge, unlike any other ever reported for calcium carbonate. Various magnifications of the crystals are shown in Figure 6.

The SEM photographs show the large size of the crystals formed when AR-980 is in solution. Typical single crystal or agglomerations of crystals are 1-10 microns in size. Figure 6 (A) shows the precipitate to be on the order of 100-200 microns in length and are thus considerably larger than those seen in Figures 4 and 5. The crystals appear to be filamentous in nature and grow from a central nucleating site which appears to be a small calcite crystal. Figure 6 (B), (C), and (D) show increasing magnification of the crystals and the "fan" like structure that seems to be the start of crystal growth. We hypothesize that calcium carbonate precipitating from solution in the presence of AR-980 grows in this manner and the growths emanating from the nucleating crystal are mechanically unstable. These crystals are then easily removed by an energy input into the system, such as mild agitation or the passing of water through a heat transfer device. It is not presently known if these crystals represent a thermodynamically stable form of calcium carbonate or if they are kinetically stabilized due to the presence of polymer in the crystal matrix. Further work is planned in this area.

CONCLUSIONS

From our work we conclude the following:

- We have developed an effective preparation for creating synthetic polymers containing a high (> 50 mole %) level of maleic acid. This preparation is especially advantageous since it is carried out in an aqueous environment.
- The synthetic scheme is versatile. We can control the polymer composition and molecular weight over wide ranges. The process is efficient in that greater than 99 % of the monomer is incorporated into the polymer.
- Aquatreat AR-980 is an effective threshold inhibitor under the conditions of our stress test. The stress test duplicates qualitative results reported from field applications.
- The polymer induces significant crystal habit modification.
- The modified crystals are mechanically unstable which accounts for the decreased adherent calcium carbonate in our stress test. Crystals precipitated with AR-980 in the system would be more easily removed.

Our approach of combining polymer synthesis with meaningful applications testing gives us the opportunity to optimize polymer composition for performance in a variety of stressed conditions. The unique crystal modifying ability of the polymer could also find uses in allied applications such as detergents and textiles.

ACKNOWLEDGMENTS

The authors would like to thank Alco Chemical for permitting us to publish this research. Also, we thank Anne Austin and Susan Fesser of Alco Chemical for their synthesis and GPC efforts, respectively. We also thank Robert Holzer of National Starch and Chemical for the SEM work.

REFERENCES

1. *Drew Principles of Industrial Water Treatment,* Drew Chemical Corporation, page 200, 1985.
2. Austin, A. B., Carrier, A. M., Belcher, J. H. and Standish, M. L., U. S. Patent applied for.
3. Feeser, Susan, Application of Aqueous GPC Using Photodiode Array for the Determination of Monomer Incorporation in Various Copolymers, *International GPC Symposium '94,* June 5-8, 1994.
4. NACE International, Laboratory Screening Tests to Determine the Ability of Scale Inhibitors to Prevent the Precipitation of Calcium Sulfate and Calcium Carbonate From Solution (For Oil and Gas Production Systems), *NACE Standard TM0374-90, Item 53023.*

7

THE DISSOLUTION OF CALCIUM CARBONATE IN THE PRESENCE OF MAGNESIUM AND INORGANIC ORTHOPHOSPHATE

Th. G. Sabbides and P. G. Koutsoukos

Institute of Chemical Engineering and High Temperature Chemical
    Processes
P.O. Box 1414
Department of Chemical Engineering
University of Patras
GR-265 00 Patras, Greece

INTRODUCTION

Damage of historical monuments due to environmental factors has been recognized over the past century and has triggered scientific research aiming at the protection of the marble structure of the monuments. Acid rain and atmospheric sulfur dioxide have been held responsible for the deterioration of the marble and limestone artwork.[1,2,3] Aside from acidity, wet precipitation by itself is believed to contribute to dissolution even at neutral pH. Calcium carbonate polymorphism perplexes the problem as the chemical composition of the various stones used is different depending on their origin. The various calcium carbonate polymorphs which are also naturally encountered include vaterite, aragonite and calcite.[4,5] The three polymorphs differ both from crystallographic point of view and also with respect to their solubilities, vaterite being the most and calcite the least soluble phase over a temperature range between 0° and 90°C.[6] Besides temperature the solubility of calcium carbonate polymorphs depends on the solution pH. Solubility increases rather sharply with pH and the aqueous medium in contact with stones may easily (particularly in the presence of acidic pollutants) become undersaturated with respect to all calcium carbonate polymorphs. In this case, dissolution may proceed. The physicochemical processes involved in the dissolution of calcium carbonate are very interesting and need to be understood in order to assess quantitatively the factors governing the dissolution of this salt in aqueous media.

From the mechanistic point of view the crystal growth and the dissolution of a solid phase in solution proceed through similar steps suggesting that they may be symmetric processes. The following steps may be distinguished in the detachment of the elementary units from the stone materials undergoing the deterioration when in contact with an

Mineral Scale Formation and Inhibition, Edited by Zahid Amjad
Plenum Press, New York, 1995

undersaturated solution (i) diffusion of the solute along a step edge from a kink (ii) migration on the surface from a step edge (iii) desorption reaction from the solid surface (iv) diffusion in the solution boundary layer adjacent to the solution[7] Any of these steps, alone or in combinations, may limit the kinetics of dissolution. When step (iv) is limiting, the kinetics are bulk diffusion controlled, when step (i) is limiting, the kinetics are controlled by surface diffusion. The dissolution of calcium carbonate has been the subject of numerous investigations over a wide pH range. Various laboratory studies have shown transport and surface controlled reactions of calcite in aqueous solutions.[8] Investigations over a wide pH range have shown that above pH 5.5 the rates are pH independent.[9,10] The problem associated with the dissolution studies reported in the literature is that they were done at conditions either of variable undersaturation or in synthetic or natural seawater, in which the presence of additional ions may complicate the picture. The present work aimed at studying the mechanism of calcium carbonate dissolution at neutral pH. The study involved experiments in which the driving force for the dissolution process, the solution undersaturation, was kept constant. The kinetics of dissolution were thus measured very precisely in high ionic strength aqueous solutions, corresponding to natural aquatic systems and the effect of the presence of Mg^{2+} and phosphate ions on the dissolution of calcium carbonate polymorphs was measured.

EXPERIMENTAL PROCEDURE

All experiments were done in a double walled Pyrex® glass reactor volume totaling approximately 230 ml, thermostated with circulating water from a constant temperature bath, at 25.0±0.1°C. Calcium chloride and sodium bicarbonate stock solutions were prepared from the respective solids (Merck. Pro Analisi). The standardization of the stock solutions was done by ion chromatography (Metrohm IC 690). The undersaturated solutions were prepared by mixing equal volumes (100 ml each) of solutions of calcium chloride and sodium carbonate made from the respective stock solutions by dilution. The ionic strength of each of the constituent solutions was adjusted to 0.62 M (3.5% w/v) with sodium chloride previously dried (Merck, Pro Analisi). Next, the solution pH was measured by a pair of glass (Radiometer G-202C) and saturated calomel electrodes (Radiometer G-202R) standardized before and after each experiment by NBS buffer solutions at pH 6.865 (0.025M KH_2PO_4 + 0.025M Na_2HPO_4) and 9.18 (0.01M Borax).[11] The pH of the undersaturated solutions was adjusted following the mixing of the components by slow addition of standard hydrochloric acid (Merck titrisol). The solution was magnetically stirred at 250 rpm and was allowed to equilibrate for a period of two hours. The criterion of equilibration was the pH constancy. After the equilibration period well characterized synthetically prepared[12] crystals of calcite, aragonite and vaterite were introduced in the undersaturated solutions. The specific surface area (multiple point dynamic B.E.T.) of the solid materials examined was 1.6 m^2/ g for calcite, 0.8 m^2/ g for aragonite and 5.1 m^2/ g for vaterite.

After the inoculation of seed crystals the dissolution process started without any induction time. Because of the dissolution process:

$$CaCO_3(s) + H^+ \Leftrightarrow Ca^{2+} + HCO_3 \qquad (1)$$

the solution pH tends to shift to more alkaline values. Change of the solution pH as small as 0.003 pH units triggered the addition of titrant solutions from two mechanically coupled burettes of a modified automatic titrator (Radiometer TT 2b) with an electric burette (ABU 12).

Assuming that the working solution contained X_1M of $CaCl_2$, X_2M Na_2CO_3 (total carbonate), X_3M HCl for the adjustment of the solution pH and X_4M of a supporting electrolyte MX (NaCl in our case) the titrant solutions were made as follows:

i. buret 1 $(2x_1-C)M$ $CaCl_2 + (2x_4+C)M$ $NaCl_2$

ii. buret 2 $(2x_2-C)M$ $Na_2CO_3 + (2x_3-C)M$ HCl

In the above equations C is a constant determined from preliminary experiments and is related with the velocity of dissolution of the calcium carbonate salt in our experimental conditions. As may be seen, in the dissolution experiments the constant C should be $2x_1$, $2x_2$, and $2x_3$. The factor 2 in front of x_1, x_2, x_3, and x_4 corresponds to the dilution anticipated by the solutions addition from the two burettes.

Samples were withdrawn during the dissolution process randomly and were filtered through membrane filters (Sleicher & Schuell 0.2 μm). The filtrate was analyzed for calcium by ion chromatography in order to confirm the constant solution composition. In all experiments the analysis showed that throughout the course of the dissolution process the calcium concentration remained constant to within $\pm 2\%$.

The solids on the membrane filters were analyzed by powder x-ray diffraction (Philips 1840/30 PW), infrared spectroscopy (FT-IR Nicolet 740) and scanning electron microscopy (SEM, JEOL JSM-5200).

The pistons of the burettes adding the titrant solutions were mechanically coupled with the recorder pen and the volume, V, added as a function of time, t, was translated into moles of calcium carbonate dissolved as a function of time. The rates of dissolution at any time were thus determined after normalization for the total surface area corresponding to the seed crystals which provided the active sites which initiated and further sustained the dissolution process. For the normalization both the reduction of the surface area because of dissolution and the area of the crystals withdrawn during sampling was taken into consideration. A laser particle counter (Spectrex ILI-1000) was used to measure the number and the size of crystallites in the undersaturated solution during dissolution .

RESULTS AND DISCUSSION

The driving force for the dissolution process:

$$CaCO_3(s) \Leftrightarrow CaCO_3(aq) \Leftrightarrow Ca^{2+}(aq) + CO_3^{2-}(aq) \tag{2}$$

is the difference between the chemical potentials of the solute in the undersaturated solutions from the corresponding values at equilibrium:

$$\Delta\mu = \mu_{CaCO_3(aq)} - \mu_{CaCO_3(\infty)} \tag{3}$$

where the subscripts aq and ∞ refer to the undersaturated solution and equilibrium respectively.

$$\Delta\mu = (\mu^{\circ}_{CaCO_3(aq)} + kT \ln a_{CaCO_3(aq)}) - (\mu^{\circ}_{CaCO_3(\infty)} + kT \ln a_{CaCO_3(\infty)}) \tag{4}$$

where k is the Boltzmann constant, T the absolute temperature, and a denote activities. Assuming $\mu^{\circ}_{CaCO_3(aq)} = \mu^{\circ}_{CaCO_3(\infty)}$:

$$\Delta\mu = kT \ln \frac{a_{CaCO_3(aq)}}{a_{CaCO_3(\infty)}} = \frac{kt}{2} \ln \frac{(a_{Ca^{2+}}a_{CO_3^{2-}})(aq)}{(a_{Ca^{2+}}a_{CO_3^{2-}})(\infty)} \qquad (5)$$

In equation 5 the denominator $(a_{Ca^{2+}}a_{CO_3^{2-}})_{(\infty)}$ is the thermodynamic solubility product of the calcium carbonate phase considered and the numerator is the activity product of the undersaturated solutions. For the calculation of the driving force the computation of the ionic activities is needed. The solution speciation was computed by the HYDRAQL program[13] written in FORTRAN 77 and tranferred to a VAX 11 / 750 computer. The extended Debye-Hückel equation corrected for ion interactions was used for the calculation of the activity coefficients.[14] The relative solution undersaturation with respect to a polymorph x was calculated from equation:

$$s_x = 1 - \left(\frac{(Ca^{2+})(CO_3^{2-})}{K_{sx}^\circ}\right)^{1/2} = 1 - (\Omega_x)^{1/2} \qquad (6)$$

The rates of dissolution of the various calcium carbonates were found to depend strongly on the solution undersaturation. The kinetics data were fitted to the semiempirical equation:

$$R_d = k_d s \sigma_x^n \qquad (7)$$

In equation 7, R_d is the dissolution rate measured from the rate of titrants addition, k_d the rate constant for dissolution, s a function of the active sites for the dissolution process and n the apparent dissolution order. The experimental data and the results obtained are summarized in Table 1. The general trend found for all polymorphs was that the dissolution rates increased with increasing undersaturation. Plots of the logarithms of the dissolution rates for the three polymorphs studied as a function of the logarithms of the respective relative undersaturation gave a satisfactory fit according to equation 7 as may be seen from the kinetics plots in Figure 1.

The least squares fit for the three polymorphs gave the following results:

Table 1. Dissolution of calcium carbonate polymorphs at constant undersaturation

Exp #	$Ca_t / 10^{-4}$ mol l^{-3}	Seed crystals	σ	$R_d / x10^{-5}$ mol min^{-1} m^{-2}
349	4.92	calcite	0.924	29.1
353	6.90	calcite	0.910	26.2
350	6.96	calcite	0.892	22.6
352	9.89	calcite	0.847	18.7
p45	5.47	aragonite	0.930	13.9
p41	5.96	aragonite	0.920	13.8
p40	7.43	aragonite	0.910	12.1
p42	9.90	aragonite	0.880	10.6
376	10.89	aragonite	0.870	10.3
p44	14.75	aragonite	0.820	6.9
p64	4.98	vaterite	0.940	22.0
p59	5.96	vaterite	0.930	21.3
p61	6.94	vaterite	0.910	20.1
p57	7.45	vaterite	0.900	19.2
p58	7-93	vaterite	0.899	17.0

25° C, pH 7.0, 3.5% NaCL, Total calcium Ca_t = total carbonate, C_t.

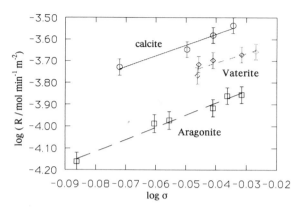

Figure 1. Kinetics of dissolution of calcium carbonate polymorphs at constant undersaturation. Plots of the logarithm of the rates of dissolution as a function of the logarithm of the relative undersaturation. pH 7.0, 3.5% NaCl, 25°C.

$$\log R_{d,calcite} = -3.38 + 5.01 \log\sigma_{calcite} \tag{8}$$

$$\log R_{d,aragonite} = -3.67 + 5.5 \log\sigma_{aragonite} \tag{9}$$

$$\log R_{d,vaterite} = -3.538 + 4.5 \log\sigma_{vaterite} \tag{10}$$

The dissolution of calcium carbonates therefore may be described by equation 11:

$$R_d = K\sigma^{5.0\pm0.5} \tag{11}$$

where K is the dissolution rate constant. Equation 11 is valid for $0 < \Omega < 1$.

Values of n>4 have been reported for the dissolution of aragonite and calcite in seawater.[15] The high apparent order together with the insensitivity of the dissolution rates on the stirring rates between 200-450 rpm suggested a surface controlled process, i.e. that the rate determining step is the diffusion of a crystal building unit from the active site to the crystal surface. Similar conclusions have been obtained for the dissolution of dicalcium phosphate dihydrate.[15]

Moreover, the high values for the apparent order of the dissolution of all calcium carbonate polymorphs may suggest control of the dissolution by a polynucleation mechanism suggested by Christoffersen.[17,18,19] In this case, the rate of dissolution may be written as:

$$R_d = k_d\sigma^{2/3} - (-\ln\Omega^{1/2})^{1/6} \exp(-A/\ln\Omega^{1/2}) \tag{12}$$

In equation 12, A is related with the surface energy, γ, of the dissolving crystallite by equation 13[20]:

$$A = \frac{\pi a^4 \gamma^2}{3kT} \tag{13}$$

where a is the linear size of a growth unit.

Rearranging equation 12 and taking the logarithms of both sides we obtained:

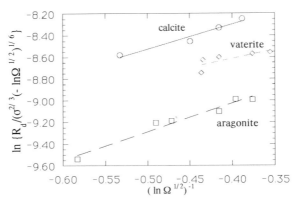

Figure 2. Kinetics of dissolution of calcium carbonate according to the polynuclear model. Plot of the left hand side of equation 14 as a function of $\ln\Omega^{1/2}$ pH 7.0, 3.5% NaCl, 25°C.

$$\ln\frac{R_d}{\sigma(-\ln\Omega^{1/2})^{1/6}} = \ln k_d - \frac{A}{\ln}\Omega^{1/2} \qquad (14)$$

Plots of the left hand side of equation 14 as a function of $[\ln\Omega^{1/2}]^{-1}$ for the various calcium carbonate polymorphs gave a satisfactory linear fit as may be seen in Figure 2. The surface energies computed from the slopes ranged between 20-30 mJ m^{-2}, values rather low but of the right order of magnitude.

The kinetics data suggested that the order of dissolution velocity is:

Figure 3. Scanning electron micrographs of calcium carbonate polymorphs: (a) calcite (b) aragonite (c) vaterite.

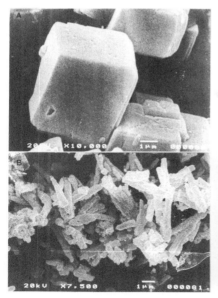

Figure 4. Dissolution of calcium carbonate polymorphs at constant supersaturation, pH 7.0, 3.5% NaCl, 25°C. Scanning electron micrographs of (a) calcite (b) aragonite and (c) vaterite after dissolution.

aragonite > vaterite > calcite

It has been suggested that the faster kinetics of dissolution of aragonite compared with calcite has direct implications for the importance of aragonite for the ocean water.[21] The series found may be correlated with the morphology of the crystals during the dissolution, shown in the scanning electron migrographs in Figures 3 and 4. Calcite and vaterite developed pits on their surface while the prismatic aragonite crystals appeared to dissolve along the c axis yielding smaller crystals.

Measurements of the number of crystallites and of their size distribution during the course of dissolution of the calcium carbonate polymorphs showed a decrease of the number of particles as a function of time, as may be seen in Figure 5a. The size evolution profile of the dissolving crystals at sustained undersaturation revealed some interesting features. A typical profile of the particle size changes during the dissolution of calcite is shown in Figure 5b. At the initial stages of dissolution (approx. up to 50 min) a sharp decrease of the crystals with sizes between 2-16 μm was observed with a simultaneous increase of the smaller sizes (1-2 μm).

Taking into consideration the decrease of the total number of crystallites because of the dissolution the increased percentage of the fraction of crystallites with the smallest size may be attributed to the rapid dissolution of the large crystals through the development of channels (as shown in the electron micrographs) which eventually yield smaller size crystals. Finally, the smaller crystallites also dissolve and the prevalence of larger sizes (4-16 μm) after extensive dissolution may be attributed to aggregation processes.

Previous work[12] has shown that aragonite may selectively grow on calcite seed crystals in supersaturated solutions containing magnesium at concentration levels 5.3 mM. We have thus examined the dissolution kinetics of the aragonite grown on calcite. The experimental conditions and the rates of dissolution measured are summarized in Table 2:

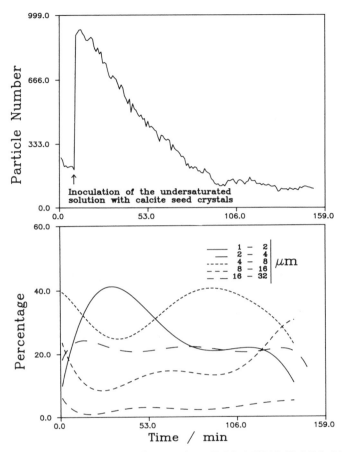

Figure 5. Dissolution of calcite at constant undersaturation, pH 7.0, 3.5% NaCl, 25°C. (a) Variation of the calcite particle number as a function of time. (b) Variation of the particle size distribution as a function of time.

The kinetics plot according to equation 7, of the logarithm of the rate of dissolution of the "composite" calcite - aragonite crystals as a function of the logarithm of the relative undersaturation gave a linear fit as may seen in Figure 6. The rate equation in this case was:

$$\log R_{d,c\text{-}a} = -4.87 + 3.6 \log\sigma_{aragonite} \tag{15}$$

Comparison of the kinetics equations 7 and 15 showed that the dissolution of aragonite seed crystals is considerably faster than the dissolution of aragonite grown on calcite. A possible explanation for this difference may be found from the morphological examination of the dissolving crystals. The scanning electron micrographs of the calcite - aragonite crystals before and after dissolution, shown in Figures 7a and 7b respectively, showed that the dissolution of the aragonite crystals takes place first, always along the c axis but from one side only (the other being attached onto the calcite surface). In the case of aragonite seed crystals both prismatic faces were dissolving.

Moreover, it should be noted that despite the fact that magnesium was not incorporated in the lattice of the aragonite crystals grown on calcite and no evidence of the presence

Table 2. Dissolution of aragonite grown on calcite at constant undersaturation; 25°C, pH 7.0 3.5% NaCl; $Ca_t = C_t$

Exp #	Ca_t / 10^{-4} M	σaragonite	R_d / x10^{-6} mol min^{-1} m^{-2}
360	10.0	0.878	8.6
361	7.0	0.914	9.7
362	5.0	0.932	11.0

of magnesium calcites was found in the x-ray diffraction spectra, analysis of the crystals showed the presence of magnesium, with a ratio of Mg:Ca = 0.3:100. It is very likely that Mg was present on the surface of the growing aragonite crystals. We thus have further investigated the interaction of magnesium ions with the various calcium carbonate polymorphs. The adsorption isotherms shown in Figure 8 suggested that the extent of adsorption followed the order: aragonite > calcite > vaterite. The magnesium probably through adsorption, blocked the active sites for dissolution thus yielding lower dissolution rates for the aragonite grown on calcite in comparison with the pure aragonite crystals.

The inhibiting role of the presence of magnesium ions in the dissolution process of calcium carbonates was confirmed with further experiments in which the dissolution of calcium carbonate was examined at conditions of constant solution composition (including constant magnesium concentration). The results obtained from these experiments are summarized in Table 3.

The kinetics plots according to the formalism of the semiempirical equation 7 are shown in Figure 9. As may be seen, the presence of magnesium resulted in a significant reduction of the dissolution rates but the reduction was higher for 1 ppm of Mg rather than in the presence of 11 ppm. In the latter case however, the concentration was higher than that needed for a surface coverage corresponding to the plateau of the adsorption isotherms.

In this case, a higher magnesium concentration in the undersaturated solution was anticipated. Measurements of the electrophoretic mobility of the calcite seed crystals have shown that calcite has a high negative charge at the experimental solution conditions. It is thus reasonable to suggest that the Magnesium - Calcite interactions are strong because they are enhanced by electrostatic interactions.

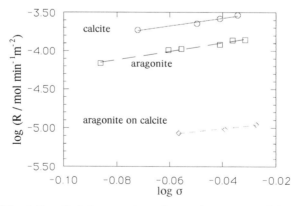

Figure 6. Kinetics of dissolution of calcite, aragonite, and aragonite grown on calcite crystals at conditions of sustained undersaturation, pH 7.0, 3.5% NaCl, 25°C. Plot of the logarithm of the rates of dissolution as a function of the relative undersaturation.

Table 3. Dissolution of calcite at constant undersaturation in the presence of magnesium; pH 7.0 25°C, 3.5% NaCl, $Ca_t = C_t$

Exp #	$Ca_t / x10^{-4}$ M	Mg ppm	$\sigma_{calcite}$	$R_d / x10^{-5}$ mol min^{-1} m^{-2}
p69	4.98	1.0	0.935	10.9
p67	5.97	1.0	0.922	10.2
p65	6.96	1.0	0.909	8.4
p66	7.94	1.0	0.896	7.8
p70	9.92	11.0	0.871	6.6
p71	8.00	11.0	0.896	13.3
p76	10.00	11.0	0.870	10.8
p72	10.92	11.0	0.857	9.9
p75	12.96	11.0	0.832	7.4

In order to further investigate the importance of electrostatic interaction of calcium carbonates with foreign ions we studied the effect of the presence inorganic orthophosphate in the undersaturated solution. At neutral pH the HPO_4^{2-} is the predominant species. Despite the high negative charge of these ions they have been found to be taken up by calcite at conditions of pH and ionic strength typically employed in our experiments.[22] The phosphate uptake may be attributed to surface complex formation with the Ca^{2+} lattice ions, since the electrical double layer is compressed by the high ionic strength of the aqueous medium (ca.

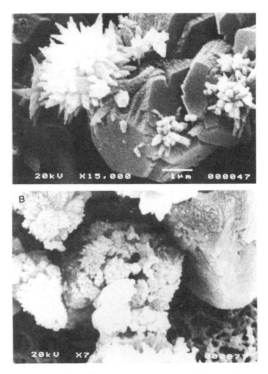

Figure 7. Scanning electron micrographs of aragonite grown on calcite seed crystals: (a) before dissolution and (b) after dissolution at constant undersaturation, pH 7.0, 3.5% NaCl, 25°C.

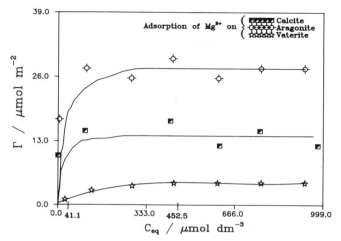

Figure 8. Adsorption isotherms for the adsorption of Mg^{2+} ions onto calcite. pH 8.0, ionic strength 0.1 M, 25°C.

0.62 M). The experimental conditions and the results obtained for the dissolution rates in the presence of 1 and 11 ppm of inorganic orthophosphate are summarized in Table 4.

The dependence of the logarithm of the rates of dissolution as a function of the relative undersaturation in the absence and in the presence of 1 and 11 ppm of inorganic orthophosphate is shown in Figure 10.

From the linear least squares fit the following relationships were found for the dissolution rates in the presence of 1 ppm, R_1 and 11 ppm, R_{11}, of inorganic orthophosphate and the rate of dissolution without phosphate R_o as a function of the undersaturation ratio $\Omega^{1/2}$ defined in equation 6:

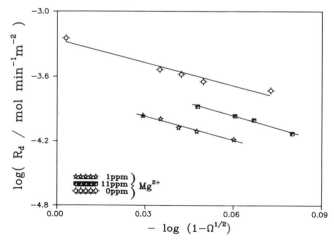

Figure 9. Kinetics of calcite dissolution at constant undersaturation in the presence of magnesium. Plots of the logarithm of the rates as a function of the logarithm of the relative undersaturation. pH 7.0, 3.5% NaCl, 25°C.

Table 4. Dissolution of calcite in the presence of inorganic orthophosphate at constant undersaturation pH 7.0, 25°C, 3.5% NaCl. Experimental conditions and dissolution rates $Ca_t = C_t$

Exp #	$Ca_t / x10^{-4}$ M	Phosphate ppm	σcalcite	$R_d / x10^{-5}$ mol min^{-1} m^{-2}
p18	5.46	1.0	0.929	4.2
p13	5.96	1.0	0.922	4.5
p14	6.46	1.0	0.916	4.1
p12	7.93	1.0	0.897	3.2
p15	9.88	1.0	0.871	2.4
p30	4.98	11.0	0.935	2.7
p29	5.47	11.0	0.929	3.1
p31	6.94	11.0	0.909	2.4
p27	7.93	11.0	0.897	2.3
p32	9.89	11.0	0.871	1.9

$$\frac{R_1}{R_o} = 0.22(1 - \Omega^{\frac{1}{2}})^{4.75} \tag{16}$$

$$\frac{R_{11}}{R_o} = 0.10(1 - \Omega^{\frac{1}{2}}) \tag{17}$$

In general, provided that $\Omega^{\frac{1}{2}} < 1$ and that Ω is a real number:

$$(1 - \Omega^{\frac{1}{2}})^a = \acute{a} \sum_{n=0}^{\infty} (\Omega^{\frac{1}{2}})/(n) \tag{18}$$

Equation 16 thus became:

$$\frac{R_1}{R_o} = 0.22 (1 - 4.75\Omega^{\frac{1}{2}} + 8.9W - 8.16\Omega^- + 3.57\Omega^2 - 0.53\Omega^{\frac{5}{2}}) \tag{19}$$

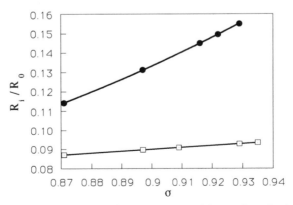

Figure 10. Kinetics of dissolution of calcite in the presence of inorganic orthophosphate at a constant undersaturation. Plot of the ratios of the rates of dissolution in the presence of phosphates over the rates measured in their absence as a function of the relative solution undersaturation. pH 7.0, 3.5% NaCl, 25°C.

factors of higher order were insignificant.

From equation 19 it may be seen that for $\Omega \to 1$, (saturation) $R_1/R_o \to 0$ (i.e. $R_1 \to 0$). That is, with decreasing undersaturation the rate of calcite dissolution in the presence of 1 ppm of inorganic orthophosphate decreased and provided that $R_o \neq 0$ the retardation effect is more inconstant at lower undersaturation. At high undersaturation, $\Omega \to 0$, $R_1/R_o \to 0.22$, there is still a definite reduction in the rate of dissolution with a limit of 80% of the uninhibited case. At higher phosphate concentration:

$$\frac{R_{11}}{R_o} = 0.10(1 - \Omega^{1/2}) \tag{20}$$

At low undersaturations ($W \to 1$) again $R_{11} \to 0$ but at high deviations from equilibrium, the retardation effect is stronger as the overall reduction of the rates of dissolution of calcite may reach 90% of the uninhibited process. The lower limiting rates predicted for the dissolution of calcite in the presence of 11 ppm of inorganic orthophosphate may be attributed to the higher extent of adsorption which was found to increase with increasing phosphate in the solution.

CONCLUSIONS

The investigation of the dissolution of calcium carbonates at constant undersaturation at neutral pH showed that the three polymorphs, calcite, aragonite and vaterite undergo dissolution via a surface controlled, polynuclear mechanism. The relative order for the dissolution was found to be aragonite > vaterite > calcite. Aragonite overgrown on calcite crystals yielded the slowest dissolution rates because of the presence of Mg^{2+} ions on its surface and because of the fact that dissolution took place along only one end of the acicular aragonite crystallites.

The presence of magnesium caused a reduction of the rates of dissolution which seemed to depend on the extent of Mg uptake on calcium carbonate. Inorganic orthophosphate on the other hand showed higher inhibition of the calcite dissolution with increasing concentrations in the undersaturated solutions. The increase in inhibition was attributed to the larger amount of adsorbed phosphate at higher concentrations. The different behaviour with respect to the influence of the rates of dissolution of the calcium carbonates of the cationic (Mg^{2+}) and anionic (HPO_4^{2-}) impurities in the undersaturated solutions should be attributed to their different behaviour at the calcium carbonate / electrolyte interface.

ACKNOWLEDGMENTS

The authors express their gratitude for partial support of this work by the commission of the Union, Contract No JOU2-CT92-108.

REFERENCES

1. Amoroso, G.G. and V. Fassina, 1983, "Stone decay and conservation" Elsevier.
2. Rosvall J., S. Aleby, O. Lindqvist, L.E. Olson (eds), 1986, "Air Pollution and Conservation", Safeguarding our Architectoral Heritage an Interdisciplinary Symposium , Oct. 15-17, Rome, Swedish Institute of Classical Studies.
3. Weaver M.E. , 1987, Acid rain versus Canada's Heritage" Heritage Canada Foundation Ottawa .

4. Carlson W.D., 1975, Reviews in Mineralogy" vol. 11, Reeder R.J., (ed.), *Mineralogical Soc. Am.*,11: 191-242.
5. Kitano V. , 1962, A Study of the Polymorphic Formation of Calcium Carbonate in Thermal Springs with an Emphasis on the Effects of Temperature, *Bull. Chem. Soc. Japan*, 35 : 1980-1985.
6. Plummer, L.N. and E. Busenberg, 1982, The solubilities of calcite, aragonite, and vaterite in CO_2-H_2O solutions between 0-90°C and an evaluation of the aqueous model for the system $CaCO_3$ - CO_2-H_2O, *Geochim. Cosmochim. Acta.*, 46:1011-1040.
7. Morel F., 1983, Principles of Aquatic Chemistry, J. Wiley, N.Y.
8. Plummer, L.N., D.L. Parkhurst, T.M.L. Wigley, 1979, Critical Review of the Kinetics of Calcite Dissolution and Precipitation, Chemical Modeling in Aqueous Systems, E.A. Jenne (Ed), ACS Washington D.C. ACS Symp. Ser. No 93:537-573.
9. Plummer, L.N., T.M.L. Wigley, D.L. Parkhurst, 1978, The Kinetics of Calcite Dissolution in CO_2 - water Systems at 5° to 60°C and 0.0 to 1.0 atm CO_2, *Am. J. Sci.*, 278:179-216.
10. Sjoberg E.L., 1978, Kinetics and Mechanism of Calcite Dissolution in Aqueous Solutions at Low Temperatures, Stockholm Contrib., *Geol.*, 32: 92p.
11. Bates R.G., 1973, pH Determination, John Wiley N.Y.
12. Sabbides, Th. G. and P.G. Koutsoukos, 1993, The crystallization of calcium carbonate in artificial seawater; Role of the substrate, *J. Crystal Growth*, 133: 13-22
13. Papelis, C., K.F. Hayes, J.O. Leckie, 1988, HYDRAQL: A Program for the Computation of the chemical Equilibrium Composition of Aqueous batch Systems Including Surface Complexation Modeling of Ion Adsorption at the oxide / solution interface, Technical Rep. No 306 (Stanford University, Stanford, CA,).
14. Truesdell, A.H., B.F. Jones, 1973, WATEQ, A Computer Program for Calculating Chemical Equilibria of Natural Waters, US Geological Survey, Washington DC.
15. Keir, R.S., 1980, The Dissolution Kinetics of Biogenic Calcium Carbonate in Sea Water, *Geochim. Cosmochim. Acta.*, 44: 241-252.
16. Zhang, J., G.H. Nancollas, 1991, Dissolution Kinetics of Calcium Phosphates Involved in Biomineralization in Advances in Industrial Crystallization, (J. Garside, R.J. Daven, A.G. Jones Eds.) Butterworth, Heinemann Oxford pp. 47-62.
17. Christoffersen, J. 1980, Kinetics of Dissolution of Calcium Hydroxyapatite. III. Nucleation - Controlled Dissolution of a Polydisperse Sample of Crystals, *J. Crystal Growth*, 49 : 29-38.
18. Hillig, W.B., 1966, A Derivation of Classical Two-Dimensional Nucleation Kinetics and the Associated Crystal Growth Ions, *Acta Metal.*, 14 :1868-1875.
19. Nielsen, A.E., 1984, Electrolyte Crystlal Growth Mechanism, *J. Crystal Growth*, 67: 289-302.
20. Christoffersen, J., Christoffersen M.R., 1981, Kinetics of Dissolution of Calcium Hydroxyapatite. IV. The Effect of some Biologically Important Inhibitors, *J. Crystal Growth*, 53 : 42-54.
21. Byrne, R.H., Acker J.G., Betzer P.R., Feely R.A., Cates M.H., 1984, Water Column Dissolution of Aragonite in the Pacific Ocean, *Nature*, 312 : 321-326.
22. Sabbides, Th. G., 1994, Investigation of the Calcium Carbonate / Water Interface Doctoral Thesis, University of Patras, Department of Chemical Engineering, Patras Greece, 200 pp.

8

PHOSPHOCITRATE

Potential To Influence Deposition of Scaling Salts and Corrosion

J. D. Sallis, W. Juckes, and M. E. Anderson

University of Tasmania
Department of Biochemistry
Box 252C G.P.O. Hobart, Tasmania, Australia 7001

INTRODUCTION

Compounds to prevent the formation of scaling salts and protect against corrosion are continuously being sought as the potential advantages of those presently in use are often restricted by secondary problems. Biodegradability, acidity and toxicity are a few key concerns which has led researchers to look for more environmentally acceptable, effective natural compounds. Polypeptides, for example, rich in acidic amino acids (particularly aspartate) have been isolated from marine organisms and demonstrated to assist in controlling nucleation and crystallization of calcium salts.[1] Such studies confirm that inhibitory power from many of the compounds in use stems from the presence of acidic groups or negatively charged ionic species. On these criteria, 2-phosphonooxy-1,2,3 propanetricarboxylic acid (phosphocitrate = PC), a natural compound reported to occur in mammalian soft tissue[2,3] and the hepatopancreas of the blue crab[2] and possessing 5 possible dissociable groups would appear to be worthy of consideration. For experimental purposes, synthetic PC has been prepared[4,5,6,7] and to date, in vivo studies with this compound have not highlighted any discernable toxicity.[8,9,10] This is not surprising as any ultimate destruction or metabolism of PC leads to phosphate and citrate, products in themselves considered as harmless and in many instances beneficial. Stability of PC under the majority of working conditions so far explored has never posed a problem.

Past research on PC has centered on its ability to control deposition of calcium phosphates and calcium oxalates either under *in vitro* controlled media conditions or *in vivo* associated with simulated pathological states.[10-15] The accumulated data leave no doubt that the compound is a powerful inhibitor of the transformation of calcium phosphate to crystalline hydroxyapatite[6,11] and that it prevents formation of calcium oxalate monohydrate (COM) arising from the conversion of calcium oxalate dihydrate (COD). Nucleation, growth and aggregegation of COM are events further strongly influenced by PC.[16,17] Not only are the calcium salts referred to above affected by the presence of PC but two well known magnesium crystalites, magnesium ammonium phosphate (struvite) and magnesium monohydrogen phosphate (newberyite), also can be prevented from forming.[18] The greater strength of PC's inhibitory power in comparison to other inhibitors derives from a powerful crystal binding affinity generated by its strong negative charge/size ratio and a more favorable stereochemistry. At pH 7.2, 3.5 of the available groups are believed to be dissociated.[19] Detailed crystallographic studies recently completed have demonstrated that PC can better penetrate and coordinate its functional groups to the calcium ions present in the (-1 0 1) face of the growing calcium oxalate monohydrate crystal than for example citrate (see Wierzbicki et al. this volume).

With this background knowledge then, it would appear that PC has potential to restrict the deposition of other common scaling salts and perhaps contribute protection against rusting and corrosion. These aspects are the basis of the present investigations whereby the action of PC on calcite and gypsum formation and deposition is discussed together with its ability to disperse iron particulates and prevent corrosion.

MATERIALS

PC was prepared and characterized by methods previously reported.[2,3,6] The neutral sodium salt was used for all studies. Where indicated, polyacrylate (Aldrich Chem Co, Milwaukee USA) of average M.W. 2,000 was trialled for inhibitory comparison purposes. All other chemicals used were reagent grade.

EXPERIMENTAL AND RESULTS

Studies Detailing the Influence of Phosphocitrate on Calcite and Gypsum Formation and Deposition

Calcite. A calcium carbonate nucleation assay was established essentially as described by Sikes et al.[20] An unstable supersaturated solution of calcium carbonate initially

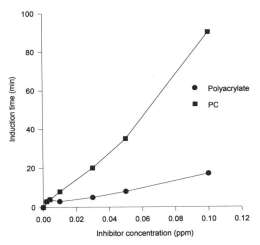

Figure 1. Comparison of the inhibitory strength of phosphocitrate and polyacrylate on the nucleation of calcium carbonate. Data represent mean ± se (n = 2).

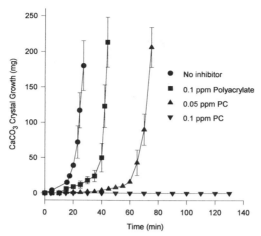

Figure 2. Inhibitory influence of phosphocitrate and polyacrylate on calcium carbonate crystal growth. Data are expressed as mean ± se. (n = 3).

was prepared to which the test inhibitor was added. Under a positive nitrogen atmosphere and at 20°C, the reaction was initiated by raising the pH to 8.1 with KOH. It remained constant until nucleation commenced at which point a decrease in pH occurred and is referenced as the induction time.

In the absence of any inhibitor (control), the induction time was approximately 1 min. Both inhibitors lengthened the induction time almost linearly with increasing concentration. On a weight basis, PC was seen to be superior to polyacrylate.

The ability of both compounds to restrict growth of calcium carbonate also was tested. Experimental conditions were modelled after those of Kazmierczac et al.[21] whereby a metastable solution of calcium carbonate was prepared and the reaction initiated by the addition of aged calcite seed crystals. The latter were prepared as described by Reddy and Nancollas.[22] The mixture was again made alkaline (pH 8.5) and constant composition was maintained by continuous automatic titration of reactants into solution.

Both PC and polyacrylate proved to be potent inhibitors. No growth was observed within 2 h in the presence of 0.1 ppm PC while at the same concentration, polyacrylate was less inhibitory with induction taking about 1 h. SEM of crystals formed in the presence of either inhibitor showed marked changes to the internal latticework and faces.

In comparison to the normal rhombohedral flat-faced calcite crystal (A), PC (B) and polyacrylate (C) both led to distorted crystal surfaces and a honey-combed interior.

Gypsum. Inhibition by PC of the formation and deposition of calcium sulphate on a heat transfer metal surface was tested using the system described by Amjad.[23] A copper U-tube (12 mm outer diameter which had been scrupulously cleaned, polished and delipi-dated) was immersed in a crystallization cell containing a supersaturated solution of calcium sulfate to which inhibitor was then added. The cell was placed in a bath held at 2°C and a temperature differential was applied by circulating hot water at 70°C through the tube. At the termination of the experiment, the deposit was removed with 1M HCl and calcium analyzed by atomic absorption spectroscopy.

Preliminary experiments in the absence of any inhibitor indicated linearity of salt deposition with time so for convenience, the data were obtained at 30 min. Strong inhibition

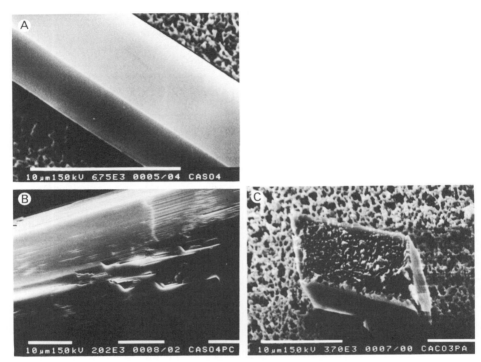

Figure 3. Scanning electron micrographs of calcium carbonate crystals grown in the absence and presence of phosphocitrate or polyacrylate. Solid bar = 10 μm.

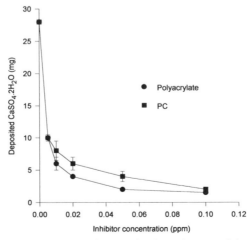

Figure 4. Comparative inhibition of phosphocitrate and polyacrylate on calcium sulfate deposition on the surface of a copper heat exchanger. Data are expressed as mean ± se. (n = 2).

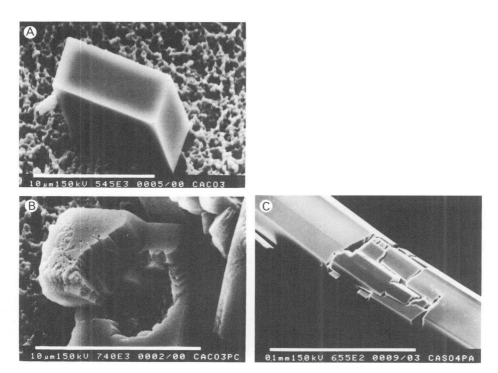

Figure 5. Scanning electron micrographs of a gypsum crystal grown in the presence or absence of inhibitor. Bar = 10 um (control, PC) and 100um (PA).

was exerted by either of the test compounds. At a concentration of 0.005 ppm (w/w), there was a 60% decrease in the amount of gypsum deposited while at a higher concentration (0.1 ppm), almost total inhibition was observed. Micrographs from SEM studies revealed the powerful nature of both inhibitors in respect to controlling formation of gypsum. There was marked breakup of the crystal faces (Figures. 5B-PC and 5C-PA) by comparison to those crystals grown in the absence of any inhibitor (Figure 5A).

The existing evidence then shows that PC does have the ability to restrict the formation and growth of all of the scaling salts tested to date, but does it have any influence over dispersion of iron particulates or prevent rusting and corrosion?

Studies on Iron Dispersion Activity

The possibility that iron salts or particulates could be kept in solution by the presence of PC was investigated as described by Garris and Sikes.[24] A 500 ml solution containing 4.5 ml NaHCO$_3$ (0.4 M) and 1.8 ml CaCl$_2$ (1.0 M) with or without PC was prepared and adjusted to pH 8.50 with NaOH. A slurry of Fe$_2$O$_3$ (50 mg) was added to initiate the reaction and the mixture stirred for predetermined experimental periods. Aliquots were then taken approximately 1.5 cm below the surface and transferred to cuvettes for reading at 500 nm. It has

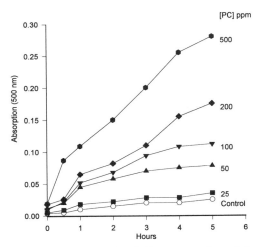

Figure 6. Iron oxide dispersion curves with variable concentrations of phosphocitrate present.

previously been established that there is a positive correlation between iron dispersion activity and absorbance.[24]

Increasing concentrations of PC clearly provided increasing dispersive activity. The mechanism of the inhibitor's action has not yet been probed but it is assumed that the contribution of the dissociation of anionic groups present in the compound would be responsible both for solubilization and dispersion of colloidal iron material.

Figure 7. Visual appearance of steel coupons immersed in a salt solution for 7 days in the presence and absence of phosphocitrate.

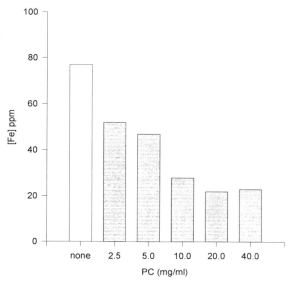

Figure 8. Concentrations of iron oxide present on plates exposed to a corrosive medium in the presence of variable concentrations of phosphocitrate.

Studies Related to Rusting and Corrosion of Mild Steel

In a series of experiments to test the effectiveness of PC to control rusting and corrosion, mild steel plates and /or coupons were used with a variety of protocols. Initial preparation of the plates or coupons consisted of washing with 20% HCl, polishing the surface to a mirror finish with progressively finer grades of abrasive grit paper (300 to 1200) and finally rinsing with ethanol to delipidate. In additional preparation of plates, an area approximately 1 cm^2 was masked off with tape, the entire plate dip-coated (3 x) in a

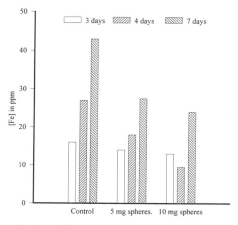

Figure 9. Prevention of iron oxide deposition on an exposed mild steel surface using phosphocitrate microspheres.

polyurethane base paint, allowing to dry between coats. The tape was removed and the exposed surface briefly cleaned again with ethanol. For routine experiments, the prepared plates or coupons were immersed in 10% NaCl and were allowed to stand at room temperature for periods up to 7 days, with or without the inclusion of PC.

A blackening of the exposed metal surface when PC was omitted from the corrosion medium (Figure 7A - control) was clearly evident and a heavy orange-brown flocculant precipitate formed rapidly within 3 days. In the presence of 0.5 mg PC / ml however (Figure 7B-PC), the coupons still retained a grey - silver appearance and the color of the medium slowly increased from pale yellow to a yellow-orange. It was evident that the inhibitor was preventing iron hydroxide flocculation and at the same time keeping the released iron in a more soluble form. This phenomenon was probably the resultant influence of both the carboxyl groups and the phosphate moiety. Millipore filtration did not reveal any significant retention of colloidal particles at a pore size of 0.22 μm.

In similar experiments with plates, at the end of 7 days, the exposed metal surface was briefly rinsed with distilled water and then allowed to air dry. Deposition of iron oxide occurred and the extent of this rusting was quantitated. The iron oxide was removed with HCl and the concentration of iron determined by a thiocyanate assay procedure.[25]

A dose - response relationship could be demonstrated with PC concentrations up to 20 mg/ml in respect to rusting.

The majority of protocols to test the inhibitory influence of PC were focussed on adding the compound directly to the bathing medium. As an alternate mode of delivery, a slow release preparation of PC was investigated for its potential to be included into a paint for coating a metal surface. Microspheres of PC were prepared as previously reported.[26] The spheres consisted of a central core containing the sodium salt of PC trapped within an albumin cross-linked glutaraldehyde vesicle which itself was coated with polymerized poly DL lactide. During the formation of the outer more impermeable shell in an aqueous environment containing Ca^{++}, the calcium salt of PC was trapped. The net result of such a preparation was to provide a mixed salt of PC efficiently trapped within a microsphere and having variable but continuous release rates of PC. The PC microspheres then were suspended in the polyurethane paint and applied to the metal surface in the manner described earlier. The percentage PC available in relation to the mass of microsphere was approximately 22%.

When the PC painted plates were exposed to 10% NaCl for 7 days, rinsed in water, allowed to dry and the rust quantitated, the following data were obtained.

It was evident that PC was releasing from the paint and offering protection against rusting as the exposed surface continued to maintain a normal appearance and the iron quantitation data supported this visual observation. While more and more iron oxide continued to form from day 3 when PC was not included in the painted surface, the opposite effect was noted in all situations where PC was present. The result was in fact more impressive than supplying the PC directly to the NaCl solution as less PC would have been present. These findings then would suggest that versatility could be introduced in any treatment regime using PC. The compound could be added directly in a soluble form to circulating water or it might be added as slow release microspheres or it could be introduced in the form of an adhesive tape or coating. More complex investigations with model pilot plant situations would obviously be mandatory to defining the roles suggested.

The power of PC to protect against pitting also was investigated, in this instance by SEM. Polyurethane painted plates (no PC in the paint) with the 1 cm² polished surface exposed were again immersed in 10% NaCl for 7 days with or without PC respectively. At the end of this period, the deposited salts were removed from the metal surface by dipping in Clarke's cleaning solution comprising concentrated HCl (1 liter), antimony trioxide (20

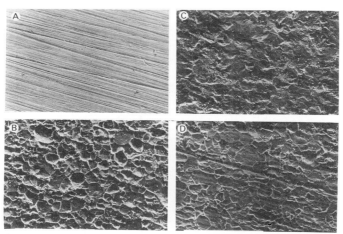

Figure 10. Scanning electron micrographs of mild steel and the pitting protection afforded by phosphocitrate. Bar = 100 μm.

g) and stannous chloride (50 g), room temperature for 10 min. A sample of metal was then mounted on a stub for SEM viewing.

In comparison to the micrograph of an unexposed normal polished mild steel plate (Figure 10A), there was evidence of extensive pitting on plates immersed in the corrosion medium (Figure 10B). With increasing amounts of PC in solution however, pitting was markedly reduced and in fact, there was little if any pitting with the highest concentration of PC used (Figures. 10C and D). An organic film appeared to reside over the metal surface, an observation consistent with that described for polyaspartate.[27]

Gravimetric and Electrochemical Testing for an Anticorrosive Effect of Phosphocitrate

Attempts to quantify an anticorrosion - PC dose-response relationship using two common approaches have so far proved elusive. A gravimetric technique initially was investigated using preweighed polished steel coupons which were then totally immersed in 10% NaCl solution in the presence or absence of PC. After 6 weeks, the coupons were removed, placed in Clarke's solution, rinsed, dried and reweighed. Although there were obvious visual changes identical to those described earlier with the plate experiments, no significant changes in metal weights were recorded. Long term experiments (up to 12 months) are normally recommended for accuracy in assessing metal loss in this type of analysis. As PC is known to hydrolyse slowly in an aqueous environment over a period of months, the technique was considered unsuitable in this instance.

Instead, electrochemical testing by polarization resistance measurements was pursued. A standard 1 l capacity 5-necked polarization cell was fitted with two platinum counter electrodes, a central working electrode with associated holder for a mild steel coupon (surface area = 4.91 cm^2) a saturated calomel reference electrode and a salt bridge probe positioned within 2 mm of the metal surface. The cell was connected to a potentiostat/galvanostat (Wenking stepping motor potentiometer model SMP 72 and associated precision potential meter PPT 70, Gerhard Bank Electronic, Göttingen, Germany). The cell was filled with artificial sea water,[28] inhibitor added and conditions allowed to stabilize for a minimum

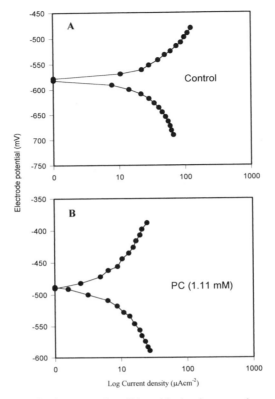

Figure 11. Potentiostat polarization curves for mild steel in the absence and presence of phosphocitrate.

of 20 min. Polarization resistance (Rp) was measured as the slope of a curve of potential (E) vs current density (i) at (E_{corr}), where i = 0.

The upper curve in each instance was compared and in the presence of PC (Figure 11B) it appeared to shift to the left and be a little steeper relative to the control (Figure 11A). Analysis of the Tafel slope drawn from curves plotted at the various concentrations of PC tested enabled the compilation of the data shown in Table 1.

Table 1. Influence of phosphocitrate on corrosion parameters relating to the exposure of mild steel to salt water. Values obtained are the mean of data derived from results of a minimum of 6 experiments for each concentration of PC

Addition PC (mg/L)	E-corr (mV)	i-corr ($\mu A/cm^2$)
none	- 676	17.0
250	- 606	14.0
500	- 525	6.2
750	- 536	8.8
1000	- 549	9.2

With PC present, there was progression to a more positive potential suggestive that the compound was acting as an anodic corrosion inhibitor. Although i_{corr} values decreased with increasing amounts of PC up to 500 mg/l , a dose response was not evident with higher concentrations of PC. This anomaly might perhaps be attributable to an increase in negative ions in solution interfering with the measurement. At 0.05% PC then, the compound could be considered a moderate corrosion inhibitor.

CONCLUSIONS

The studies confirm that PC does have the ability to control scaling salt formation and deposition. Further, the compound can prevent the precipitation of iron salts from solution and offers protection against rusting and corrosion. The demonstrated flexibility to provide PC where a situation might arise for a soluble additive or alternately, from a slow release microsphere form seems useful. It means that the inhibitor can be incorporated into inert materials for coating surfaces or provided as colloidal particles. There is also the advantage that phosphocitrate could be used as either the acid or neutral salt and from the point of view of degradation products, it should totally meet environmental guidelines. Of course, many other important aspects such as thermostability, long-term viability in aqueous alkaline environments and performance under complex mixture conditions have to be investigated before the true potential of phosphocitrate can be established.

Research support was provided by the University of Tasmania through the ARC Small Grants Scheme. The SEM services of the University Central Science Laboratory is acknowledged. The authors also are grateful to Mr. W. McEwan and Mr. G. Betts of the Hydro-Electric Commission, Moonah, for technical advice and the provision of electrochemical testing equipment.

REFERENCES

1. Wierzbicki, A., Sikes, C.S., Madura, J.D., and Drake, B., 1994, Atomic force microscopy and molecular modeling of protein and peptide binding to calcite, *Calcif. Tissue Int.* 54: 133-141
2. Lehninger, A.L., 1977, Mitochondria and biological mineralization processes: An exploration, *Horiz. Biochem. Biophys.* 4: 1-30.
3. Williams, G., and Sallis, J.D., 1981, The sources of phosphocitrate and its influential role in inhibiting calcium phosphate and calcium oxalate crystallization. In: *Urolithiasis, Clinical and Basic Research*, Smith, L.H., Robertson, W.G., and Finlayson, B., (eds). pp569-577 Plenum Press, N.Y.
4. Meyer, J., Bolen, R.J., and Stakelum, J.J., 1959, The synthesis of citric acid phosphate: *J.A.C.S.* 81: 2094 - 2096.
5. Williams, G., and Sallis, J.D., 1980, The synthesis of un-labelled and ^{32}P-labelled phosphocitrate and analytical systems for its identification, *Anal. Biochem.* 102: 365-373.
6. Tew, W.P., Mahle, C., Benavides, J., Howard, J.E., and Lehninger, A.L., 1980, Synthesis and characterization of phosphocitrate acid, a potent inhibitor of hydroxylapatite crystal growth, *Biochemistry* 19: 1983-1988.
7. Pankowski, A.H., Meehan, J.M., and Sallis, J.D., 1994, Synthesis via a cyclic dioxatrichlorophosphorane of 1,3-dibenzyl-2-phosphonooxy citrate, *Tetrahedron Letters* 35: 927-930.
8. Tew, W.P., Malis, C.D., Howard, J.E., and Lehninger, A.L., 1981, Phosphocitrate inhibits mitochondrial and cytosolic accumulation of calcium in kidney cells *in vivo, Proc. Natl. Acad. Sci.* 78: 5528-5532.
9. Tsao, J.W., Schoen, F.J., Shankar, R., Sallis, J.D., and Levy, R.J., 1988, Retardation of calcification of bovine pericardium used in bioprosthetic heart valves by phosphocitrate and a synthetic analogue, *Biomaterials* 9: 393-397.
10. Krug, H.E., Mahowald, M.L., Halverson, P.B., Sallis, J.D., and Cheung, H.S., 1993, Phosphocitrate prevents disease progression in murine progressive ankylosis, *Arthritis Rheum.* 36: 1603-1611.

11. Williams, G., and Sallis, J.D., 1982, Structural factors influencing the ability of compounds to inhibit hydroxyapatite formation, *Calcif. Tissue Int.* 34: 169-177.

12. Shankar, R., Crowden, S., and Sallis, J.D., 1984, Phosphocitrate and its analogue N-sulpho-2-amino tricarballylate inhibit aortic calcification, *Atherosclerosis* 52: 191-198.

13. Sallis, J.D., Brown, M., and Parker, N.M., 1990, Phosphorylated and non-phosphorylated carboxylic acids. Influence of group substitutions and comparison of compounds to phosphocitrate with respect to inhibition of calcium salt crystallization. In: *Surface Reactive Peptides and Polymers: Discovery and Commercialization.* Sikes, C.S., and Wheeler, A.P., (eds) ACS Books, Washington. pp.149-160.

14. Johnsson, M., Richardson, C.F., Sallis, J.D., and Nancollas, G.H., 1991, Adsorption and mineralization effects of citrate and phosphocitrate on hydroxyapatite, *Calcif. Tissue Int.* 49: 134-137.

15. Cooper, C.M., and Sallis, J.D., 1993, Studies on the gastrointestinal absorption of phosphocitrate, a powerful controller of hydroxyapatite formation, *Int. J. Pharm.* 98: 165-172.

16. Richardson, C.F., Johnsson, M., Bangash, F.K., Sharma, V.K., Sallis, J.D., and Nancollas, G.H., 1990, The effects of citrate and phosphocitrate on the kinetics of mineralization of calcium oxalate monohydrate. In: *Materials Synthesis Utilizing Biological Processes,* Rieke, P.C., Calvert, P.D., and Alper, M., (eds;) 174: 87-92. Materials Research Society Series, Pittsburgh, Pa.

17. Wierzbicki, A., Sikes, C.S., Sallis, J.D., Madura, J.D., Stevens, E.D., and Martin, K.L., Scanning electron microscopy and molecular modelling of calcium oxalate monohydrate crystal growth inhibition by citrate and phosphocitrate, *Calcif. Tissue Int.* (in press).

18. Sallis, J.D., Thomson, R., Rees, B., and Shankar, R., 1988, Reduction of infection stones in rats by combined antibiotic and phosphocitrate therapy, *J. Urol.* 140: 1063-1066.

19. Ward, L.C., Shankar, R., Sallis, J.D., A possible antiatherogenic role for phosphocitrate through modulation of low density lipoprotein uptake and degradation in aortic smooth muscle cells. *Atherosclerosis* 65: 117-124

20. Sikes, C.S., Yeung, M.L., and Wheeler, A.P., 1991, Inhibition of calcium carbonate and phosphate crystallization by peptides enriched in aspartic acid and phosphoserine. In: *Surface Reactive Peptides and Polymers: Discovery and Commercialization,* Sikes, C.S., and Wheeler, A.P., (eds.), ACS Books, Washington, DC, pp.50-71.

21. Kazmierczac, T.F., Tomson, M.B., and Nancollas, G.H., 1982, Crystal growth of calcium carbonate. A controlled composition kinetic study, *J. Phys. Chem.* 86: 103-107.

22. Reddy, M.M., and Nancollas, G.H., 1971, The crystallization of calcium carbonate. 1. Isotopic exchange and kinetics, *J. Colloid Interface Sci.* 36: 166-172.

23. Amjad, Z., 1988, Calcium sulfate dihydrate (gypsum) scale formation on heat exchanger surfaces: the influence of scale inhibitors, *J. Colloid Interface Sci.* 123: 523-536.

24. Garris, J.P., and Sikes, C.S., 1993, Use of polyamino acid analogs of biomineralizing proteins in dispersion of inorganic particulates important to water treatment, *Colloids and Surfaces A* 2-3: 103-112.

25. Vogel, A.I., 1961, *Quantitative Inorganic Analysis.* 3rd. edn. pp.786-787. Longmans, London.

26. Sallis, J.D., Meehan, J.D., Kamperman, H., and Anderson, M.E., 1993, Chemically modified phosphocitrate and entrapment in microparticles for sustained inhibition of undesirable biomineralization, *Phosphorus, Sulfur and Silicon* 76: 281-284.

27. Little, B.J., and Sikes, C.S., 1991, Corrosion inhibition by thermal polyaspartate. In: *Surface Reactive Peptides and Polymers: Discovery and Commercialization,* Sikes, C.S., and Wheeler, A.P., (eds), ACS Books, Washington, DC. pp 263-279.

28. Dawson, R.M.C., Elliott, D.C., Elliott, W.H., and Jones, K.M., 1986, *Data for biochemical research.* 3rd. Edition. p.448. Oxford University Press, N.Y.

9

ADSORPTION BEHAVIOUR OF POLYELECTROLYTES IN RELATION TO THE CRYSTAL GROWTH KINETICS OF BARIUM SULFATE

M. C. van der Leeden[1] and G. M. van Rosmalen[2]

[1] Delft University of Technology
Laboratory of Physical Chemistry
Julianalaan 136, 2628 BL Delft, The Netherlands
[2] Delft University of Technology
Laboratory for Process Equipment
Leeghwaterstraat 44, 2628 CA Delft, The Netherlands

ABSTRACT

For two polyphosphinoacrylates PPAA-I and PPAA-II and a copolymer of maleic acid and vinylsulfonic acid PMA-PVS, the adsorption levels on $BaSO_4$ were determined as a function of time and compared with the growth inhibiting effectiveness of these compounds in $BaSO_4$ crystallization. Only slightly different values are found for the equilibrium adsorption plateau of the best growth retarder PMA-PVS and the moderate inhibitor PPAA-II, but the time needed to establish the plateau coverage appears to be considerably shorter in the case of PMA-PVS. An indication is found that the adsorption rate of PPAA-II interferes with the growth rate of $BaSO_4$ in the absence of an inhibitor under the given experimental conditions, which has a negative influence on its growth inhibiting performance.

INTRODUCTION

Barium sulfate precipitation forms a severe problem in especially the off-shore oil and gas production, where seawater is injected into the oil bearing strata in order to squeeze the oil through the porous rock layers into the production well, the so-called secondary recovery method. The water, originally present in the pores of the formation layers, often contains a relatively high concentration of barium ions which is, for instance, the case in many North Sea production fields. When this water comes into contact with the seawater, which contains a considerable concentration of sulfate ions, barium sulfate precipitation in

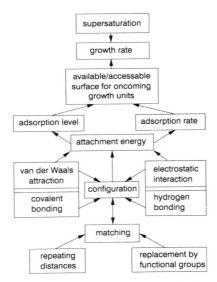

Figure 1. Factors influencing the adsorption process and, consequently, the growth kinetics of a mineral salt in the presence of an inhibitor.

the downhole equipment and in the pores of the rock layers is inevitable, resulting in a reduced productivity of the well. The treatment of this barium sulfate scale problem is mainly focussed on its prevention by the addition of scale-control additives, generally indicated as inhibitors. Most frequently applied inhibitors are phosphonate and polyelectrolyte compounds, because of their high effectiveness, even when dosed in trace amounts, and their extremely good thermal stability. Polyelectrolyte-type inhibitors are primarily designed for their ability to form a relatively large number of coordinative bonds with the cations of the mineral surface.

Many parameters together, however, compose the effectiveness of a growth inhibitor. And although it is generally accepted that preferential adsorption of a growth inhibitor at the crystal surface is an essential step in its specific performance, the precise nature and role of the various factors influencing this step are still not fully disclosed. The often interrelated factors and parameters, which influence the adsorption process and thus the growth kinetics of a mineral salt in the presence of an inhibitor, are presented in Figure 1.

In the past decade, many of these aspects of inhibitor adsorption in relation to the inhibitor performance have been studied for different types of inhibitors. In the next section, a brief overview of the achieved knowledge will be presented.

INHIBITOR PERFORMANCE IN RELATION TO ITS ADSORPTION BEHAVIOUR

The normal crystal growth kinetics will be disturbed by the adsorption of inhibitor molecules in competition with the regular growth units. The degree of disturbance will be determined by the adsorption level and the adsorption rate, the last being mostly ignored in the description of the growth inhibition process. These factors are determined by the attachment energy which is influenced by parameters as type and strength of the bonds between surface ions and functional groups of the inhibitor, their electrostatic interaction, the inhibitor configuration, and the matching of the interatomic distances and orientation of the lattice ions in the crystal surface layer with respect to the functional groups of the inhibitor landing upon the crystal surface.

Adsorption Level

The adsorption level that can be reached by an inhibitor will be determined by its affinity towards the crystal surface. The attraction can be an electrical interaction, for which at least some ionization of the functional acidic groups of the inhibitor is needed.[1] Total ionization, however, prevents the additive from adsorption because the attendant entropy loss cannot be compensated by the energy gain obtained by bondage with the surface.[2] For the bondage with the crystal surface, van der Waals interaction is essential. In addition, functional groups are needed which are able to form coordinative bonds with the cations at the crystal surface. Also functional groups which are able to form hydrogen bridges can contribute to the bonding.

The surface coverage needed to obtain growth blockage will depend on the type of inhibitor, the pH-value, the type of mineral and its surface relief related to its growth mechanism. For small phosphonate molecules, surface coverages of only 5% and 12% in the case of $CaSO_4 \bullet 2H_2O$ (gypsum)[3] and $BaSO_4$ crystallization[4], respectively, appeared to be sufficient to attain total growth blockage at pH-values of 5. This coverage is considerably lower than the maximum equilibrium surface coverage in both cases and may be explained by complete adsorption of the small molecules along the steps.[5,6,7]

In general, the relatively large and less flexible polyelectrolyte inhibitors will not primarily adsorb along the steps but onto the crystal terraces, thus providing obstacles for the advancing growth steps.[8]

Inhibitor Configuration

For an effective use of both attraction and bonding possibilities by the inhibitor, a flat configuration of the inhibitor upon the crystal surface is desired. Hydrophobic groups, for instance, appeared to be have a negative influence on the inhibitor performance.[1] Such groups incline to extend from the surface, resulting in a loopy configuration of the inhibitor molecule, they increase the molecular weight of the inhibitor without adding extra bonding possibilities and they may even cause steric hindrance by masking functional groups or by hampering their access to the crystal surface. The result is a relatively high surface coverage, but a low attachment strength between inhibitor molecules and the surface. Non-functional loops and tails may also be present in the case of a low ionization degree of the inhibitor molecule, depending on the pH of the solution and on the ionization constants of the functional groups. With only a few ionized functional groups, also the attraction by electrostatic interaction between inhibitor molecule and the surface will be weak.

Adsorption Rate

Besides the degree of surface coverage and the inhibitor configuration, also the adsorption rate of the inhibitor molecules may play an important role. The surface coverage by an inhibitor is usually determined through measurement of its equilibrium adsorption isotherm. During growth from a supersaturated solution in the presence of an inhibitor, the adsorption equilibrium is, however, seldom reached owing to the competition between inhibitor molecules and $BaSO_4$ growth units on their way to reach lattice positions at the surface. The rate at which bonding with the surface can be established, will primarily depend on the electrostatic interaction between the crystal surface and the inhibitor ions. Thereafter, the strength of the chemical bonds formed between functional groups and crystal surface and the intrinsic diffusity of the inhibitor molecules in the immobilized Nernst layer surrounding the crystal will determine the time needed to reach the final position of a small inhibitor molecule on the surface. For relatively large polyelectrolyte inhibitors an additional process of relaxation towards the equilibrium conformation of the molecules upon the crystal surface can be expected, while little time is available between the deposition of subsequent growth layers.

Information about adsorption rates of growth inhibitors is scarcely found in the literature. Adsorption of (5×10^{-7} M) nitrilotri(methylene phosphonic acid) on $BaSO_4$ crystals was reported to be completed for 80-90% within 1-2 minutes[9], while reaching the adsorption equilibrium required 20-30 minutes. For the adsorption of a much larger polyphosphonate (phosphonylated polyphenylene oxide, $M_w \approx 30,000$) onto hydroxyapatite, equilibrium was reached after 3 hours or more, depending on the additive concentration.[10] For these relatively large polyelectrolyte inhibitors, relaxation towards their equilibrium conformation after diffusion to the surface can occur on a timescale of hours.[11]

Only a few studies on the competition for attachment to the surface between the regular growth units and polyelectrolyte molecules are known in the literature. The growth inhibiting effect of anthocyan on potassium bitartrate was interpreted by a time-dependent factor by Laguerie, et al.[12], expressing the surface poisoning of the crystal terraces by the additive. Contrarily, almost instantaneous adsorption of some low molecular weight polyelectrolytes on gypsum crystals was concluded by Weijnen and van Rosmalen.[13] It can be expected that, if the time needed to reach equilibrium surface coverage is shorter than the time between the deposition of two growth layers, the adsorption rate will not influence the inhibitor performance. If, however, the adsorption process is relatively slow and especially if a relatively high surface coverage is needed to hamper the growth, the adsorption rate may become an important parameter in the inhibition process. The formation of macrosteps was explained by a time-dependent impurity adsorption model by van der Eerden and Müller-Krumbhaar.[14] In case of a slow impurity adsorption process, a step with a height of one monolayer, following shortly after a preceding step, will meet a terrace swept relatively clean from impurities and will hence be less retarded than its predecessor. As a consequence, this step will overtake the first one, and this process will be repeated for the next oncoming steps. This results in coalescence of monosteps leading to a macrostep, which propagates considerably slower along the crystal surface than a monolayer step, thus causing growth retardation.

Matching

Appropriate matching between the acidic groups and the cations at the crystal surface may facilitate the adsorption process.[15]

In the matching process not only interatomic distances, but also the orientation of structural units in the crystal surface layer and the charge distribution over the inhibitor molecules have to be taken into account. Matching of additives has comprehensively been studied for organic-type crystals, which are more suitable for designing structurally specific additives.[16-19] Only a few examples of such "tailor-made" additives are known for inorganic and mineral salts.[20-24] Rawls and Cabasso studied the adsorption of a polyphosphonate onto hydroxyapatite and observed the exchange of phosphonate groups with phosphate groups in the crystal lattice.[10] Replacement of lattice ions by functional groups may also happen in the case of the adsorption of sulfonate containing compounds on $BaSO_4$. For example, a random copolymer of maleic acid and vinyl sulfonic acid has proved to be a very effective growth inhibitor for $BaSO_4$ crystallization.[25] Its good performance could be attributed to the relatively high concentration of carboxylic acid groups (owing to the maleic acid building

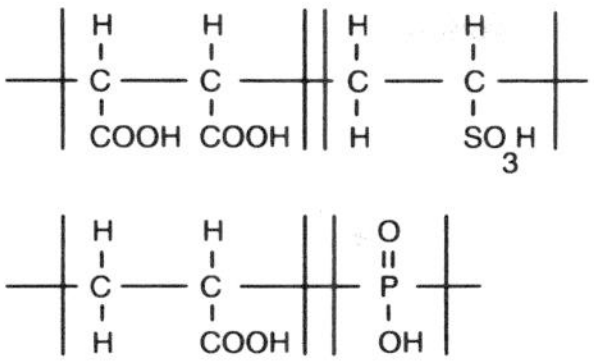

Figure 2. The generic formulas, the molecular weights M_w, and the polydispersities M_n/M_w of the investigated compounds.

units), as well as to the presence of sulfonic acid groups directly attached to the backbone of the molecule. The positive influence of the sulfonic acid group may be twofold: Apart from its potential to form strong bonds with Ba^{2+}-ions at the surface, it may also replace a sulfate lattice ion. If matching of the sulfonate groups plays a role, the inhibitor may be preferentially adsorbed on a crystal face where the sulfate lattice ions have an oxygen-suflur bond orientated perpendicular to the crystal surface and thus easy replacement by the inhibitor-bound sulfonate groups can occur. After prolonged growth in the presence of the inhibitor such a face will become more pronounced. This has indeed been observed.[26] $BaSO_4$ crystals, which were originally bounded by {002}, {210} and less frequently by {211} faces, developed {011} and {101} faces after growth in the presence of the maleic acid-vinylsulfonic acid copolymer. Only these two crystal faces have lattice sulfate ions with one S-bond directed normally out of the surface. A flatter adsorption of the inhibitor molecule may take place if the repeating distance between its functional groups corresponds with that of the lattice ions. The polymer backbone can then act as a fence at the crystal surface, thus forming a solid obstacle for the propagating steps.[24]

AIM OF THE STUDY

In this paper we compare the growth inhibiting performance of three polyelectrolytes in barium sulfate crystallization with their adsorption behaviour, with an emphasis on the effect of inhibitor adsorption rate. The three polyelectolytes are a random copolymer of maleic acid and vinyl sulfonic acid, denoted as PMA-PVS, and two polyphosphino-acrylates with different molecular weights, indicated by PPAA-I and PPAA-II. The generic formulas of these compounds, as well as their molecular weights M_w and their polydispersities M_n/M_w, are given in Figure 2.

EXPERIMENTAL

To determine the growth inhibiting performance of the three compounds, free-drift seeded suspension growth experiments were performed at $25.00 \pm 0.01°C$.[1,27] The growth process was initiated by the addition of 1 ml $BaSO_4$ suspension, containing about 50 mg well-defined seed crystals with a specific surface area of 0.30 ± 0.03 m^{2}g^{-1}, to 200 ml of a well-stirred solution with an initial relative supersaturation σ_i of 4.6, a pH-value of 5.0 ± 0.01, and, if applied, 0.1 ppm inhibitor. The growth process was followed by recording the suspension conductivity as a function of time. Some additional growth experiments were performed at pH = 3.8±0.01 and σ_i = 4.9 in the presence of 0.1 ppm PMA-PVS or PPAA-II, in which the added seeds were pretreated with an 0.1 ppm inhibitor solution, saturated with $BaSO_4$, before initiation of the growth process.

For the adsorption experiments dry $BaSO_4$ crystals purchased from Baker were used. The specific surface area of these crystals was 6.3±0.1 m^2g^{-1}, determined by BET (N_2) gas adsorption. To measure the adsorption as a function of time, an accurately weighted amount of 200 mg $BaSO_4$ crystals was added to 40.00 ml of a saturated $BaSO_4$ solution with a pH-value adjusted to 5.0, containing an inhibitor concentration ranging from 10 to 50 ppm. Immediately after addition of the crystals (t = 0), the suspension was well-shaken during a certain registered period and subsequently filtered over a 0.22 μm pore size filter. The moment of complete separation of crystals and solution was taken as the adsorption time. The residual inhibitor concentration in the filtrate was determined by titration with 5 X 10^{-3} M NaOH solution. In the case of the polyphosphinoacrylate compounds, the solutions were titrated in a nitrogen atmosphere from pH = 4.000 to 7.000 and the sample concentration was determined from a calibration line.[28] Within this pH range, about 70% of the carboxylic acid groups are ionized. For PMA-PVS, the pH range had to be extended to pH = 9.000 to obtain a sufficiently accurate calibration line.

RESULTS AND DISCUSSION

Growth Experiments

Sequence in Growth Retarding Performance. First, the growth curves of $BaSO_4$ crystals in the absence (blank) and presence of 0.1 ppm of the tested compounds were measured. Figure 3 shows the corresponding mean linear growth rates $\overline{R}$ as a function of the

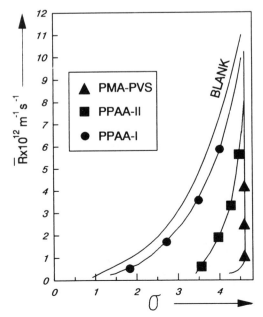

Figure 3. Plots of the mean linear growth rate $\overline{R}$ versus the relative supersaturation σ, in the presence and absence of 0.1 ppm PMA-PVS, PPAA-I and PPAA-II, at pH = 5 and 25°C.

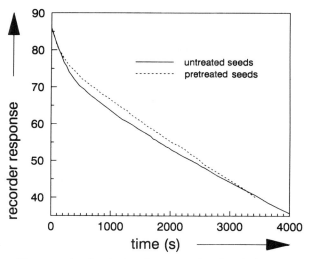

Figure 4. Dependence of the supersaturation (α recorder response) on time, in the presence of 0.1 ppm PPAA-II at pH = 3.8 and 25°C, after addition of untreated seeds and of seeds pretreated in 0.1 ppm PPAA-II.

relative supersaturation σ, starting from an initial supersaturation σ_i of 4.6. $\bar{R}$ is defined as the time derivative of the concentration ($\sim$ conductivity) versus time curve divided by the total surface area of the crystals at time t, as described in detail.[1] The total increase in surface area of the crystals for the blank amounted only 4%. The relative super-saturation σ of the bulk solution is given by ($c-c_{eq}/c_{eq}$), with c_{eq} = equilibrium concentration. From these curves, the following sequence in effectiveness at pH = 5 and 25°C can be concluded:

$$PMA\text{-}PVS > PPAA\text{-}4 > PPAA\text{-}1.$$

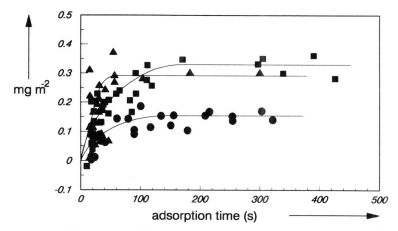

Figure 5. Adsorption of 50 ppm PMA-PVS (closed triangle), PPAA-II (closed square) and PPAA-I (closed circle) at pH = 5 and 25°C. Surface coverage (mg m^{-2}) versus time (s).

In the presence of PMA-PVS, the mean linear growth rate $\overline{R}$ approaches zero while σ hardly decreases. The $\overline{R}$ - σ growth curve obtained in the presence of PPAA-1 approaches that of the blank. The same sequence was found at pH = 3.8 and 25°C.

In addition, growth experiments were performed at higher concentrations of PPAA-II and PMA-PVS to determine the concentration where total growth blockage occurs. No detectable growth could be measured anymore in the presence of 0.2 ppm PMA-PVS and 0.5 ppm PPAA-II at pH = 5 and 25°C. The concentration where the hardly effective inhibitor PPAA-I causes total growth blockage was not determined, but it will anyhow be much higher than for PPAA-II, in view of its poor effectiveness.

Growth with Pretreated Seed Crystals. To check whether any measurable effect of the inhibitor adsorption rate on the growth curve can be measured, growth experiments were performed at σ_i = 4.9 and pH = 3.8 in the presence of 0.1 ppm PMA-PVS or PPAA-4, using either seed crystals without special treatment or seed crystals previously immerged in an inhibitor solution (saturated with BaSO$_4$) of the same concentration for at least 6 hours before being applied. Within this time, adsorption should be completed. A pH-value of 3.8 instead of 5 was selected for these experiments because at pH = 3.8 adsorption of both inhibitors proceeds less easily than at pH = 5 owing to a lower degree of dissociation, resulting in a poorer effectiveness, and differences will become more pronounced. Also, the slightly higher initial supersaturation of 4.9 instead of 4.6 will negatively influence the inhibitor performance. The growth rate of the blank is independent of the solution pH and initial supersaturation.[1]

In the presence of 0.1 ppm PPAA-II, different supersaturation-vs.-time curves were found, indeed, for the two different types of seed crystals. Growth is retarded more effectively when seeds, pretreated with 0.1 ppm PPAA-II, were added. This is shown in Figure 4 where the recorder response, which is directly related to the bulk concentration and hence to the supersaturation, is plotted versus time. Addition of the untreated seeds results in a faster decrease in supersaturation, indicating a less effective growth inhibition owing to the inhibitor's time-dependent adsorption. Both curves coincide after ≈3200 seconds (s) growth time, when the supersaturation has decreased until σ ≈ 3.0. The largest measured difference in supersaturation between both curves is ≈13%.

For PMA-PVS, no significant difference was found between the supersaturation-vs.-time curves measured with normal seeds and with pretreated seeds, which is an indication that the PMA-PVS adsorption rate plays no role in the inhibition process. This will be verified by measuring time dependent adsorption isotherms.

Time Dependent Adsorption

Time dependent adsorption measurements were performed in order to find out if the three tested compounds behave differently. Formerly determined equilibrium adsorption isotherms of PPAA-I, PPAA-II and PMA-PVS at pH = 5, revealed that the equilibrium coverage is maximal, i.e. at the plateau level, for PPAA-I and PPAA-II at an initial concentration from ≈ 50 ppm, and for PMA-PVS from ≈ 20 ppm.[26] Therefore the first series of time dependent adsorption experiments were done with 50 ppm inhibitor. The results are presented in Figure 5. The minimum feasible measuring time was ≈8 seconds. Below this time the curves were extrapolated to the 0,0 origin. Despite the substantial spread in the measured points, it can be concluded that the best growth retarder PMA-PVS reaches its equilibrium adsorption plateau value of 0.29 mg m^{-2} in a significantly shorter period (≈40 s) than the two polyphosphinoacrylates (100-140 s). PPAA-I seems to reach its equilibrium plateau value (0.15 mg m^{-2}) in a slightly shorter time than PPAA-II (0.33 mg m^{-2}), probably because of its lower M_w.

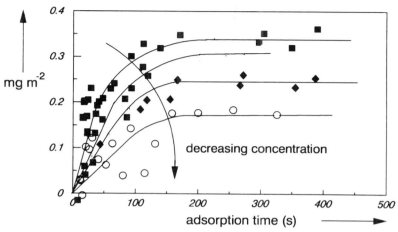

Figure 6. Adsorption of 10 ppm (open circle), 20 ppm (closed diamond), 25 ppm (star) and 50 ppm (closed square) at pH = 5 and 25°C versus time.

Additional experiments were performed with PPAA-II and PMA-PVS concentrations varying from 10 to 50 ppm to study the influence of inhibitor concentration on the adsorption time. Since for PPAA-II the maximum equilibrium surface coverage of 0.33 mg m^{-2} is reached at concentrations ≥50 ppm[26], its equilibrium surface coverages will be lower at concentrations <50 ppm. Two opposite effects will influence the adsorption time with decreasing concentration: the driving force for adsorption will become smaller, leading to a longer adsorption time, and a lower surface coverage has to be established, requiring a shorter adsorption time. In Figure 6 some of these curves (50, 25, 20 and 10 ppm PPAA-II) are presented. It can be seen that the initial slope of the curves decreases with decreasing initial concentration and that the time needed to reach the equilibrium surface coverage remains between 100 and 140 seconds. Plotting of the initial slopes versus the initial concentrations, revealed a linear relation according to:

$$\text{slope (mgm}^{-2}\text{s}^{-1}) = (1.11 \pm 0.02) \times 10^{-4} \times C \text{ (ppm)}.$$

PMA-PVS reaches its maximal equilibrium surface coverage at an initial concentration of ≈20 ppm.[26] Additional time dependent adsorption experiments with concentrations of 25 and 10 ppm resulted in curves which coincided with the curve measured for an initial PMA-PVS concentration of 50 ppm. Determination of the change of the initial slopes with decreasing concentration was therefore not possible for PMA-PVS, but it can anyway be concluded that the PMA-PVS adsorption time is about the same for initial concentrations varying from 10-50 ppm.

Inhibitor Effectiveness versus Adsorption Behaviour

Knowing at what inhibitor concentrations total growth blockage occurs at pH = 5, the corresponding surface coverage can be estimated from the extrapolated part of the equilibrium adsorption isotherm of the inhibitor. This value is only reached if the adsorption is completed before the surface is covered by the next growth layer. A first indication that this is not the case for PPAA-II, but probably true for PMA-PVS, was obtained from the growth experiments with pretreated seed crystals. When the time needed to establish the inhibitor surface coverage causing growth blockage is known, we can find out whether competition between $BaSO_4$ growth units and inhibitor molecules for a position at the crystal surface may occur by comparing this adsorption time with the times needed to grow one $BaSO_4$ layer at the prevailing supersaturation.

At pH = 5, total growth blockage of well-defined $BaSO_4$ crystals with a specific surface area of 0.30 ± 0.03 m^2 g^{-1} was found to occur in the presence of either 0.2 ppm PMA-PVS or 0.5 ppm PPAA-II. The corresponding inhibitor surface coverages, derived from the extrapolated equilibrium adsorption isotherms, are 3.8×10^{-3} mg m^{-2} and 6.6×10^{-3} mg m^{-2}, respectively. These values will, however, not be reached when the adsorption proceeds slower than the deposition of the next growth layer.

Equating concentrations measured with smooth crystals (growth experiments) and very rough crystals (adsorption experiments) seems not entirily justified. In a former study, however, we found similar values for the maximum surface coverage needed for total growth blockage by the phosphonate inhibitor HEDP (1-hydroxy-ethane-1,1 bisphosphonic acid) for very rough and for well-defined $BaSO_4$ crystals. We therefore assume that the effect of surface roughness is also not too dominant for the compounds investigated in this study.

The times needed for the establishment of the calculated surface coverages can unfortunately not be measured, because the current detection techniques do not allow determination of PPAA-II and PMA-PVS at concentrations below 10 ppm.

Hence the adsorption time of 0.5 ppm PPAA-II could only be estimated from the initial slope of its time dependent adsorption curve, using the found relation of slope vs. initial concentration for inhibitor concentrations ≥10 ppm (= 5.55×10^{-5} mg m^{-2} s^{-1}). Substitution of the surface coverage of 6.6×10^{-3} mg m^{-2} then leads to an adsorption time of 118 s.

Since the measured times needed to establish equilibrium surface coverage by PMA-PVS at concentrations varying from 10 to 50 ppm remained 40 s, we assume that also for the very low concentration of 0.2 ppm, with an equilibrium surface coverage of $3.8'10^{-3}$ mg m^{-2}, the adsorption time is ≈40 s.

The increase in $BaSO_4$ surface during a growth experiment is almost negligible, and will therefore play no role in the extrapolation of the adsorption characteristics determined in the range of 10 to 50 ppm to the very low concentrations as used in the growth experiments. During a growth experiment in the absence of an inhibitor, when growth of the $BaSO_4$ seed crystals continues until the supersaturated $BaSO_4$ solution has reached its equilibrium value, ≈2 mg $BaSO_4$ is formed on ≈50 mg seed crystals with a specific surface area of 0.30 m^2 g^{-1}. During an experiment in the presence of an inhibitor even less $BaSO_4$ is formed, because

Table 1. Mean time t (s) needed to grow one $BaSO_4$ monolayer in the absence of an inhibitor for various relative supersaturations σ

Relative supersaturation (σ)	Mean linear growth rate $\bar{R} \times 10^{12}$ m s^{-1}	Mean layer growth time τ (s)
4.6	11.4	38.7
4.5	10.6	41.6
4.0	7.2	60.9
3.5	4.8	92.2
3.0	3.1	143.9
2.5	1.9	229.8
2.0	1.2	375.5
1.5	0.7	668.0
1.0	0.2	2133.3

the saturation value of the barium sulfate solution is usually not reached owing to the growth retardation. Although the $BaSO_4$ surface area hardly increases during a growth experiment, it is of course constantly refreshed.

The obtained inhibitor adsorption times of 118 s and 40 s for PPAA-II and PMA-PVS, respectively, were subsequently compared with the $BaSO_4$ growth rate in the absence of an inhibitor. The mean time needed for the deposition of one $BaSO_4$ growth layer was calculated for relative supersaturations ranging from 4.6 ($= \sigma_i$) to $\sigma = 1.0$. For the height of one growth layer d_{hkl} a value of $(8.612 \times 10^{-29} \text{ m}^{-3})^{1/3} = 4.416 \times 10^{-10} \text{ m}^{29}$ was taken, and with the help of the determined mean linear growth rate values $\overline{R}$ (m s^{-1}) for the various relative supersaturation values, a mean layer growth time t for each supersaturation was obtained. Some of the calculated values are listed in Table 1.

It can be seen, that the estimated value for the time needed to reach PPAA-II equilibrium surface coverage interferes with the mean times needed to grow a barium sulfate monolayer at σ-values of 4.6 until ≈ 3.0. It can thus be expected that the PPAA-II performance is reduced owing to the relatively slow adsorption rate of the inhibitor within this supersaturation range. Once the supersaturation is decreased by the growth of the seed crystals, the growth rate decreases and the inhibitor molecules have sufficient time to adsorb and block the attachment of oncoming barium sulfate growth units. This supersaturation range of 4.6 until ≈ 3.0 agrees with the supersaturation range where a difference was found between the growth curves measured with normal seeds and pretreated seed crystals.

For PMA-PVS, the adsorption time of 40 s interferes with $BaSO_4$ monolayer growth times only at supersaturations higher than 4.5-4.6. It can therefore be understood that no adsorption time effect was measured for this compound in growth experiments with this initial supersaturation.

CONCLUSIONS

- The sequence in growth retarding effectiveness of the three investigated polyelectrolytes at pH = 5 and at 25°C is:

  $$PMA\text{-}PVS > PPAA\text{-}II > PPAA\text{-}I.$$

- Total blockage of $BaSO_4$ growth at pH = 5, 25°C and an initial relative supersaturation σ_i = 4.6 occurs at a concentration of ≈ 0.2 ppm PMA-PVS or ≈ 0.5 ppm PPAA-II.
- Pretreatment of $BaSO_4$ seed crystals in 0.1 ppm PPAA-II results in a better growth retardation of $BaSO_4$ at pH = 3.8, σ_i = 4.9 and 25°C in the presence of 0.1 ppm PPAA-II than when non-treated seeds are added. For $BaSO_4$ growth in the presence of 0.1 ppm PMA-PVS under the same growth conditions, pretreatment of the seeds has no measurable effect.
- The better growth retarding effectiveness of 0.1 ppm PMA-PVS at pH = 5 and 25°C compared to 0.1 ppm PPAA-II can be related to a faster adsorption of PMA-PVS.
- The adsorption rate of PPAA-II at pH = 5 and 25°C interferes with the growth rate of $BaSO_4$ in the σ-range of 4.6 to 3.0, leading to a diminished growth-inhibiting effectiveness of PPAA-II.

ACKNOWLEDGMENT

The authors are indebted to P.A. de Jong and S. Nathoenie for their technical assistance.

REFERENCES

1. M.C. van der Leeden, J. Reedijk, and G.M. van Rosmalen, *Estudios Geol.*, 38, 279.
2. M.C. van der Leeden and G.M. van Rosmalen, 1987, *Desalination*, 66, 185.
3. Weijnen, M.P.C., M.C. van der Leeden, and G.M. van Rosmalen, 1987, *Geol. Chemistry and Mineral Formation in the Earth Surface (Eds. R. Rodriquez Clemente and Y. Tardy)*, 753.
4. M.P.C. Weijnen, G.M. van Rosmalen, P. Bennema, and J.J.M. Rijpkema, 1987, *J. Crystal Growth*, 82, 509-527.
5. M.P.C. Weijnen, G.M. van Rosmalen, and P. Bennema, 1987, *J. Crystal Growth*, 82, 528-542.
6. J. Christoffersen, M.R. Christoffersen, S.B. Christensen, and G.H. Nancollas, 1983, *J. Crystal Growth*, 62, 254.
7. N. Cabrera, and D.A. Vermileya, 1958, *Growth and Perfection of Crystals (Eds. R.H. Doremus, B.W. Roberts and D. Turnbull)*, Wiley, New York, 393.
8. R.J. Davey, 1976, *J. Crystal Growth*, 34, 109.
9. Leung, W.H. and Nancollas, G.H., 1978, *J. Crystal Growth*, 44, 163.
10. H.R. Rawls and I. Cabasso, 1984, *Proc., Symp. Adsorption and Surface Chemistry of Hydroxyapatite (Eds. D. Misra, N. Dwarika)*, Plenum Press, New York, 115-128.
11. M.A. Cohen Stuart, 1986, *Advances in Colloid and Interface Science*, 24, 143.
12. C. Laguerie, B. Ratsimba, and C. Frances, 1990, *BIWIC 1990, Bremer International Workshop for Industrial Crystallization (Ed. J. Ulrich)*, 10.
13. M.P.C. Weijnen and G.M. van Rosmalen, 1985, *Desalination*, 54, 239.
14. J.P. van der Eerden and H. Müller-Krumbhaar, 1986, *Electrochimica Acta*, 31, no. 8, 1007.
15. J.S. Gill and R.G. Varsanik, 1986, *J. Crystal Growth* 76, 57.
16. L. Addadi, Z.Berkovitch-Yellin, I. Weissbuch, J. van Mil, L.J.W. Shimon, M. Lahav, and L. Leiserowitz, 1985, *Angew. Chem.* 24, 466.
17. L. Addadi, Z.Berkovitch-Yellin, I. Weissbuch, M. Lahav, and L. Leiserowitz, 1986, *Topics in Stereochemistry, vol.16 (eds. E.L. Ehil et al.)*, John Wiley & Sons Inc., 1.
18. M. Vaida, L.J.W. Shimon, Y. Weisinger-Lewin, F. Frolow, M. Lahav, L. Leiserowitz, and R.K. Mc Mullin, 1988, *Science*, 241, 1475.
19. G.M. van Rosmalen and P.Bennema, 1990, *J. Crystal Growth* 99, 1053-1060.
20. M.A. van Damme-van Weele, thesis, 1965, Twente University of Technology, The Netherlands.
21. S. Sarig, A. Glasner, and J.A. Epstein, 1975, *J. Crystal Growth* 28, 295.
22. R.J Davey, S.N. Black, L.A. Bromley, D. Cottier, and B. Dobbs, 1992, *Nature*, 253, (6344), 549.
23. S.N. Black, L.A. Bromley, D. Cottier, and R.J. Davey, 1991, J. Chem. Soc., *Faraday Trans* 87 (20), 3409.
24. L.A. Bromley, D. Cottier, R.J. Davey, and B. Dobbs, 1993, *Langmuir*, 9 (12), 3594.
25. M.C. van der Leeden and G.M. van Rosmalen, 1988, *Proc., Third Intl. Symposium on Chemicals in the Oil Industry, Royal Soc. of Chemistry London*, 68.
26. M.C. van der Leeden and G.M. van Rosmalen, 1995, *J. Colloid and Interface Sci., 171, 142.*
27. S.T Liu and G.H. Nancollas, 1970, *J. Crystal Growth* 6, 281.
28. J. Blaakmeer, M.A. Cohen-Stuart, and G.J. Fleer, 1990, *Macromolecules*, 23, 230.
29. A.E. Nielsen, 1969, *Kinetics of Precipitation*, Pergamon, Oxford.

10

DEVELOPMENT OF NEW PHOSPHONIC ACID BASED SCALE INHIBITORS AND EVALUATION OF THEIR PERFORMANCE IN RO APPLICATIONS

Alexei G. Pervov and Galina Y. Rudakova

NII VODGEO
42, Komsomolsky pr., Moscow, 119826, Russia
NPO IREA
4, Bogorodsky val, Moscow, 107258, Russia

INTRODUCTION

The significance of thorough pretreatment of Reverse Osmosis (RO) feed water can not be overestimated. The neglect of these processes has resulted in shutdowns in a number of RO facilities. The problem of fouling and scaling in membrane equipment is extensively investigated and successfully avoided in majority of cases due to advances in research and operational practices. The problems that do arise are reflected in professional literature.[1-5] Many new advanced chemicals are being used for scaling control[3,4] and cleaning procedures.[2]

However, a number of research programs lack a rational approach to identifying the causes of the problems and to understanding the mechanisms of flux and rejection decline in the unsuccessful operation. Laboratory test simulation techniques do not always correspond to real conditions in membrane units. Generally, the results of new investigations are presented after a long term course of pilot plant and field testing.[3] This could be valid for certain conditions at certain facilities, however, the results of such studies cannot be applied to variety of other feed water conditions.

The mechanisms of scaling and/or fouling layer formation are not fully understood. Long term operational experience of RO facilities demonstrates that the process of foulant accumulation and even its control techniques are determined by the type of the unit construction.[5] Introduction of improved types of membrane modules could be viewed as one possible solution of fouling and scaling problems.

The present study is devoted to resolution of scaling problems and their control in the applications of Russian desalination facilities furnished with spiral wound modules manufactured by POLIMERSINTEZ (Russia). The studies were conducted in NII VODGEO to understand the scaling and fouling mechanisms and to provide industrial facilities with

Mineral Scale Formation and Inhibition, Edited by Zahid Amjad
Plenum Press, New York, 1995

operational guidelines. As a result of these, a series of chemicals produced by NPO IREA (Russia) were recommended as antiscalants and cleaning agents. Once the mechanisms of scaling and fouling are understood, the tendencies of foulant accumulation could be observed and methods for its control could be successfully devised and evaluated.

In this study, new test procedure is offered which enables us to evaluate pretreatment qualities for various possible foulants. This provides a quantitative comparison of the effectiveness of different pretreatment techniques, such as additives and/or acid dosing, and permits its rapid testing in variety of feed water conditions (i.e., a wide range of total dissolved solids, TDS, and pH values and scale-forming species concentrations). Pretreatment requirements can also be successfully determined through the application of the suggested test procedure.

AN OVERVIEW OF SCALING MECHANISMS FROM PREVIOUS STUDIES

The investigation of the sparingly soluble salt scaling in RO systems has shown that the main cause of crystal formation in spiral wound modules is attributed to the contact between spacer (mesh) bundles and the outer membrane surface.[9] During the module manufacturing, pressure is applied to spacer while the membrane leaves are being wound around the product tube. The contact sites provide additional local resistance to the flow above the membrane surface which lowers flow rates and results in an increase in concentration. The degree of supersaturation reached often appears to be sufficient to initiate crystal nucleation. The results of experimental study suggests that crystals are brought out of the sites and subsequently precipitate on the outer membrane surface according to suspended particles fouling mechanism. Microscopic observation of membrane surfaces during autopsies of scaled-up modules confirms this crystallization mechanism.[5]

The influence of the membrane material on calcium sulfate and calcium carbonate scaling propensities also should be mentioned. High concentrations attained in deadspots substantially depend on pressures in membrane channel. Low pressure composite membrane based modules demonstrated no susceptibility to calcium sulphate scaling despite severe brine conditions during testing. The decline of product flux of the fouled module is apparently caused, not by pressure drop across the foulant layer, but by the increase of feed solution concentration due to the polarization from the fouling. Otherwise, the reduction of the product flux during foulant accumulation would not be accompanied with substantial decrease in rejection. An increase in the differential pressure across the fouled modules in the vessel should be recorded as it reduces the active applied pressure (net driving pressure) and thus lowers the flux. Delta pressure increase due to fouling is also attributed to the presence of the mesh spacer in the channel. Thus the influence of module construction and the type of membranes should be accounted during the consideration of pretreatment requirements and cleaning schedules. A special test procedure has been developed that will determine the most likely foulants that should be removed from feed water and to evaluate fouling or scaling rates directly for the type of modules employed in industrial installation.

As the influence of mesh spacer on fouling is investigated, it was found that the problem could be overcome by the introduction of new module construction without contact between spacer and membrane surface. For conventional RO systems, the influence of module construction should be accounted during the additive evaluation tests.

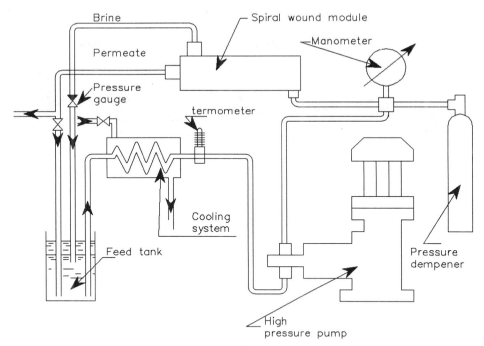

Figure 1. Schematic diagram of spiral wound module test system.

EXPERIMENTAL

A test study was carried out focusing mainly on calcium carbonate scale control in RO desalination systems built around spiral wound modules with low pressure TFC membranes. The experiments were conducted in the test array shown in Figure 1. The system contained 4" spiral wound modules. The test procedure was conducted in the batch concentration mode, where reject was recycled to recirculation tank and permeate was withdrawn from the system. The amount of accumulated scalant (calcium carbonate) in this case was calculated as the difference between the predicted amount of calcium in concentrated feed solution (initial concentration multiplied by concentration factor) and actual amount of calcium in the circulating solution. We were thus able to calculate scaling rates for any type of the feed water entering the membrane unit.

To embrace possible feedwater compositions, we added calcium chloride to natural well water to increase hardness, sodium chloride to change the TDS, and sodium hydroxide for pH adjustment. Using the test procedure described above, calcium carbonate scaling tendencies for various feed water conditions as well as effectiveness of various types of inhibitors have already been extensively investigated in relation to different spiral wound modules. Figure 2 shows the steps of calculation of scaling rates in the 4" module (model 4021) with the well water used as the feed source.

The graph in Figure 2a shows the dependency of calcium content in the recirculating feed solution on the volume concentration factor. Figure 2b shows the amount of accumulated scale as a function of the concentration factor (CF). Every value of CF corresponds to a certain amount of time elapsed from the beginning of the experiment. Thus the amount of accumulated scale could be presented as mass of scale versus time. The scaling rate for every

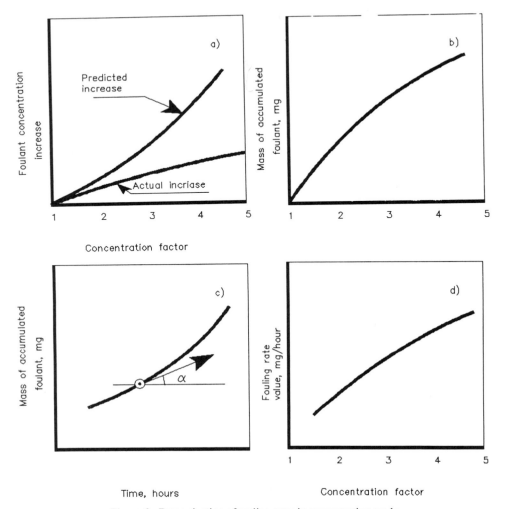

Figure 2. Determination of scaling rates in concentration mode.

moment of accumulation could be determined as the tangents to the slope of scale amount versus time curve (Figure 2c). And, finally, the graph shown in Figure 2d yields scaling rate versus CF dependency. The results of this testing enables us to evaluate scaling rates in every module in every pressure vessels of an RO assembly for any feed water composition and or system configuration (Figure 3).

The developed test procedure provides a rapid quantitative comparison of effectiveness of various antiscalants by direct calculation of scaling rates. Figure 4 shows the results of testings conducted with well water using different samples of antiscalants. The widely used sodium hexametaphosphate, sodium tripolyphosphate, hydroxyethylidene diphosphonic acid (HEDP), nitrilo *tris*-methylene phosphonic acid (NTP), and several newly developed formulations were tested.

It may be understood from Figure 4, by the slope of the curves, "Phosphanol" and "IOMS" could be recognized as the strongest antiscalants. Among these formulations only "Phosphanol" is used on an industrial scale and is permitted for the use in water supply practice. Phosphanol is a sodium salt of dioxydiaminopropanol tetra metal phosphonic acid, developed by IREA. All IOMS's are salts of oxypropanephosphonic acid. Figure 5 shows the results of phosphanol testings using well water as a feed. Variation in the feed composition is accomplished by addition of different chemicals to the well water and further experimental data processing enable us to calculate scaling curves for different feed water composition (Figure 6). Figure 7 shows the guidelines of how to identify the scaling curve (scaling rate versus CF at the inlet to RO module) for certain actual values of pH, TDS, Calcium and Alkalinity.

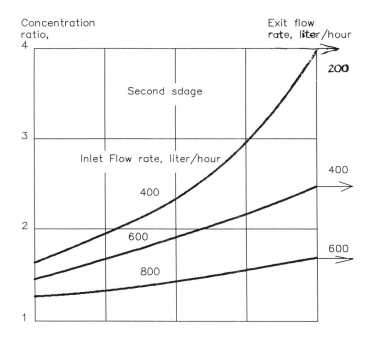

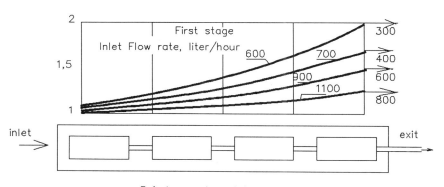

Spiral wound modules

Figure 3. Concentration ratio increase in the pressure vessels of RO unit.

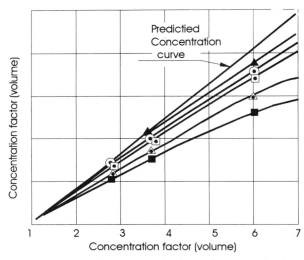

Figure 4. Feedwater concentration curves: evaluation of different antiscalant performance.

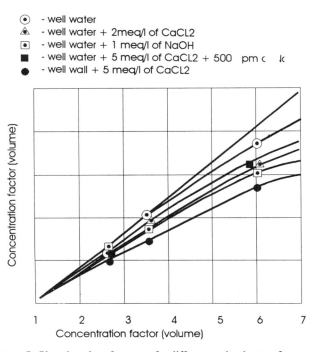

Figure 5. Phosphanol performance for different antiscalant performance.

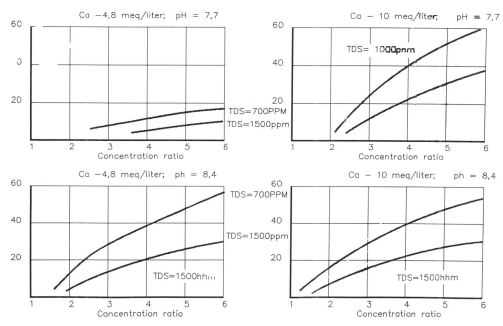

Figure 6. Results of calcium carbonate scaling rates determination under different feed water conditions and concentration ratio increase.

PRACTICAL APPLICATIONS

Tables 1 & 2 present the results of experimental data processed to recommend optimum cleaning schedules (operational period between cleanings and volume of cleaning solution per one vessel) for various cases of feed water composition and RO unit recovery when "Phosphanol" is dosed.

Table 1. Identification of feed water composition

pH	TDS (PPM)	Alkalinity, meq/l					No. of lines
		4–6			6–8		
		Calcium, meq/l			Calcium, meq/l		
		4–6	6–8	8–10	6-8	8-10	
7.2–7.7	500–800	A1	B1	C1	D1	E1	1
	800–1200	A2	B2	C2	D2	E2	2
	1200–1600	A3	B3	C3	D3	E3	3
7.8–8.4	500–800	A4	B4	C4	D4	E4	4
	800–1200	A5	B5	C5	D5	E5	5
	1200–1600	A6	B6	C6	D6	E6	6
No. of columns		A	B	C	D	E	

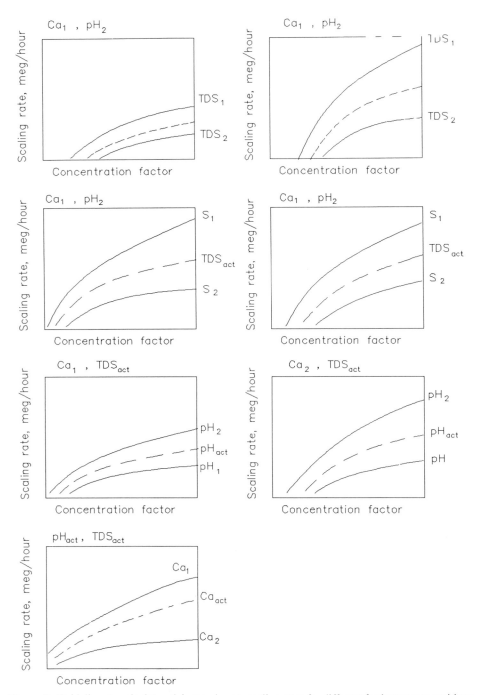

Figure 7. Guidelines to calculate calcium carbonate scaling rates for different feedwater compositions.

Table 2. Determination of operational conditions

Water composition	Two-stage scheme multiplicity of concentration - 1,72	two-stage scheme multiplicity of concentration - 2,5	Inhinitor dosage
	Interval between cleaning procedures, h		
A1	2000	1000	1
A2	2000	1500	1
A3	2000	2000	1
A4	500	—	2
A5	1000	—	2
A6	2000	300	2
B1	2000	400	2
B2	2000	500	2
B3	2000	1500	2
B4	*	—	3
B5	250	—	3
B6	800	—	3
C1	2000	—	3
C2	2000	300	3
C3	2000	500	3
C4	*	*	3
C5	*	*	3
C6	*	*	3
D1	500	—	2
D2	800	250	2
D3	2000	400	2
E1	300	—	3
E2	500	—	3
E3	700	—	3

Experimental methods for the optimum operational period between cleanings are based on precipitate accumulation testing and are described previously.[5] Antiscalant addition doesn't necessarily prevent scale formation in RO modules, though it may reduce the rate of scaling by a factor of 10-20 times. The present guidelines prescribe roughly the interval between cleanings for wide range of brackish well water compositions. To increase the RO unit durability and to reduce the consumption of cleaning chemicals, for certain feed water compositions it would be necessary to operate the unit at lower recoveries.

HOW TO USE THE DEVELOPED GUIDELINES

1. To select the optimum operational mode, first it is necessary to identify the feed water composition according to Table 1. The identification indices are found in Table 1 according to certain columns and lines which correspond to the concentrations of calcium and bicarbonate, as well as the pH and TDS values. Each column of Table 1 has its own letter symbol. Thus, feed water composition can be identified by means of Table 1 and each composition will have its own index. The most common ground water compositions are found within the lines 1-3 and columns A-B.
2. Table 2 offers optimum operational conditions according to water compositions presented in Table 1. Intervals between cleanings are calculated basing on scaling rates in the RO modules with brine concentration factor values of 2.5 and 1.7. The optimum operational modes are developed for the pretreatment with 1 ppm of

"Phosphanol" added to the feed. The chemical composition of cleaning solution may include citric acid, EDTA, HEDP or other chelating agents at 2 percent concentrations. All listed solutions exhibit similar dissolution abilities for calcium carbonate. The intervals between cleanings listed in Table 2 are selected to provide the most efficient removal of calcium carbonate for current conditions (100 liter of cleaning solution per one vessel of two elements and 150l for three elements). WATERLAB has developed RO units that are operated with simplified pretreatment. The only measure to control scale formation is inhibitor injection at 1-3 ppm concentrations. This enables us to ensure reliable performance during 1000-3000 hours of constant operation. The developed dosing technique utilizes solenoid valves and injectors at the inlet of the high-pressure pumps. The use of concentrated liquid antiscalant formulations (25-50 per cent) resulted in the development of automatic RO machines with built-in antiscalant dosing units. The inhibitor tank volume is sufficient for 1000-3000 hours of operation.

REFERENCES

1. R.Larry Reitz, Development of a Broad Spectrum Antiscalant for Reverse Osmosis Systems, *Technical Proceedings* 12th Annual Conference of the Water Supply Improvement Association, Orlando, FL, May 13-18, 1984. Vol. 1: Sess. 1-6, Topsfield, Mass., s.a. F 1-26.
2. S.I. Gracham, R.L. Reitz, C.E. Hickman, 1989, Improving Reverse Osmosis Performance Through Periodic Cleaning, *Desalination*, 74, 113-124.
3. C.K. Evans and M.A. Finan, Comparison of the Use of Different Antiscalant Treatment in a Pilot Reverse Osmosis Plant at Ras Abu Jarjur, Bahrain, *Technical Proceedings*, IDA World Conference on Desalination and Water Reuse, Aug 25-29, 1991, Washington DC, Volume 2, Session: Advanced Materials and Chemicals.
4. Z.Amjad, 1985, Applications of Antiscalants to Control Calcium Sulphate Scaling in Reverse Osmosis Systems, *Desalination*, 54, 263-276.
5. A.G.Pervov, 1991, Scale Formation Prognosis and Cleaning Schedules in RO Systems Operation, *Desalination*, 83, 77-118.

11

THE INFLUENCE OF SOME PHOSPHONIC ACIDS ON THE CRYSTAL GROWTH OF CALCIUM FLUORIDE

Abbas Abdul-Rahman,[1] M. Salem,[2] and G. H. Nancollas

[1] Organon Inc.
   375 Mt. Pleasant Ave., West Orange, New Jersey 07052
[2] Chemistry Department, Faculty of Science
   Menoufia, University, Egypt
[3] State University of New York at Buffalo
   Buffalo, New York 14214

ABSTRACT

The influence of several phosphonate species on the kinetics of crystal growth of calcium fluoride has been studied using the Constant Composition (CC) method. The proportionality of the rate of crystal growth to the square of the relative supersaturation was consistent with a spiral growth mechanism and the reaction was significantly retarded in the presence of micromolar concentrations of phosphono-formic, -acetic, and -propionic acids. Two other additives, ethylenediamine tetra (methylene phosphonate) (ENTMP) and a polyacrylate, (GOOD-RITE® K-732 from The BFGoodrich Company, molecular weight = 5100) were also effective inhibitors, but the combined retarding effect of their mixtures was less than the sums of those of the components. The results of equilibrium adsorption experiments of phosphonoformic, phosphonoacetic and phosphonopropionic acids on calcium fluoride crystal surfaces were consistent with those calculated from the crystallization experimental data.

INTRODUCTION

In view of the widespread use and application of the alkaline earth metal fluorides, an understanding of their crystal growth mechanisms is of prime importance. Calcium fluoride growth has been shown to be sensitive to impurities even when these are present at very low levels.[1] To gain a better understanding of the influence of additives on the rate of crystallization, the Constant Composition, CC, technique has been used in the present work since it provides kinetic information even at supersaturations that are unattainable by more conventional free drift methods.[2] In general, the addition of foreign substances to a growing crystal medium may result in rate reductions due to complexation with lattice ions and/or by the blocking of growth sites through adsorption. In the latter case, rate retardation would be more sensitive to low additive concentrations, while extensive complexation would be expected to lead to variations in the stoichiometric ratios of the lattice ions and lower the degree of supersaturation of the crystallizing medium.

Application of the CC method, made possible by using a fluoride ion-selective electrode, has been shown to satisfactorily maintain, potentiostatically, the activity of the ions of interest.[3]

EXPERIMENTAL

Supersaturated solutions of calcium fluoride were prepared using calcium nitrate (or chloride) and sodium (or potassium) fluoride solutions prepared from Ultrapure (Alfa Chemical Company) and Reagent Grade (J. T. Baker Company) chemicals in deionized, triply distilled water. Cation concentrations were determined by atomic absorption and/or by exchanging metal ions for protons on an ion exchange resin column (Dowex-50) followed by acid base titrations. Calcium fluoride seed crystals were prepared in polyethylene containers under nitrogen by mixing 50 ml aliquots of 0.25 mol l^{-1} potassium fluoride and 0.125 mol l^{-1} calcium chloride solutions at 25°C. The crystals were washed free of chloride ions using saturated solutions of calcium fluoride and were aged for at least one month before use. The composition was confirmed by X-ray powder diffraction (Phillips XRG 3000 X-ray diffractometer, Ni filter and CuKα radiation) and the specific surface area, 6.30 m^2g^{-1}, was determined by BET nitrogen adsorption (Quantasorb II, Quantachrome, Greenvale NY, 30% N_2/He).

Crystal growth experiments were made in a polyethylene lined 300 ml double-walled Pyrex glass reaction cell maintained at 25.0±0.1°C. The reaction media were stirred magnetically and bubbled with nitrogen gas saturated with respect to the cell solution. The fluoride electrodes (Orion, Model 94-09) were calibrated using standard potassium fluoride solutions both before and at the end of each experiment. Supersaturated solutions were prepared by mixing solutions of calcium nitrate (or chloride), potassium (or sodium) fluoride and potassium (or sodium) nitrate (or chloride) in the reaction cell. Crystal growth was then initiated by the addition of calcium fluoride seed crystals and the fluoride activity was monitored using a fluoride ion specific electrode in conjunction with a thermal-electrolytic silver-silver chloride reference electrode. Any change in the fluoride ion activity triggered the addition of calcium and fluoride titrants controlled by means of a pH-Stat (Metrohm Combitrator, Model 3D, Brinkman Instrument Co.). During the crystallization experiments, the constancy (±1%) of the concentration of calcium ions was also verified by filtering aliquots of the growth media through 0.2 mm Millipore filters and analyzing the filtrates by atomic absorption spectrophotometry (Perkin Elmer, Model 503). The filtered solid phases

Table 1. Thermodynamic data for speciation calculations

Reaction	Equilibrium constant
$HF \rightarrow H^+ + F^-$	6.61×10^{-4}
$HF_2^- \rightarrow HF + F^-$	2.95×10^{-1}
$H_2O \rightarrow H^+ + OH^-$	1.00×10^{-14}
$CaF^+ \rightarrow Ca^{2+} + F^-$	9.12×10^{-2}
$CaOH^+ \rightarrow Ca^{2+} + OH^-$	4.27×10^{-2}
$CaF_2^{(s)} \rightarrow Ca^{2+} + 2F^-$	3.47×10^{-11}

Table 2. Crystallization of calcium fluoride at 25°C, T(Ca):T(F) = 1:2, ionic strength = 0.15 mol l^{-1}, 200 rpm stirring speed

Experiment no.	T(Ca)/10^{-4} mol L^{-1}	s	CaF$_2$ seeds/mg	Rate /10^{-6} mol min^{-1}
20	5.97	0.79	17.6	11.0
24	5.97	0.79	48.2	33.0
25	6.35	0.90	22.3	16.5
26	6.16	0.84	22.3	14.3
27*	5.97	0.79	22.3	11.5
28	5.77	0.73	24.2	10.6
29	5.58	0.67	26.1	9.4
30	5.39	0.61	30.2	7.9
31*	5.97	0.79	22.3	11.6

*Stirring rate was 300 rpm.

were investigated by X-ray diffraction and scanning electron microscopy (ISI Model II Scanning Electron Microscope).

RESULTS AND DISCUSSIONS

The kinetics of the crystal growth reactions was analyzed in terms of the concentrations of the free calcium and fluoride ionic species. These were calculated using mass balance, electroneutrality and ionic strength expressions together with thermodynamic equilibrium constants of the various ion pairs and complexes summarized in Table 1.[4,5,6] The speciation computations were made by successive approximations of the ionic strength as described previously.[3,7]

Activity coefficients were calculated from the extended form of the Debye-Hückel equation proposed by Davies.[8] The relative supersaturations, σ, were expressed by Eq. 1

$$\sigma = \{[(Ca^{2+})(F^-)]^{1/3} - (K_{sp})^{1/3}\}/(K_{sp})^{1/3} \qquad (1)$$

in terms of the molar concentrations of the free calcium and fluoride lattice ions in the supersaturated solution, (Ca^{2+}) and (F^-), and the solubility product, K_{sp}, at the ionic strength of the experiment.

During the CC experiments, summarized in Table 2, the supersaturation was maintained constant by the addition of titrants containing the lattice ions.

Typical plots of titrant volume shown in Figure 1, were used to calculate the rates of crystal growth[9] expressed in Eq. (2)

$$\text{Rate of Growth}, R_g = ks(K_{sp})^{1/v} \sigma^n = k'\sigma^n \qquad (2)$$

In Eq. 2, k and k' are rate constants, s is related to the number of growth sites available on the surface of the seed crystals, and v (= 3) is the number of ions in the formula unit. K_{sp} is the solubility product at the ionic strength of the experiment (0.15 mol l^{-1}) and n is the effective order of the growth reaction. It can be seen in Table 2 that the rate of titrant addition (mol min^{-1}) at each supersaturation increased proportionally to the mass of CaF$_2$ seed crystals used to initiate the growth reactions.

A. Abdul-Rahman et al.

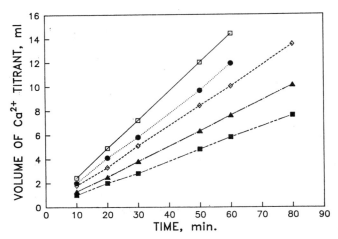

Figure 1. Typical plots of the volume of calcium nitrate (or chloride) titrant added during the crystallization experiments. Experiment 25 (open squares), 26 (closed circles), 27 (open diamonds), 29 (closed triangles), and 30 (closed squares).

It was evident from the relatively large increase in the size of the cubic seed crystals and the absence of secondarily nucleated particles, that growth took place exclusively on the surface of the added seed crystals. The effective order of the growth reactions was obtained from the slopes of the logarithmic plots of Eq. 2 shown in Figure 2.

The n value, 1.94 ± 0.1, showed that the rate of crystallization followed a rate law second order with respect to relative supersaturation, consistent with a spiral growth mechanism. Further evidence was provided by the insensitivity of the growth rates to changes in fluid dynamics in experiments 27 and 31 shown in Table 2.

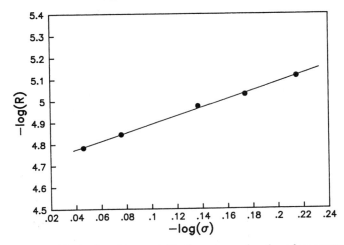

Figure 2. Logarithmic plot of the crystallization rate as a function of supersaturation.

Table 3. The influence of phosphonoformic (PF), phosphonoacetic (PA), and phosphonopropionic (PP) acid additives on the crystallization rate of calcium fluoride at 25°C, T(Ca):T(F) = 1:2, T_{Ca} = 5.96 x 10^{-4} molar, ionic strength = 0.15 molar, 200 rpm stirring speed

Experiment no.	Additive	[Additive] /10^{-6} mol L^{-1}	Rate/10^{-6}mol min^{-1} m^{-2}
27	—	0.0	11.5
35	PF	0.50	5.8
36	PF	1.00	4.0
37	PF	2.00	3.2
38	PF	3.00	3.0
39	PF	4.00	2.9
40	PA	0.50	8.0
41	PA	1.00	6.7
42	PA	2.00	6.0
43	PA	3.00	5.7
44	PA	4.00	5.5
45	PP	0.50	8.8
46	PP	1.00	7.8
47	PP	2.00	7.0
48	PP	3.00	6.7
49	PP	4.00	6.5

Crystallization experiments in the presence of phosphonate additives made at a constant supersaturation, σ = 0.79, are summarized in Table 3 and the data are plotted in Figures 3 and 4.

It can be seen that the rate of crystallization was markedly reduced even at very low additive concentrations. As the latter increased (Figure 3), the retardation reached maxima of 75%, 51% and 43% for phosphonoformic (PF), phosphonoacetic (PA) and phosphono-propionic acids (PP), respectively.

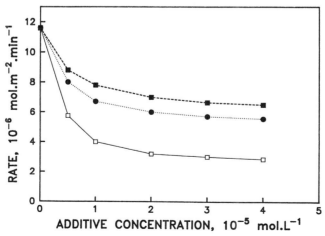

Figure 3. Crystal growth of calcium fluoride in the presence of phosphonoformic (open squares), phosphonoacetic (closed circles), and phosphonopropionic (closed squares) acids.

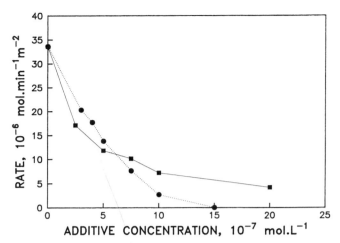

Figure 4. Crystal growth of calcium fluoride in the presence of ENTMP (closed squares) and K-732 (closed circles).

CC crystal growth experiments were also made in the presence of ethylenediamine tetra(methylene phosphonate), ENTMP, and a polyacrylic acid, K-732. The growth curves, plotted in Figure 4 showed that both were effective growth inhibitors. Complete inhibition of CaF_2 growth was achieved by the addition of only 1.5×10^{-6} mol l^{-1} of K-732. It is interesting to note (Figure 5) that when mixtures of ENTMP and K-732, at a total concentration of 1.0×10^{-6} mol l^{-1}, were added to the supersaturated solutions prior to seeding, the resulting growth rate was greater than the sum of the rates with the individual components at the same concentration. This suggests molecular interference between the adsorbates facilitating diffusive access of calcium and fluoride ions to growth sites through the disrupted adsorption layer.

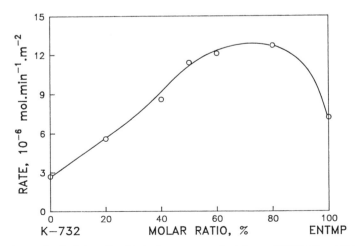

Figure 5. Crystal growth of calcium fluoride in the presence of mixtures of ENTMP and K-732. Plot of rate against molar ratio at a constant total concentration of 1.00×10^{-6} mol l^{-1}.

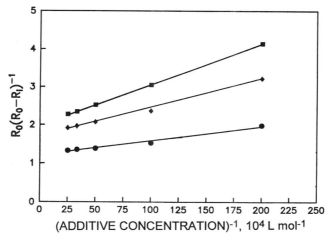

Figure 6. "Kinetic" Langmuir plots of the influence on calcium fluoride crystal growth rate of phosphonoformic (closed circles), phosphonoacetic (closed diamonds), and phosphonopropionic (closed squares) acids.

Crystal growth retardation in the presence of additives can be interpreted in terms of the elimination or blocking of growth sites on the crystal surfaces. In addition, the inhibitor molecules may compete with fluoride ions in the Stern layer[10] thereby preventing their integration into the growing crystals. It can be seen in Figure 3, that even at maximum retardation by the three phosphonate inhibitors, crystal growth continued at lower rates indicating that crystal lattice ions were still able to reach the surface between the adsorbed additive molecules. At maximum surface coverage, repulsive forces between the adsorbed molecules probably limited their surface concentrations so that the crystal surfaces retained their accessibility to Ca^{++} and F^- ions from solution. Assuming that the reduction in crystal

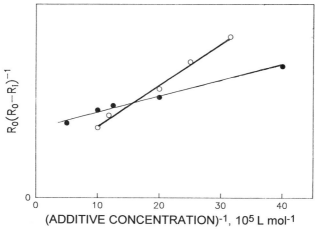

Figure 7. Langmuir-type plots of relative crystal growth rate reduction in the presence of ENTMP (closed circles) and K-732 (open circles).

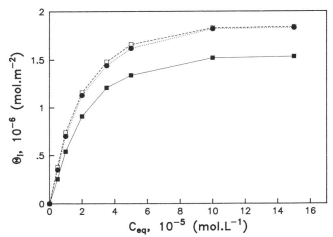

Figure 8. Equilibrium adsorption isotherms of phosphonoformic (closed squares), phosphonoacetic (closed circles), phosphonopropionic (open squares) acids.

growth rate reflected the increased coverage on the CaF_2 crystal surface, it can be described in terms of a simple Langmuir-type isotherm.[11] The growth rates in the absence (R_o) and presence (R_i) of inhibitors, can be represented by such a model using Eq. 3.

$$R_0/(R_0 - R_i) = 1 + (KC_{add})^{-1} \qquad (3)$$

Typical plots of $R_0/(R_0 - R_i)$ against the reciprocal of the inhibitor concentration, C_{add} are shown in Figures 6 and 7.

The linearity confirms the applicability of this model. Values of the resulting "kinetic" adsorption constants, K, are 0.93×10^6, 1.62×10^6, 1.90×10^6, 2.5×10^6 and 3.8×10^6 l mol^{-1} for phosphonoformic acid, K-732, phosphonoacetic acid, phosphonopropionic acid and ENTMP additives, respectively.

Equilibrium adsorption experiments of the phosphonoic acids on calcium fluoride crystals were made at pH = 6.0±0.2 and 25.0±0.2 °C. It was shown, in timed experiments, that equilibrium was reached within the first few minutes but the suspensions were allowed to come to equilibrium for 24 hours. Adsorption isotherms are shown in Figure 8. Although it was not possible to fit the data using the linear form of Langmuir equation, it can be seen that the extent of adsorption decreases in the order

PA >> PP>PF in good agreement with the results of the crystal growth experiments.

ACKNOWLEDGMENT

We thank the National Institute of Health for a grant, DE03223, in support of this work.

REFERENCES

1. G. H. Nancollas, R. A. Bochner, E. Liolios, L. J. Shyu, Y. Yoshikawa, J. P. Barone and D. Svrjek, 1982, The Kinetics of Crystal Growth of Divalent Metal Fluorides, Am. Inst. Chem. Eng., Symp. Ser., 215:26.

2. M. B. Tomson and G. H. Nancollas, 1978, Mineralization Kinetics: A Constant Composition Approach, *Science*, 200:1059.

3. L. J. Shu, and G. H. Nancollas, 1980, The Kinetics of Crystallization of Calcium Fluoride. A New Constant Composition Method, *Croatica Chem. Acta*, 53:281.

4. A. J. Ellis, 1963, The Effect of Temperature on the Ionization of Hydrofluoric Acid, *J. Chem. Soc.*, 4300.

5. R. E. Connick and M. S. Tsao, 1954, Complexing of Magnesium Ion by Fluoride Ion, *J. Am. Chem. Soc.*, 76:5311.

6. F. G. Gimblett and C. B. Monk, 1954, E.M.F. Studies of Electrolytic Dissolution Part 7. Some Alkali and Alkaline Earth Metal Hydroxides in Water, *Faraday Soc. Trans.*, 50:965.

7. G. H. Nancollas, 1966, Interactions in Electrolyte Solutions, Elsever, Amsterdam.

8. C. W. Davies, 1960, Ion Association, Butterworths, London.

9. G.H. Nancollas, 1979, The Growth of Crystals in Solution, *Adv. Colloid Inter. Sci.*, 10:215.

10. H. R. Rawls, T. Bartels and J. Arends, 1982, Binding of Polyphosphonates at the Water/Hydroxyapatite Interface, *J. Colloid Inter. Sci.*, 87:339.

11. P. Koutsoukos, Z. Amjad and G. H. Nancollas, 1981, The Influence of Phytate and Phosphonate on the Crystal Growth of Fluorapatite and Hydroxyapatite, *J. Colloid Inter. Sci.*, 83:599.

12

THE EFFECT OF MINERAL DEPOSITS ON STAINLESS STEEL

Pavlos G. Klepetsanis, Nikolaos Lampeas, Nikolaos Kioupis, and
Petros G. Koutsoukos

The Institute of Chemical Engineering and High Temperature Chemical
    Processes
P.O.Box 1414
Department of Chemical Engineering
University of Patras
GR-265 00 Patras, Greece

INTRODUCTION

The occurrence of various types of scales in both recirculation and once through cooling systems is very common due to the composition of the make-up water used to replace the losses due to evaporation. Calcium sulfate, mainly as gypsum, and the much less insoluble calcium carbonate polymorphs are the salts most frequently encountered.[1,2] The deposits that are formed greatly reduce heat transfer causing energy losses or material damage especially when coupled with corrosion. In all cases in which scale is encountered several steps precede the appearance of the solid phase. These include nucleation (primary and secondary), crystal growth and secondary changes such as recrystallization, aging, aggregation, etc. The solution supersaturation is a very important factor for the control of the formation of solid precipitates in the aqueous phase but it should be examined in conjunction with other factors such as the nature of the substrate on which scale deposits are formed. This is important in cases in which more than one type of salt is formed. It has been suggested by Newkirk and Turnbull[3] that the lattice misfit between two crystalline solids may serve as a criterion for the prediction of the overgrowth of one crystalline deposit on the matrix provided by another. In the present work we have investigated the formation of calcium sulfate dihydrate (CSD) in aqueous solutions supersaturated both with respect to this phase and with respect to all calcium carbonate polymorphs. More specifically, the deposition of CSD on metallic surfaces both clean and covered with calcium carbonate scale was studied in order to investigate the relationship of these two salts in terms of heterogeneous precipitation. Investigation of the the relationship of CSD with the various calcium carbonate polymorphs was done in a batch reactor at low (25°C) and at higher temperatures (120 and 140°C).

Mineral Scale Formation and Inhibition, Edited by Zahid Amjad
Plenum Press, New York, 1995

The presence of reducing conditions may yield conditions favorable for the formation of sulfide ions from the reduction of sulfate. Sulfide ions may participate in the corrosion process of the steel parts in contact with the fluid. Corrosion is responsible for the partial dissolution of iron which may trigger the precipitation of iron sulfide. The sulfide deposits, depending on the solution pH, may promote further corrosion.[4,5] Calcium carbonate deposits, on the other hand, are known to provide protection against corrosion. The correlation between various types of scale formation (carbonate and iron sulfide) on steel substrates and corrosion has also been investigated with electrochemical measurements.

EXPERIMENTAL

Crystallization / Deposition Experiments

The experiments described in the present work were done in batch reactors, magnetically stirred (ca. 400 rpm) in order to insure homogeneity of the solution. Thermostatting of the reactors was done by circulating water through the glass jacket surrounding the reactor. The experiments at 120 and 140°C were done in an autoclave with a glass lining (Parr). The supersaturated solutions were prepared in the reactors by mixing equal volumes of solutions containing calcium and a salt of the anion of the solid being investigated, respectively. The deposition proceeded either in the bulk of the solution or on the surface of the stainless steel (SS 304) U tubes through which 60.0 ± 0.5°C thermostatted water was flowing. Immediately following the mixing of the solutions, physicochemical properties of the supersaturated solution such as conductivity and/or calcium concentration were monitored. In the former case, a conductivity probe (Radiometer PP1042) with a conductivity meter (Metrohm 660 Conductometer), appropriately calibrated as described in Klepetsanis and Koutsoukos[6], was employed. In experiments in which the precipitation process was monitored through the measurement of the calcium ion concentration in the solutions, aliquots of the aqueous phase were withdrawn, filtered through membrane filters (Millipore 0.22 μm) and the filtrates were analyzed for calcium and sulfate ions by ion chromatography (Metrohm IC 690 with a conductivity and a spectrophotometric detector and a Shimadzu electronic signal integrator). The deposited solid phases were characterized by physicochemical methods including powder x-ray diffraction spectroscopy (Philips 1840/30), scanning electron microscopy

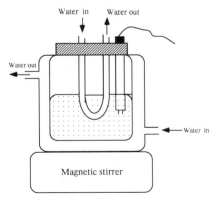

Figure 1. Experimental setup for the measurement of the kinetics of deposition of insoluble salts on steel surfaces provided by a U-shaped tube immersed in solutions supersaturated in the salt deposited.

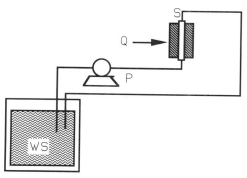

Figure 2. Experimental setup for the deposition of calcium carbonate in stainless steel specimens. WS: working solution, S: specimen, P: peristalic pump for the circulation of the supersaturated solution, Q: external heating of the specimen tube.

(SEM, JEOL JSM 5200), infrared spectroscopy (Nicolet FTIR 740), and specific surface area measurements (multiple point nitrogen adsorption BET). The experimental setup for the rate measurements of the deposition of minerals on the steel U-tubes is shown in figure 1.

Layers of calcium carbonate deposited on steel specimens were obtained by the continuous flow of supersaturated solutions, keeping the supersaturation of the circulating solution constant by the addition of the appropriate titrant solutions[7] The experimental set-up is shown in figure 2.

Electrochemical Measurements

The electrochemical investigation of the behavior of stainless steel SS 304 in corrosive media of Sodium Chloride in the presence (and in the absence) of sulfide ions (and Iron Sulfide deposits) includes two kinds of experiments:

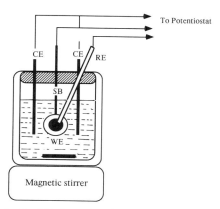

Figure 3. Experimental setup for the DC electrochemical measurements. CE: counter electrodes (graphite), RE: reference electrode with salt bridge, WE: working electrode.

- DC-measurements: Open Circuit Potential (OCP) measurements and Potentiodynamic Scans.
- AC-measurements: Impedance measurements

The DC-measurements (Open Potential Potential and Potentiodynamic Scans) were done in an electrochemical cell consisting of a 0.5l thermostatted glass vessel, magnetically stirred, and equipped with an appropriate electrode system. The cell was sealed with a polyamide lid with ports accomodating a graphite counter electrode, a reference (Ag/AgCl/KCl (sat.)) electrode with a salt bridge, and the working electrode. The sample of steel examined (stainless steel) was embedded in an otherwise chemically inactive resin matrix in the form of rods.

The specimens used as working electrodes were shaped as plates. The electrodes were connected to the Potentiostat (INTERTECH PGS-151). The function of potentiostat (Open Circuit Potential and Potentiodynamic Scans Measurements) was selected by appropriate switch. The experimental setup is shown in figure 3:

The potentiostat employed for the measurements has two outputs - one for measuring voltage, and the other one for the measurements of the current - and one input for external scan voltage. The potentiostat has also the ability for control from digital Input/Output ports of computer data acquisition cards using the appropriate software. The total surface area exposed in the solution was 1 cm^2 and the surface was polished with a series of silicon carbide papers up to 1000 grit (except for the measurements in which the specimen surface was covered by deposits). The polished surface was next washed well with triply distilled water and dried with distilled acetone. After the immersion of specimen in the corrosive media the open circuit potential (OCP) with respect to the reference electrode was measured for at least two hours. The potentiodynamic scans were done after the measurement of the OCP starting from the cathodic region, at potentials about E_{OCP} = -250mV and ended to the anode region at potentials E_{OCP} = +400mV with a scan rate 0.167 mV/s (ASTM-standard procedure). The measurements were done by a combination of data acquisition cards (Advantech PCL-812 PG and PCL-726) and an XT personal computer with the appropriate connections with the potentiostat (PGS-151). The data were collected by the computer and were stored in files for further treatment. The measurements of the OCP of the steel specimens were done in aerated, stirred solutions in:

  a. The absence of sulfides in the solution and any deposits on the specimen surface over a range of sodium chloride concentrations,
  b. The presence of various concentrations of sulfides in the solution but in the absence of deposits on the surface of the specimen, and
  c. The absence of sulfides in solution but in the presence of an iron monosulfide deposit on the surface of the specimen.

The pH of above experiments was adjusted to approximately 6.5 by the addition of standard 0.1M HCl or 0.1M NaOH solutions as needed.

The deposition of the iron monosulfide on the stainless steel surfaces was done by immersion of specimen plates in an iron monosulfide suspension which was prepared by a rapid mixing of stoichiometric mixtures of ammonium sulfide and ferrous ammonium sulfate. The speciments were removed from the suspension at different time intervals and were next placed in a 0.4 M NaCl solution for investigation (OCP and Potentiodynamic measurements) of the effect of the various deposition times in the electrochemical behavior of stainless steel. It should be noted that it was previously found that the FeS deposition on stainless steel specimens suspended in stirred suspensions increased with time.

The AC- impedance measurements were done in the same thermostatted vessel using the Electrochemical Impedance Analyzer (Model 6310,EG&G). The Electrochemical Im-

Table 1. Precipitation of calcium sulfate on various substrates. Initial conditions and kinetics results

Exp. #	$Ca_t = (SO_4)_t$ * 10^{-2} mol l^{-1}	Substrate	θ °C	Ind. time, τ min	Rates 10^{-5} mol l^{-1}min^{-1}
1	5.50	calcite	25	250	-
2	5.20	calcite	25	600	-
3	5.00	calcite	25	840	-
4	4.70	calcite	25	1020	-
5	4.50	calcite	25	1250	-
6	4.00	SS304	60	5	57.4
7	3.50	SS304	60	21	33.5
8	3.25	SS304	60	25	9.5
9	2.75	SS304	60	77	1.0
10	2.40	SS304	60	-	No deposition
11	3.50	SS304+calcite	60	5	14.7
12	2.80	SS304+calcite	60	33	7.7
13	2.60	SS304+calcite	60	79	6.6
14	2.40	SS304+calcite	60	111	1.4
15	2.00	SS304+calcite	60	stable, no precipitation	
16	5.1	SS316$^+$	80	24	50
17	4.9	SS316$^+$	80	32	658
18	4.7	SS316$^+$	80	68	541
19	4.5	SS316$^+$	80	89	357
20	4.3	SS316$^+$	80	128	255

* Ca_t: total calcium $(SO_4)_t$ = total sulfate
$\Leftrightarrow$ experiments in which the calcium sulfate supersaturated solutions were flowing through the tube specimens at a flow rate of 500 ml/ h, pH=7.0

pedance Analyzer was connected to the computer via GPIB-Card (PC-AT of National Instruments) and the impedance measurements were done with the suitable software.

The AC - measurements include impedance measurements of stainless steel specimens in aerated, stirred solutions from an initial frequency of 10 mHz to a final frequency of 100 KHz with AC-amplitude 5.00 mV RMS with respect to the OCP and:

a. The absence of sulfide ions at various times and
b. The presence of sulfide ions at various times.

RESULTS AND DISCUSSION

The driving force for the formation of the various calcium sulfates from the supersaturated solutions is the difference between the chemical potentials, $\Delta\mu$, of the salt in the supersaturated solutions, μ_s, and those corresponding to the equilibrium, μ_∞:

$$\Delta\mu = \mu_s - \mu_\infty = [\mu_s^O + RT \ln(\alpha_{CaSO_4})_s] - [\mu_\infty^O + RT \ln(\alpha_{CaSO_4})_\infty] \tag{1}$$

provided that $\mu°_s = \mu°_\infty$ and taking mean activities:

$$a_{CaSO_4} = (a_{Ca^{2+}} \, a_{SO_4^{2-}})^{1/2} \tag{2}$$

equation 1 becomes:

$$\Delta\mu = -\frac{RT}{2} \ln \frac{(\alpha_{Ca^{2+}} \alpha_{SO_4^{2-}})}{(\alpha_{Ca^{2+}} \alpha_{SO_4^{2-}})_\infty} \qquad (3)$$

or

$$\Delta\mu = -\frac{RT}{2} \ln \frac{IP}{K_s^O} = -\frac{RT}{2} \ln \Omega \qquad (4)$$

In equation 4 IP and K_s^O are the activity product and the thermodynamic solubility product of the salt formed respectively. The ratio IP/K_s^O is defined as the supersaturation Ω. The computation of the ionic activities of all species was done by taking into account all equilibria, mass and charge balance equations with successive approximations for the ionic strength as described earlier by Klepetsanis and Koutsoukos.[8]

The experimental conditions and the results obtained for the deposition of calcium sulfate on various substrates are summarized in Table 1.

In all experiments included in Table 1, the thermodynamically stable calcium sulfate dihydrate (CSD) phase was formed. The supersaturated solutions used were tested before for stability and it was verified that they were stable in the absence of other foreign surfaces. The deposition therefore of CSD on the metallic surfaces or the seed crystals introduced in the supersaturated solutions was selective on the foreign substrate. At low temperatures and in solutions seeded with calcite seed crystals the formation of CSD was very slow initially and the uncertainty for the measurement of the rates of growth upon increasing the amount of seed crystals introduced by a factor of 2 or 3 did not show any significant change in the long induction times while the measurements of the rates which were done by monitoring the specific conductivity of the suspension, were influenced by the slurry densities and thus are not included in Table 1. The deposition, however, on stainless steel was found to be faster and was preceded by short induction times that decreased with increasing supersaturation. The rates, on the other hand, decreased with decreasing supersaturation and were considerably slower in comparison with the rates of spontaneous precipitation of CSD at similar conditions (as may be seen in Table 2.)

Moreover, the presence of a calcium carbonate coating on the 304 stainless steel surfaces facilitated the deposition of CSD as may be seen from the shorter induction times. The rates of deposition were, however, slower. It should be noted that the U-shaped tubes used in these experiments were previously immersed in supersaturated calcium carbonate solutions at 60°C. This treatment resulted in the deposition of a calcium carbonate layer consisting mainly of vaterite which slowly converted to yield the more stable calcite, as shown by Dalas and Koutsoukos.[7] A comparison with experiments 16 through 20 in which CSD was deposited at 80°C at rates comparable to those of spontaneous precipitation, suggested also that in the experiments from 25-60°C, the deposition of CSD was done

Table 2. Spontaneous precipitation of calcium sulfate dihydrate at 60°C

Experiment #	$(Ca)_t = (SO_4)_t$ $\times 10^{-2}$ mol l^{-1}	Ind. Time, τ min	Rate x 10^{-3} mol l^{-1} min^{-1}
110	3.80	36	83.8
111	3.65	62	57.0
112	3.50	112	35.7
113	3.35	188	13.0

Table 3. Comparison of lattice compatibilities between calcium sulfate dihydrate and the various calcium carbonate polymorphs

Phase	Crystal system	Face parameters	Unit cell lattice a x b (Å^2)	Misfit %
$CaSO_4 \cdot 2H_2O$, gypsum	Monoclinic	010	5.75x15.15	-
Calcite	Orthorhombic	1010	4.99x17.06	5
Aragonite	Hexagonal	100	5.72x15.88	0.5
Vaterite	Hexagonal	1010	33.96x14.32	1.5

selectively on the substrates introduced (i.e. calcite and SS 304). The estimated lattice misfit for the (1010) face of CSD and the three calcium carbonate polymorphs shown in projected dimensions in Table 3, is better than 5% (i.e. sufficient to account for the observed overgrowth). The dependence of the induction times on the solution supersaturation is shown in Table1.

It has been suggested by Mullin[9] that the change in the slope of the linear dependence of log t_{ind} vs $(\log\sigma)^{-2}$ may be considered as a change between homogeneous and heterogeneous nucleation. In the case of the calcium carbonate lined stainless steel the limit has been shifted from a value of σ about 10 for the uncovered to 40. The limit for the spontaneous precipitation case is at about $\sigma = 30$. The transition limit for the deposition of the flowing supersaturated solution was around $\sigma = 15$ (i.e. the same as that for the U tubes as may be seen in Figure 5.)

The CSD deposition rates, R_P, were plotted as a function of the relative solution supersaturation according to the semiemperical equation:

$$R_P = k_P\sigma^n \qquad (5)$$

Where k_P is the rate constant, σ the relative solution supersaturation defined as:

$$\sigma = \Omega^{\frac{1}{2}} - 1 \qquad (6)$$

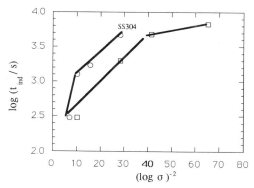

Figure 4. Dependence of the induction time on the relative solution supersaturation for the deposition of calcium sulfate dihydrate on: (open circle) 304 stainless steel surfaces; (open square) 304 stainless steel coated with calcium carbonate deposits.

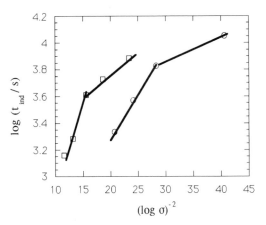

Figure 5. Induction period for the formation of calcium sulfate dihydrate as a function of the relative supersaturation. Plots of the logarithm of the measured induction times as a function of the inverse square of the logarithm of the relative supersaturation: (open square) spontaneous; (open circle) on heated 316 SS tubes at a flow rate of 0.5 l/h, pH 7.0, 60°C.

and n the apparent order of the process. Values of n = 1-2 correspond to spiral growth and n > 2 to polynuclear growth. In the case n = 1, the rate may also be diffusion controlled.[10] Log-log plots according to equation 5 are shown in Figure 6.

The apparent order found for the deposition of CSD on the 304 SS surfaces and in the bulk supersaturated solution was found to be 5 suggesting a surface controlled polynuclear mechanism. In the case of the calcite coated SS 304, however, the apparent order was found to be 2, a value typical for spiral growth on seed crystals as suggested by Liu and Nancollas[11], Nancollas et al.[12], Christoffersen et al.[13], Rosmalen[14], etc. In the scanning electron micrographs shown in Figure 7(a and b), the outgrowth of prismatic CSD from the vaterite and calcite crystals present on the surface of the heated SS 304 surface is shown.

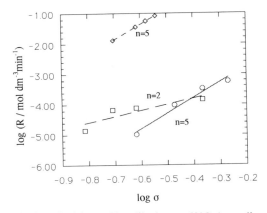

Figure 6. Kinetics of precipitation of calcium sulfate dihydrate at 60°C. (open diamond) Spontaneous precipitation. Plots of logarithm of the rates as a function of the relative solution supersaturation: (open circle) Precipitation on U-shaped tubes of 304 SS; (open square) Precipitation on U-shaped tubes of 304 SS covered with calcium carbonate.

Table 4. Initial conditions, induction times, and rates of formation of calcium carbonates from supersaturated solutions on calcium sulfate dihydrate seed crystals.

Exp. #	$Ca_t = (SO_4)_t$ $\times 10^{-3}$ mol l^{-1}	θ °C	Ind. time min	Rates $\times 10^{-10}$ mol min^{-1}m^{-2}
*200	1.13	120	stable	No precipitation
201	1.13	120	70	2.7
202	1.08	120	110	2.4
203	1.00	120	140	1.9
204	0.95	120	195	1.6
205	1.47	140	60	0.5
206	1.36	140	70	0.3
207	1.02	140	90	0.2
208	0.95	140	100	0.1

*In the absence of calcium sulfate dihydrate seed crystals

The selectivity of the CSD overgrowth on the calcium carbonate polymorphic deposits was further demonstrated in experiments in which calcium carbonate supersaturated solutions were seeded with CSD seed crystals. The experimental conditions and the results obtained are shown in Table 4.

Although the subject was not addressed, we have found that the surface roughness played an important role in the deposition of the minerals. In the experiments done we have taken care so that the pretreatment of the steel surfaces prior to immersion in the supersaturated solutions was reproduced precisely and it was verified by microscopic examination of the steel surfaces. It was found that the deposition of calcium carbonate in the interior walls

Figure 7. Scanning electron micrographs of gypsum crystals forming on vaterite crystals deposited on stainless steel.

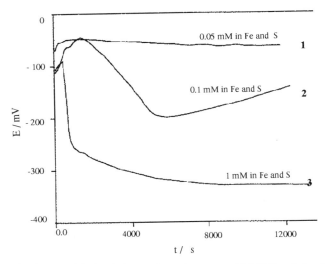

Figure 8. Open circuit potential of stainless steel specimens on which FeS scale has been deposited.

of the specimens improved their behavior towards corrosion in corrosive media containing high chloride concentrations (3.5% NaCl). The presence of calcium carbonate deposited in the inner surface of the tube showed a passive region in the anodic current range. It is interesting however to note that the corrosion potential did not change significantly suggesting that there was no tendency of affecting the propensity of the material for corrosion. In our study we have used non-uniform calcium carbonate layers.

Moreover, the presence of iron and sulfide in the aqueous medium flowing through the specimens and containing 3% sodium chloride was investigated. Experiments were done using various concentrations of iron (as $(NH_4)_2Fe(SO_4)_2$) and sulfide (as $(NH_4)_2S$) ions. The solutions contained 3% NaCl and were prepared in a reservoir thermostatted to $35.0 \pm 0.1°C$. The solutions were supersaturated with respect to iron sulfide. The precipitation was monitored with a calibrated conductivity probe and conductivity meter which allowed for

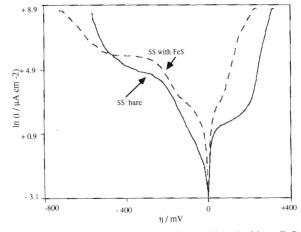

Figure 9. Polarization curves of stainless steel specimens with and without FeS scale deposits.

electronic compensation. Thus, conductivity changes as small as 1 μS could be detected. The fluid was circulated by a peristalic pump from the reservoir through the stainless steel specimen. Before entering the SS specimen, the fluid was preheated by a condenser heated by circulating water thermostatted to 90°C. The specimen was a stainless steel with smooth, shiny internal walls, 5 cm long and 6.35 mm ID. The specimen was rotated about the x axis by a precision motor (M), on glass joints. Throughout the measurement, the open circuit potential was measured with a potentiostat between the specimen and a saturated calomel electrode with a bridge filled with a 3% NaCl solution. The open circuit potential is a measure of the tendency of the system to corrode. Thus, a shift of the potential towards the more negative may be interpreted as an enhanced tendency towards corrosion. All experiments were done in the presence of air (solutions at equilibrium with atmospheric air) and the pH was adjusted to 4.7 with a standard base solution. When the concentrations of the working solution (contained in the reservoir) were 5×10^{-5}M in both total iron and total sulfide, no precipitate was formed for more than 24 hours. During this time period, the open circuit potential (after approximately 200 s) maintained a constant value of -65 mV as may be seen in Figure 8 (curve 1).

Upon raising the solution concentration so that it was 0.1mM in both total iron and total sulfide, spontaneous precipitation of iron sulfide occurred inside the stainless steel specimen, leading to a decrease of the open circuit potential to more cathodic values as low as -200 mV, after approximately 4000s, past the attainment of the previous value of the -65 mV, which corresponded to the situation of deposit-free specimen. However, upon return of the solution into the reservoir, the solid particles were redissolved and subsequently the film deposited became thinner and the potential shifted again to more anodic potentials with a tendency to reach the deposit-free specimen (curve 3). Further increase of the solution

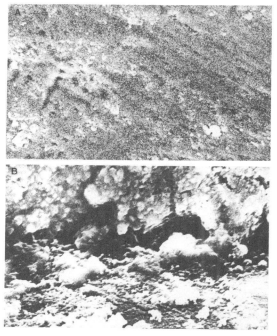

Figure 10. Scanning electron micrographs of (a) polished 304 stainless steel surface before and (b) after amorphous FeS deposition.

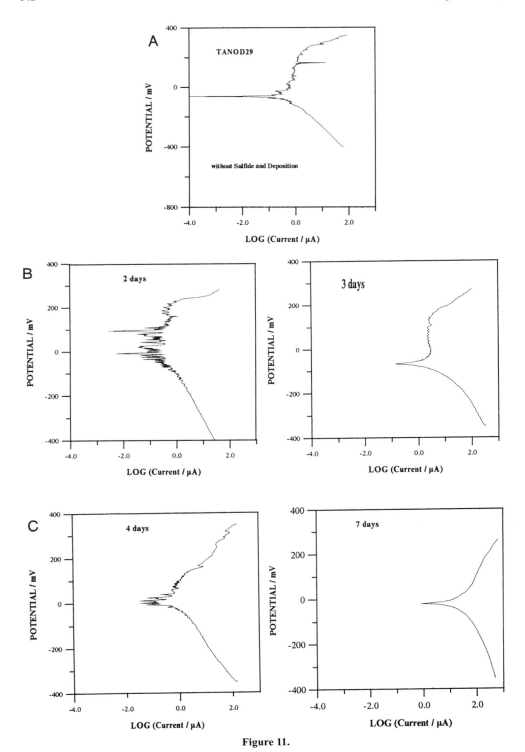

Figure 11.

supersaturation however to 1mM for total iron and total sulfide, resulted in the spontaneous precipitation of iron sulfide which persisted in the reservoir and was deposited on the walls of the specimen. Following the formation of the deposit, a sharp drop of the open circuit potential was observed to a significantly more cathodic potential, which remained constant at -315 mV. The constant potential in this case was in line with the stability of the deposit formed on the walls of the specimen (curve 2). The results of the present work showed that the formation of iron sulfide deposits formed in the presence of ferrous and sulfide ions in geothermal brines tend to favor the corrosion of the metallic surfaces, since corrosion may start at more cathodic potentials. Typical polarization curves obtained for specimens on which FeS scale had been deposited are shown in Figure 9.

It is interesting to note that the deposited amorphous FeS reduced considerably the resistance of steel by almost eliminating the passive region. It should be noted however that the time needed for the formation of the FeS deposits on steel is also an important parameter for the behavior of stainless steel. By allowing stainless steel in contact with FeS suspensions the deposit thickness increased with time, apparently due to both growth and adhesion on the steel surface. The morphology of the deposits is shown in Figure 10.

The polarization curves of the specimens with various deposits of FeS are shown in figure 11. As may be seen, the formation of the first deposit layers did not affect the passive region over a period of two days (Figure 11b) but the noise increased in comparison with the bare SS 304 (Figure 11a), possibly due to the enhancement of the surface roughness. Further deposition resulted in relatively smoother surfaces as in three days the surface coverage was complete, and microscopic examination of the deposits showed lack of bare patches. It was however after four days that the specimens showed enhanced corrosion which eventually led to the elimination of the passive region (Figure 11c), suggesting enhanced corrosion.

CONCLUSIONS

The results of the present work have shown that the deposition of calcium sulfate dihydrate takes place at higher degrees of supersaturation spontaneously and at lower supersaturations on substrates such as stainless steel 304 and 316 and also on calcium carbonate deposits consisting mainly of vaterite. In all cases a surface controlled mechanism was found to be prevalent, while the deposition on calcium carbonate substrates was found to be selective both at 25 and 60°C, probably because of the favorable lattice matching between calcium sulfate dihydrate and the calcium carbonate polymorphs. At high temperatures (120 and 140°C) vaterite was formed on calcium sulfate dihydrate substrates, supporting the lattice matching suggestion. The deposition of sparingly soluble salts on steel surfaces is related with the corrosion behavior of the metal in aggressive media containing chloride, sulfide and carbonate ions. The deposition of FeS enhanced corrosion phenomena while the deposition of calcium carbonate (calcite) yielded resistivity to the metallic surfaces towards corrosion.

ACKNOWLEDGMENT

The authors wish to acknowledge the support of this work through the CEU, JOULE II program, contract No. JOU2-CT-92-0108.

REFERENCES

1. Cowan, J.C. and D.J. Weintritt, 1976, *Water formed scale deposits*, Gulf Publ. Co., Houston TX.
2. *BETZ Handbook of industrial Water Conditioning*, (1980).
3. Newkirk, J.B., D.J. Turnbull, 1955, Nucleation of ammonium iodide crystals from aqueous solutions, *J. Appl. Physics*, 26: 579-593.
4. Sardisco, J.B., W.B. Wright, and E.C. Greco, 1963, Corrosion of iron in an $H_2S-CO_2-H_2O$ system: Corrosion film properties on pure iron, 19: 354t-359t.
5. Vedage, H., T.A. Ramanarayanan, J.D. Mumford, and S.N. Smith, 1993, Electrochemical growth of iron sulfide films in H_2S saturated chloride media, 49: 114-121.
6. Klepetsanis, P.G., and P.G. Koutsoukos, 1991, Spontaneous precipitation of calcium sulfate at conditions of sustained supersaturation, *J. Colloid & Interface Science*, 143: 299-308.
7. Dalas, E. and P.G. Koutsoukos, 1990, Calcium carbonate scale formation and prevention in a flow through system at various temperatures, *Desalination*, 78: 403-416.
8. Klepetsanis, P.G. and P.G. Koutsoukos, 1989, Spontaneous precipitation of calcium sulfate at constant calcium activity. *J. Crystal Growth*, 98: 480-485.
9. Mullin, J.W., 1993, *Crystallization*, Butterworth- Heinemann, 3rded., Oxford.
10. Söhnel, D. and J. Garside, 1992, *Precipitation. Basic Principles and industrial applications*, Butterworth - Heinemann, Oxford.
11. Liu, S.T. and G.H. Nancollas, 1973, The crystal growth of calcium sulfate dihydrate in the presence of additives, *J. Colloid & Interface Science*, 44: 422-427.
12. Nancollas, G.H., A.E. Eralp, and J.S. Gill, 1978, Calcium sulfate scale formation: A kinetic approach, *Soc. Petrol. Eng. J.*, 133-139.
13. Christoffersen, M.R., J. Christoffersen, M.P.C. Weijnen, and G.M. van Rosmalen, 1982, Crystal growth of calcium sulfate dihydrate at low supersaturation, *J. Crystal Growth*, 58: 585-595.
14. Brandse, W.P., and G.M. van Rosmalen, 1977, The influence of sodium chloride on the crystallization rate of gypsum, 39: 2007-2010.

13

THE USE OF CHELANTS AND DISPERSANTS FOR PREVENTION AND REMOVAL OF RUST SCALE

Robert P. Kreh and Wayne L. Henry,[1] John Richardson,[2] and
Vincent R. Kuhn[2]

[1] W. R. Grace & Co
7379 Route 32, Columbia, Maryland 21044
[2] W. R. Grace & Co
300 Genesee Street, Lake Zurich, Illinois 60047

ABSTRACT

A series of laboratory and pilot rig experiments are discussed to illustrate the effectiveness of various chelants and polymers in controlling or removing iron deposits in cooling systems. The organic molecules were chosen for strong interactions with iron(III) centers. Substituents have been included to increase the solubility and dispersion once these chelants and polymers are attached to ferric sites. Removal of iron oxide becomes more difficult as it ages and crystallizes, but this can be accomplished through proper selection of conditions and additives.

INTRODUCTION

One approach to controlling corrosion in cooling water systems is the use of chromate-containing programs.[1] Typically, these are operated near neutral pH, thus minimizing fouling problems due to inorganic salts of calcium (e.g., $CaCO_3$) and iron (e.g., Fe_2O_3), which are less soluble at higher pH values. Because of its toxicity, the use of chromate has been declining in recent years.[2] To control corrosion without the aid of chromate, cooling towers can be operated at higher pH values, and corrosion and scale inhibitors such as phosphate, phosphonates and polymers are employed. However, the elevated pH's have exacerbated fouling problems from calcium and iron-containing inorganic deposits, and there is a need for improved control of this fouling. In addition, low pH excursions can lead to increased corrosion and concomitant deposition of iron scale, again demanding good control of iron chemistry.

Deposit prevention and removal of iron scale can be achieved by mechanical or chemical means.

Mechanical Cleaning and Deposit Prevention

Mechanical cleaning methods commonly employed include techniques for both deposit prevention and deposit removal.

Deposit Prevention:
- Side stream filtering[3]
- Control of concentration cycles

Deposit Removal:
- Hydroblasting
- Abrasives[4]
- Brushing

In general, mechanical approaches to deposit control are expensive and not totally effective in the absence of chemical treatments. Methods for deposit removal require the equipment to be off-line and are labor intensive. In many cases, freshly cleaned surfaces are subject to flash corrosion if corrosion inhibitors are not rapidly employed. Accessibility to all fouled surfaces is also an issue, which can lead to incomplete cleaning.

Chemical Cleaning and Deposit Prevention

Chemical technology is available for both the prevention and removal of iron scale, these include:

Deposit prevention:
- Polymeric dispersants[5]

Deposit removal:
- Acids
- Chelants
- Phosphonates/neutral pH cleaners

The following section details the main principals behind each of these chemical technologies.

Polymers. Polymers have been used extensively in cooling water as dispersants for foulants such as silt and mud and for controlling scale-forming salts such as calcium carbonate,[6] iron oxide[7] and calcium phosphonate.[8] Generally, anionic polymers are used and are believed to function by increasing the negative charge on solid particles. Polyacrylate and polymaleate polymers have been found effective for calcium carbonate (threshold) inhibition.[6] Similar to the phosphonates, the performance of these polymeric threshold inhibitors can be adversely affected by soluble iron. The inclusion of sulfonate groups onto the carboxylate polymers generally improves the dispersion of precipitates of iron oxide[7] and calcium phosphate.[8] Very recently, it has been shown that hydrophobic substituents (e.g., t-butyl groups) also enhance the ability of carboxylate-containing polymers to disperse iron and calcium-containing solids.[9]

The tannins are a class of polymers which has been investigated for use in on-line removal of iron deposits.[10] These polymers are attractive because they contain hydroxyaromatic substituents which are available to bind to ferric ions. However, the amount of hydroxyaromatic substituents as a percentage of the entire polymer is relatively low, and these polymers often give insoluble iron complexes. Two approaches that we have taken to increase the solubility and the density of ferric binding sites are amino acid/hydroxyaromatic chelants and sulfonated hydroxyaromatics.

Chelants. Chelants are molecules which form strong complexes with metal ions (such as Ca(II) and Fe(III)) because they are able to coordinate to the metal ion through more than one atom. Each coordinating atom generally donates an electron pair, and 4-7 member rings are formed which include the metal ion and the chelant. An example of a chelant which has been used for cleaning of iron and calcium scales is citric acid.[11] It contains four potential electron-donating sites (three carboxyl oxygens and one hydroxyl oxygen). Generally citric acid has been used for cleaning at low pH because of the increased solubility of inorganic scales at low pH.

Cleaning at low pH has the disadvantages of increased rates of corrosion and large amounts of acid needed to adjust the pH of the entire system. Low pH cleaning is generally done off line. It is more cost effective to perform on-line cleaning at or near the normal operating pH of the system. At these higher pH values (neutral and above), the chelation of citric acid is not great enough to dissolve iron oxide scale.

Polydentate amino acid chelants such as the hexdentate ethylenediaminetetraacetic acid (EDTA) and the tetradentate nitrilo-tris(acetic acid) (NTA) are known to be stronger chelants for metal ions such as Ca(II) and Fe(III).[12] Even these chelants have difficulty dissolving iron oxide at pH values which are neutral or above. These chelants are effective in dissolving $CaCO_3$ solids at high pH,[13] but they must be used in very large concentrations (500-5,000 mg/l) to overwhelm the soluble Ca present in typical cooling water. These chelants can be corrosive to steel surfaces, particularly when used in high concentrations.[14] Recently, new chelants have been developed with increased specificity for iron by combining the chelating ability of amino acids with the strong iron affinity of hydroxyaromatic groups.[15,16] These chelants have led to the development of formulations with better control of iron scale at alkaline pH values. In this paper, we will show further advances in this approach, including the use of hydroxyaromatic ligands for strong iron chelation and sulfonic acid groups for enhanced solubility and dispersion of iron oxide particles.

Phosphonates. Organophosphonates have been introduced to replace chromates for corrosion inhibition of mild steel.[1] These inhibitors operate best under more alkaline conditions (pH 8-9) where $CaCO_3$ scale formation becomes a significant problem. Fortunately, these phosphonates can also inhibit $CaCO_3$ scale.[1] However, soluble iron ions can complex and/or precipitate with these chemicals causing them to lose their ability to inhibit corrosion and $CaCO_3$ scale.[17] Certain phosphonates have been used to remove iron oxide deposits and leave a surface which is passivated against further corrosion.[18] This cleaning method is effective at neutral pH, thereby minimizing environmental impact and eliminating the hazards associated with the handling of acid. This phosphonate approach can also passivate the metal surface preventing "flash rusting" from occurring after the cleaning solution is removed.[18]

The results discussed below will provide insight into the mechanisms by which polymers, chelants and phosphonates can prevent and remove iron oxide scales. We will also describe several new chemical approaches to improved iron control.

EXPERIMENTAL

The laboratory procedures are described below and they are correlated with the Tables and Figures which are based on each procedure.

Maintaining Soluble Iron (Table 1)

The water used for this test was of low hardness: 99 mg/l $CaSO_4$ 13 mg/l $CaCl_2$, 55 mg/l $MgSO_4$ and 176 mg/l $NaHCO_3$. To a stirred solution of 50 mg/l additive in this water at pH 8.1, was added 10 mg/l Fe (as $FeCl_3$). The pH was re-adjusted to 8.1, and the solution was mixed at 300 rpm, 130 °F for 17 hours. It was then filtered through a 0.2 micron membrane and analyzed for Fe by inductively coupled plasma (ICP).

Fe_2O_3 Dispersion (Table 2)

The water used for this test was either the low hardness water described above, or a high hardness water containing 735 mg/l $CaCl_2 \times 2H_2O$, 250 mg/l $MgSO_4 \times 7H_2O$ and 845 mg/l $NaHCO_3$. The reagents (50 mg/l) were dissolved in the water and the pH was adjusted to 8.0 ± 0.2. Anhydrous hematite (Fe_2O_3, 1-5 µm particle size, 1000 mg/l) was added, and the mixtures were shaken at 300 rpm for 17 hours at 54°C. The mixtures were then allowed to settle for 30 minutes. and samples were withdrawn at 50% depth for iron analysis (by ICP).

Dissolving Precipitated Rust Scale (Table 3)

The additive (50 mg/l) was combined with 19 mg/l of "FeO(OH)" (precipitated from $FeCl_3$) in the low hardness water at pH = 8.1. The mixture was shaken at 300 rpm at 54°C for 17 hours, followed by filtration (0.1 µm), acidification, and analysis by ICP.

Iron Dispersion/Solubilization from Rusting Coupons (Figure 1)

Two clean, mild steel coupons were exposed to the low hardness water, treated with and without a dispersant. The initial pH of this water was adjusted to 8.2, the temperature was controlled at 60°C, and the water was aerated by means of passing air through a dispersion tube. At the end of 24 hours, the coupons were removed and the water was analyzed for soluble and dispersed iron.

Iron analyzed from water filtered through a 0.2 µ filter represents soluble iron. Dispersed iron is represented as the difference in iron concentration of unfiltered and filtered iron.

Iron Deposit Removal (Figures 2-5)

A clean preweighed $\frac{1}{2}''$ O.D., mild steel tube was installed in the fouling rig. Ten liters of high hardness water containing a phosphonate treatment was pumped over the tube at 6-8 liters/min. The water temperature was maintained at about 55°C by heating the inside of the mild steel tube. A simulated pH upset was initiated by adding dilute sulfuric acid to a pH of 5. The test was then continued for 24 hours at this pH.

The fouled tube was removed and placed in an oven for drying. After drying at about 105°C for a minimum of 2 hours, the tube plus the deposit was weighed. Next, the deposit from the tube was scraped and weighed. The tube was then acid cleaned, dried and reweighed. The deposit weight and the tube weight loss was used in comparing the affect of the cleaner on iron deposit.

After a tube was fouled in accordance with the above procedure, the fouled tube section was installed in the test rig. For the cleaning evaluation, 15 l of treated water was transferred into the basin of the test rig. The pH of this water was adjusted to within the range of the test and recirculated at a rate of 2 gpm at 44°C. At the end of 24 hours, a sample of water was removed for iron analysis. The heat transfer tube was removed and dried in an oven. The tube and deposit were weighed together. The deposit was scraped from the tube and weighed. The tube was acid cleaned and weighed.

Table 1. Maintaining Soluble Iron, Following
the Addition of Soluble Ferric Ion

Line No.	1 Additive (50 mg/L)	2 Soluble Fe (mg/L)
1	None	0.0
2	EDTA	0.8
3	HPA	0.9
4	NTMP	3.6
5	HEDPA	4.0
6	EDDHMA	4.4
7	SHA-1	4.8
8	SHA-2	5.2
9	Copolymer	5.0

The test for removal at 10 mg/l additive was performed in a medium hardness water at pH 8.5-9.0 for 15 days. The tests for removal at 500 and 1500 mg/l ppm were performed in low hardness water at pH 7.0-7.5 for one day. Cleaner activity is based on comparing deposit and tube weight loss before and after cleaning. Deposit weight and tube weight loss before cleaning was determined from seven deposit formation runs. Results from these (prior to cleaning) runs show an average deposit weight of 0.74 ± 0.18 gms. and an average tube weight loss of 0.89 ± 0.18 gms.

RESULTS AND DISCUSSION

Iron Scale Prevention

Cooling water is generally contaminated with various forms of oxidized iron due to corrosion of steel equipment and/or its introduction with the feedwater. Maintaining this oxidized iron in soluble or dispersed forms can prevent the fouling of surfaces. If the iron enters the water in a soluble (e.g., ferrous) form, it is possible to maintain the iron in a soluble form which will be carried out of the system via blowdown. This is described below under "Maintaining Soluble Iron(III)". If the iron enters the system in an insoluble form (e.g., Fe_2O_3), maintaining these solids in dispersion will permit its removal. This is described below under "Dispersion of Iron Oxide".

Maintaining Soluble Iron(III)

The addition of 10 mg/l of soluble ferric ions (in the form of $FeCl_3$) to a medium hardness cooling water resulted in no soluble iron in the absence of additives (Table 1). The presence of 10 mg/l of EDTA resulted in only a small amount of soluble Fe (0.8 mg/l). The corrosion inhibitor, hydroxyphosphonoacetic acid (HPA) performed similarly. The stronger chelating phosphonates, nitrilo-tris(methylenephosphonic acid) (NTMP) and 2-hydrodxyethylidene-diphosphonic acid (HEDPA) maintained high amounts of soluble iron, as did the non-P chelating agents ethylenediamine,N,N′-(2-hydroxy-5-methylphenylacetic acid) (EDDHMA), and two sulfonated hydroxyaromatic compounds (SHA-1 and SHA-2). These results show that the stronger chelants maintain higher levels of soluble iron.

The sulfonate/carboxylate copolymer (line 9) also gives high levels of soluble iron. Considering the available ligands and the large distances between them on the polymeric

Table 2. Dispersion of 1000 mg/l Fe_2O_3 with Various Additives

Line No.	1 Additive	2 Dispersion (mg/L) in low hardness water	3 Dispersion (mg/L) in high hardness water
1	None	≤30	â30
2	EDTA	≤30	—
3	EDDHMA	≤30	—
4	ATMP	≤30	—
5	HEDPA	420	100
6	SHA-1	460	125
7	SHA-2	580	290
8	PAA	500	200
9	Copolymer	690	530

chain, the copolymer can not be a strong chelant in a thermodynamic sense (i.e., the binding constant for Fe will be low). Thus, the polymer maintains soluble iron by inhibiting its precipitation, probably by dispersing the iron oxide in very small particles which are inhibited from further crystal growth and precipitation.

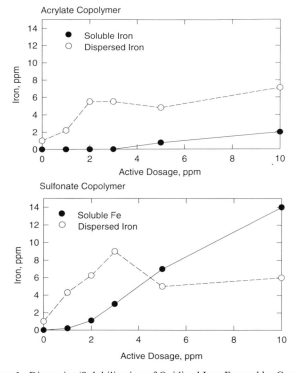

Figure 1. Dispersion/Solubilization of Oxidized Iron Formed by Corrosion.

Dispersion of Iron Oxide

Table 2 shows the dispersion of Fe_2O_3 (hematite) by various additives in waters of different hardness. The "wrap-around" hexadentate chelating agents, EDTA and EDDHMA do not disperse Fe_2O_3. This is probably due to the attachment of all anionic groups to the available ferric sites on the particles, leaving no "free" anionic charges to repel other particles. In contrast, the sulfonated hydroxyaromatics contain hydroxyl "anchors" and isolated anionic groups (sulfonates) for charge repulsion. The isolated sulfonic acid groups are "free" because they can not form a chelate with the hydroxyaromatic and a metal center.

Table 2 shows that the copolymer is the best iron oxide dispersant, especially in hard water. Note that the dispersion obtained with all of the small molecules and with polyacrylic acid (PAA) diminishes as the water hardness increases. This is likely due to complexation by solution calcium which ties up active sites which would otherwise be available for complexation and/or dispersion. With the hydroxyaromatics, the calcium probably reduces the affinity of the hydroxyaromatic ligands for surface iron sites. In the case of PAA, the number of "free" anionic carboxylates is reduced, leading to less anionic repulsion which is necessary for good dispersion. The sulfonate groups of the copolymer are less affected by calcium and probably provide the anionic repulsion while the carboxylates retain their strong affinity for the surface iron sites.

Figure 1 shows the effect of polymer concentration on the degree of dispersed and soluble iron formed during the corrosion of mild steel coupons. Two polymers are compared in this figure, a polyacrylate and a sulfonated carboxylate copolymer. At low concentrations (<3 mg/l) the amount of dispersed iron increases with increasing polymer concentration. This observation is consistent with an increase in anionic charge on the solid with increasing availability of polymer.

With both polymers the amount of dispersed iron reaches a maximum and then a plateau. The maxima observed could be due to a particle surface coverage phenomena. At higher concentrations the behavior of the two polymers differs substantially. The sulfonated copolymer maintains higher levels of soluble iron than the polyacrylate. This can be attributed to either the ability of the sulfonated copolymer to chelate iron directly or an ability to promote and stabilize the formation of colloidal iron particles.

Rust Scale Removal

If iron scale becomes attached to the internal surfaces of a cooling water system, it must be removed. The high iron affinity of chelants and certain phosphonates suggests that they may be useful for removal of rust scale. Below we show the type of cleaning which can

Table 3. Dissolved Iron From Precipitated Rust Scale

Line No	1 Additive (50 mg/L)	2 Dissolved iron (mg/L)
1	None	0.0
2	EDTA	0.25
3	HEDPA	0.37
4	NTMP	0.38
5	SHA-2	6.8
6	SHA-1	4.8
7	EDDHMA	4.4

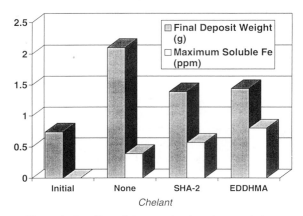

Figure 2. Iron Deposit Removal Using 10 ppm Chelant.

be achieved at low (10 mg/l), medium (500-1500 mg/l) and high (10,000 mg/l) concentration of chelant or phosphonte.

At low concentration (50 mg/l) SHA-1, SHA-2 and EDDHMA were able to dissolve iron oxide which had been formed by precipitation from $FeCl_3$ under alkaline conditions (Table 3). Note that the amino acid chelant (EDTA) and organophosphonates (HEDPA and NTMP) dissolved very little of the precipitated iron oxide. In contrast, none of these chemicals were able to dissolve dehydrated crystalline iron oxides such as a-Fe_2O_3, g-Fe_2O_3 and Fe_3O_4. This suggests that it is best to attempt removal of rust scale soon after it has formed, before it has time to crystallize and dehydrate.

Rust scale, which had been formed by corroding a mild steel tube, proved difficult to remove at low levels of chelant. Figure 2 shows that the amount of iron oxide actually increased during the 15-day " cleaning period". This was due to further corrosion during this time, and the level of chelants was not sufficient to complex all of the oxidized iron which was formed by new corrosion. Note the low levels of soluble iron during the 15-day period.

Increasing the level of additives to 500 mg/l resulted in scale removal with some additives (Figure 3). Best performance was obtained with HEDPA and EDDHMA, but removal was incomplete. Increasing the level of additive to 1500 mg/l resulted in complete

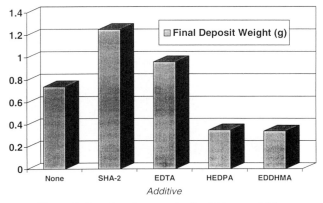

Figure 3. Iron deposit removal using 500 ppm additive.

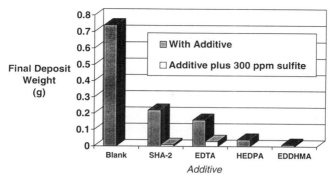

Figure 4. Iron deposit removal using 1500 ppm additive.

removal with HEDPA and EDDHMA (Figure 4). SHA-2 and EDTA gave ~75% removal, and this removal was enhanced by the presence of sulfite at 300 mg/l.

The product of this rust scale removal included finely dispersed iron oxide solids as well as soluble iron. Thus, the chelants function by dissolving freshly formed ferric solids, which have not yet crystallized to insoluble Fe_2O_3. The solubilization of freshly formed ferrous or ferric ions weakens the surface attachment of the more crystalline rust scale, which is then dispersed. The sulfite reducing agent probably helps maintain the under-deposit iron in the ferrous state, which is more soluble and more available to chelants.

The key to removal of iron oxide scale is the formation of significant concentrations of soluble iron. This is shown in Figure 5 where we plot the final deposit weight vs. soluble iron for over 100 experiments which used the same procedure as in Figures 3 and 4. On the average, a concentration of at least 30 mg/l of soluble iron was required to obtain some deposit removal and at a concentration of over 80 mg/l, removal was essentially complete.

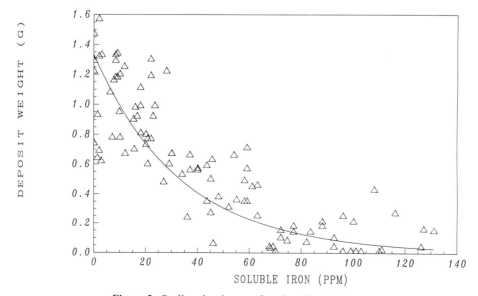

Figure 5. On-line cleaning as a function of soluble iron.

At higher concentration of additive, e.g., 10,000 mg/l HEDPA, complete solubilization of the rust scale can be obtained. This is the subject of earlier papers and patents.[18]

CONCLUSIONS

In cooling water systems, oxidized iron is sequentially transformed from soluble to precipitated amorphous and then to crystalline forms. Interaction with chemical additives at an early stage in this transformation provides the best control of iron scale. This can be accomplished by maintaining soluble iron, dispersing iron oxide solids or re-dissolving scale which has attached to metal surfaces. The form of iron contaminants present in a particular system should determine the appropriate selection of chemical additives: polymer, chelant and/or phosphonate.

Polymers

Copolymers containing sulfonate *and* carboxylate functionality are ideal for on-line prevention of rust scale by maintaining soluble ferric ions and dispersing any precipitated iron oxide.

Chelants

The strongest ferric chelants contain amino acid/hydroxyaromatic or sulfonated hydroxyaromatic ligands. These are best used for maintaining soluble iron (low level continuous dosage) or for removing rust scale at high levels of chelant.

Phosphonates

Strong iron interaction can be provided by multiple phosphonate and hydroxyl ligands in a molecule such as HEDPA. It can maintain soluble iron (for continuous on-line prevention), and it is an excellent choice for rust scale removal at high concentrations because it simultaneously passivates mild steel surfaces.

REFERENCES

1 Straus, S. D. and Puckorius, P. R., Power, June 1984, S-1.
2 Federal Register Vol 58, Number 154, 40, CFR part 63, 1993 "National Emissions Standards for Hazardous Air Pollutants for Chromium Emissions from Industrial Process Cooling Towers".
3 Gray, J. A. "Sidestream Filtering and Particulate Removal".
4 Holland, T. E. and Harding, P. J. Power, 104, June 1978.
5 Dubin, L. and Fulks, K. E. "The Role of Water Chemistry on Iron Dispersant Performance" Paper no. 118, CORROSION 84, NACE, New Orleans, 1984.
6 Binglin, Y., "Scale Inhibition by Polyacrylic Acid Fractions" Water Treatment 4257-4265, 1989, China Ocean Press - Printed in Beijing.
7 Lange, U.S. Patent 3,898,037, August, 1975.
8 Zuhl, R. W., Masler, W. F., "A Novel Material for Use in Minimizing Calcium Phosphate Fouling in Industrial Cooling Water Systems", Cooling Tower Institute Annual Meeting, New Orleans, LA, February, 1987.
9 Masler, W. F., Amjad, Z. U.S. Patent 4,566,973, Jan. 29, 1986.
10 Kaplan, R. I., and Ekis Jr., E. W. "The On-line Removal of Iron Deposits from Cooling Water Systems", paper no. 8, NACE, CORROSION 84, New Orleans, LA. April, 1984.

11 Arrington, S. and Bradley, G. "Service Water System Cleaning with Ammoniated Citric Acid" Paper no. 387, NACE, San Francisco, CA. March 9-13, 1987.

12 Bersworth, F. C., U. S. Patent 2,396, 938, March 19, 1946.

13 Burroughs, J. E. and Nowak, J. A., U.S. Patent 3,951,827.

14 Hausler, R. H., CORROSION/82, Paper No. 30, National Association of Corrosion Engineers, Houston, Texas, 1982.

15 Engstrom, G. G. and Kuhn, V. R., 1985 International Water Conference, Paper No. IWC-85-34.

16 Kuhn, V. R., Engelhardt, P. R., Mitchell, W. A., U.S. Patent 4,721,532, Jan. 26, 1988.

17 U.S. Patent 5,022,096, June 11, 1991.

18 Gray, J. A. and Heinz, K., "A Neutral pH Process for On-line Cleaning and Passivation of Hot and Chilled Water Systems", Paper no. 446, NACE CORROSION 91, Cincinnati, OH., March 1991.

14

FORMATION OF HYDROXYAPATITE IN THE PRESENCE OF PHOSPHORYLATED AND SULFATED POLYMER IN AN AQUEOUS PHASE

S. Shimabayashi, N. Hashimoto, H. Kawamura, and T. Uno

Faculty of Pharmaceutical Sciences
The University of Tokushima
Sho-machi 1-78-1, Tokushima 770, Japan

INTRODUCTION

Hydroxyapatite (HAP) is known as the main inorganic component of hard tissues such as bones and teeth. It directly crystallizes and grows in the solution when the degree of supersaturation is low. On the other hand, it is formed via amorphous calcium phosphate (ACP), octacalcium phosphate (OCP), or dicalcium phosphate dihydrate (DCPD) when the degree of supersaturation in the mother solution is high. Formation of hard tissues and crystallization of HAP in the human body are affected physicochemically by many kinds of proteins, especially phosphoproteins. These proteins are called regulator proteins which play an important role in the regulation of mineral deposition and HAP growth on various organic matrices in animal/human body. The function of phosphoproteins, however, has been still complicated. It has been suggested that both matrix-bound and soluble phosphoproteins may serve dual roles; making the initial mineral deposition on the collagen matrix and inhibiting the calcification of soft tissues.

In the present paper, phosphorylated polyvinylalcohol (Phos. PVA) was prepared as a polymeric model compound for phosphoproteins. Phosphoserine (PSer) was assumed to be a monomeric model compound because phosphoproteins usually contain PSer residues. The effects of these model compounds on the formation of HAP and on the transformation of ACP to HAP was discussed, taking into consideration the effects of unphosphorylated polyvinylalcohol (PVA) and serine (Ser). This makes it possible to examine the differences in the effect between the phosphorylated and unphosphorylated compounds of a given molecular size (i.e., PSer vs. Ser, and Phos. PVA vs. PVA), and between the low and high molecular compounds at a given ester phosphate concentration (i.e., PSer vs. Phos. PVA).[1]

Effects of phosphorylated PVA on the formation of and transformation to HAP were also compared with those of sulfated polyvinylalcohol (Sulf. PVA). Both phosphate and sulfate groups are oxoacidic and tetrahedral, while the interatomic distance of P-O (= 0.154

Mineral Scale Formation and Inhibition, Edited by Zahid Amjad
Plenum Press, New York, 1995

nm) is slightly longer than that of S-O (= 0.149 nm). That is, the size of phosphate group is a little larger than that of sulfate group. Ester phosphate is bivalent in charge, while ester sulfate univalent. The influences of these factors would become clear by making a comparison between the effect of phosphorylated PVA and that of sulfated PVA on the formation of HAP.[2]

The effects of phosphorylated cellulose (Phos. Cell.), which is insoluble in water, were also studied. Phosphorylated cellulose could be regarded as a model compound for the insoluble matrix containing phosphate groups. Its effect on the transformation to and formation of HAP was fairly different from that of water-soluble phosphorylated PVA, even though phosphate groups were contained in either compound.[3]

EXPERIMENTAL

Materials

Phosphorylated PVA was prepared through phosphorylation of PVA according to the method mentioned elsewhere[1], where the degree of polymerization of the original PVA (NL-05, Nippon Gohsei Kagaku Co., Ltd., Osaka, Japan) was about 450. The degree of phosphorylation of phosphorylated PVA was around 10 %. Phosphorylated PVA was used as Na-salt in the present paper. Sulfated PVA (K-salt) was purchased from Wako Pure Chemical Industries, Ltd. (Osaka, Japan). The degree of sulfation was 92 %, and the degree of polymerization of PVA as a starting material was about 1500. Therefore, as a result, the degree of esterification, charge density along the polymer chain, and polymer length along the chain were less in phosphorylated PVA than in sulfated PVA.

Phosphorylated cellulose was purchased from James River Company (Berlin, N.H., USA). It was used after rinsing with enough amount of water on a funnel. The content of phosphate groups was determined as 0.2 mole per mole of repeating glucose unit in average by means of chemical analysis of the residue on ignition. That is, the degree of phosphorylation was ca. 7 % with respect to the total OH-group. Phosphorylated cellulose of low phosphate content (= 0.1 mole/mole of glucose unit) was obtained by means of partial hydrolysis through heating the original phosphorylated cellulose (0.3 g) with 0.6 mol/l H_2SO_4 for 20 minutes. Cotton fiber (i.e., unphosphorylated cellulose), as a reference material for the phosphorylated cellulose, was of the grade of Japanese Pharmacopoeia XII. It was used, as is.

Sodium chondroitin-6-sulfate (Na_2Chs) was the product of Nakarai Chemicals, Ltd.(Kyoto, Japan). Its viscosity-average molecular weight was ca. 30,000. Other reagents used in the present paper were of the analytical grade. Those were used without further purification.

Methods

Calcium ion activity was determined at a given temperature using an Orion ion-sensitive electrode connected to an Orion expandable ion analyzer (model EA940). Prior to the measurement on the sample solution, the calcium electrode was calibrated with an aqueous solution of $CaCl_2$ in the presence of 0.9 % NaCl, taking activity coefficient into consideration. The suspension/solution pH was measured by a pH-meter (Hitachi-Horiba, model M-1). The pH and calcium ion activity in a mother solution, which is approximately equal to the concentration of free calcium ion ($[Ca^{2+}]$), were traced for about 100 minutes from immediately before mixing $CaCl_2$ with K_2HPO_4 in the presence of a given concentration of an additive in 0.9 % NaCl aqueous solution in order to determine the induction time from ACP to HAP. The induction time, T, from ACP to HAP was determined from the intersection of the tangents drawn to the time-courses of pH and $[Ca^{2+}]$ just before and after the second steep decrease. The induction periods, thus obtained from these two curves, were in fair agreement with each other.

The X-ray powder diffraction ($CuK\alpha$ radiation at 35 kV and 10 mA) was examined with a diffractometer (Toshiba ADO-301) at the angle $2\theta = 31.8$ degrees, which is specific for HAP, at room temperature. Samples for the diffraction measurements were obtained from time to time by filtration of a precipitate prepared by mixing $CaCl_2$ with K_2HPO_4 in the presence of a known concentration of additive at 25°C. The precipitate was jelly-like, which was thoroughly dehydrated and kept in acetone.

Mean diameter (d) of secondary particles of the precipitate was measured by a Coulter counter (type TA-II, aperture size 100 μm) at 25°C and at 100 minute after the mixing of 1.25 mmol/l $CaCl_2$ with 2.50 mmol/l K_2HPO_4 in the presence of 0.9 % NaCl containing a known concentration of various additives.

The surface of phosphorylated cellulose, where calcium phosphates (ACP and HAP) were formed or not, was observed by means of a SEM (scanning electron microscope, model Hitachi S-430).

Total calcium concentration was determined by EDTA (ethylenediaminetetraacetic acid) chelatometry at pH 13 with 1-(2-hydroxy-4-sulfo-1-napthylazo)-2-hydroxy-3-naphthoic acid. Total concentration of inorganic phosphate was determined by colorimetry at 720 nm after reduction of the complex of phosphate ammonium molybdate by stannous chloride, according to the method of Gee et al.[4]

RESULTS AND DISCUSSION

The Effects of Phosphorylated PVA

The Mean Diameter of Secondary Particles of Calcium Phosphates. Figure 1 shows the relationships between mean diameter(d) and polymer concentration (c_p), and between d/d_0 and total phosphorus concentration ($[P]$) of added phosphorylated PVA or PSer, where $d_0 = d$ at $[P]= 0$. After attaining a maximum, d decreased with c_p. The increase in d was due to the effect of interparticle bridging by the polymer adsorbed on the precipitate, while the decrease in d was caused by the effect of interparticle repulsion through the polymer adsorbed on the surface of each particle. The d-value was barely affected by Ser, while it monotonously decreased with an increase in the concentration of PSer. Consequently, the d-value decreased

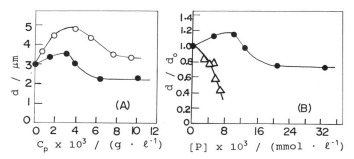

Figure 1. Mean diameter of precipitate particles. Additive was PVA (open circle), phosphorylated PVA (closed circle), or PSer (triangle). Solution pH was 7.6 - 7.9. The degree of phosphorylation of phosphorylated PVA was 19 %.

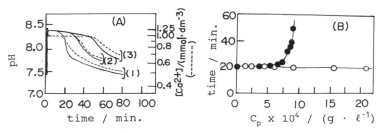

Figure 2. Time courses of pH and $[Ca^{2+}]$ (A), and induction time as a function of polymer concentration (B) at 35°C. (A) concentration of phosphorylated PVA x 10^4/(g/l) = 0 (curve 1), 9.75 (2), and 10.8 (3). (B) Added polymer: PVA (open circle) and phosphorylated PVA (closed circle). The degree of phosphorization of phosphorylated PVA was 8.17 %.

in the order of PVA > phosphorylated PVA > Ser. > PSer. It is interesting that phosphorylated compounds are more effective than unphosphorylated ones in decreasing the d-value, which is owing mainly to the adsorption via esterified phosphate group.

Transformation of Amorphous Calcium Phosphate to HAP. Figure 2 (A) shows the time courses of pH and $[Ca^{2+}]$ in the presence and absence of phosphorylated PVA. The pH increased while $[Ca^{2+}]$ decreased steeply almost along the ordinate immediately after addition of K_2HPO_4 to an aqueous solution of $CaCl_2$. The decrease in $[Ca^{2+}]$ is attributable to the formation of ACP, and the increase in pH is due to the protonation of the excess phosphate ion remaining in the mother solution. After an induction period, pH and $[Ca^{2+}]$ steeply decreased again; that is, OH⁻ and Ca^{2+} were simultaneously consumed. The decreases reflect the transformation of ACP to HAP, because the solubility of HAP is lower than that of ACP and, therefore, HAP consumes OH⁻, Ca^{2+}, and PO_4^{3-} in the mother solution as the lattice ions when it is formed from ACP. The induction time (T), obtained from Figure 2 (A), is shown in Figure 2 (B) as a function of polymer concentration, c_p. The induction time began to increase at $c_p = 7 \times 10^{-4}$ g/l in phosphorylated PVA, while no significant increase in the induction time was observed up to $c_p = 2.2 \times 10^{-3}$ g/l in PVA.

Similar measurements were done in the presence of Ser and PSer. Little specific effect of Ser was found, whereas PSer showed the retardation effect on the transformation of ACP to HAP in the concentration range higher than 3×10^{-2} mmol/l. Figure 3 shows the relationship between the induction time (T) and phosphorus concentration ([P]) of the ester

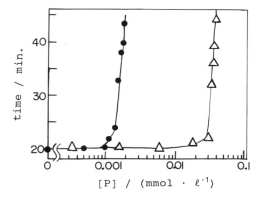

Figure 3. Relationship between the induction time and concentration of added ester phosphate at 35°C. The additive was PSer (triangle), and phosphorylated PVA (closed circle). The degree of phosphorization of phosphorylated PVA was 8.17 %.

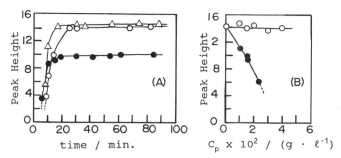

Figure 4. Diffraction strength as a function of time (A) and polymer concentration (B). Additives were PVA (open circle), phosphorylated PVA (closed circle), and none (open triangle). Phosphorylated PVA was the same sample as that used in Figures 2 and 3. Concentration of polymer in (A) was 1.6×10^{-2} g/l in common. 5 mmol/l $CaCl_2$ was mixed with 10 mmol/l K_2HPO_4 at pH 6.8 - 7.5 at 25°C.

phosphates. According to these results, the concentration of PSer should be ca. 20 times higher than that of phosphorylated PVA to show the same induction time.

X-Ray Powder Diffractometry of the Precipitate. No diffraction peaks were detected on a precipitate immediately after precipitation. However, specific diffraction peak developed at diffraction angle $2\theta = 31.8$ degrees, and the diffraction strength increased with time. It levelled off at a certain time after the precipitate formation, which depended on the experimental condition. This result means that the initial precipitate was ACP, which subsequently crystallizes to HAP.

Figure 4 shows the diffraction strength as a function of elapsed time after the precipitate formation (A), and polymer concentration (B). The strength after the levelling-off (at 30 min.) depended on species and concentration of the added polymer. The peak height was almost constant irrespective of the concentration of PVA, whereas it decreased with the concentration of phosphorylated PVA.

Similar measurements were done in the presence of Ser and PSer to study their effects on the crystallinity of the precipitate. The diffraction strength at $2\theta = 31.8$ degrees decreased with an increase in the concentration of PSer, while Ser showed no effect. That is, unphosphorylated compounds such as Ser and PVA hardly affect the degree of crystallinity, but phosphorylated compounds such as Pser and phosphorylated PVA inhibit crystallization to HAP. To compare the effect of phosphorylated PVA with that of PSer on a common scale, the relationship between [P] and relative peak height (h/h_0) was shown in Figure 5, where [P] is a concentration of the added ester phosphate, and h and h_0 are the peak heights in the presence and absence of phosphorylated compounds at 30 minutes after precipitate forma-

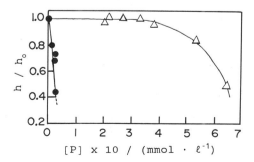

Figure 5. Relative diffraction strength as a function of the phosphorus concentration of the added ester phosphate. Experimental conditions were the same as in Figure 4. Additives were PSer (open triangle) and phosphorylated PVA (closed circle).

tion. According to Figure 5, the concentration of PSer should be about 20 times higher than that of phosphorylated PVA to achieve the same effect on h/h_0. It is interesting that the value obtained here is quite the same as that obtained through the induction time method (see Figure 3).

Though both phosphorylated (PSer, phosphorylated PVA) and unphosphorylated compounds (Ser, PVA) were assumed to be adsorbed by HAP[5,6], the former thwarted crystallization but the latter not. This is because the affinity of the ester phosphate group for the active site of HAP crystal growth is higher than that of the -OH groups of Ser and PVA. Thus, the ester phosphate groups of these organic compounds as well as condensed phosphates[7] have exhibited a significant role in the regulation of the crystal growth of HAP. It has been known that these phosphorylated organic compounds inhibit the adsorption of anionic proteins on HAP also.[8]

Some of the phosphate groups along the polymer chain of the adsorbed phosphorylated PVA participate in the adsorption in contact with HAP, whereas others remain on the polymer loops or tails protruding from the HAP surface. Negative charges of the ester phosphate groups in the thick adsorption layer effectively repel the inorganic phosphate ions approaching toward the growth sites of HAP, resulting in strong inhibition of the crystal growth. Therefore, the effect of phosphorylated PVA is more remarkable than that of PSer which is adsorbed separately on the surface of HAP. As for the formation of secondary particles of calcium phosphates (see Figure 1), adsorbed PSer simply inhibits the aggregation through steric and electrostatic repulsion. Phosphorylated PVA, in contrast, showed both dispersing and flocculating effects, depending on the polymer concentration or the amount of adsorbed polymer. This is another specific character of the polymer. These facts suggest that polymeric phosphorylated compounds in the animal body (i.e., phosphoproteins) are more important than monomeric and low-molecular phosphorylated compounds in the regulation of the crystal growth of HAP and the formation of hard tissues.

The Effects of Sulfated PVA and Related Compounds

Sulfated PVA retarded the transformation of ACP to HAP, and weakened the X-ray diffraction intensity of the precipitate after the transformation with an increase in its concentration. On the other hand, Na_2SO_4 and Na_2Chs did not show any effects on the induction time and the degree of crystallization irrespective of their concentrations, although these compounds contain sulfate group in their molecular structure.

Figure 6 shows the relationship between T/T_0 and concentration of sulfate or phosphate group of the additive, where T and T_0 are the induction time in the presence and absence of the additive. The effect of an additive on T/T_0 was in the order phosphorylated PVA > sulfated PVA > Na_2Chs = Na_2SO_4. Quite similar sequence was obtained from the

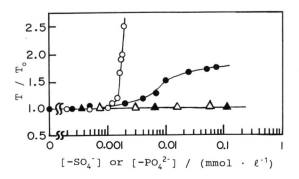

Figure 6. T/T_0 as a function of concentration of sulfate and phosphate group of the additives. The additives are phosphorylated PVA (open circle), sulfated PVA (closed circle), Na_2Chs (open triangle), and Na_2SO_4 (closed triangle). The data for phosphorylated PVA were quoted from Figure 3. Measurement was done at 30°C.

measurement of an X-ray diffraction intensity in the presence of these additives (data not shown here).

Mechanism for the retardation and inhibition by sulfated PVA seems similar to that by phosphorylated PVA. Some of the sulfate groups occupy the active growth sites for inorganic phosphate ion on the embryos/nuclei of HAP through the isomorphous substitution, and the others of the same polymer chain are unattached but close to the surface, forming a negatively charged adsorption layer. The negative charges repel inorganic phosphate ion approaching to its site on the surface, resulting in the inhibition of the crystal growth and the retardation of the transformation to HAP. However, the affinity of the sulfate groups for the sites of phosphate ion on the nucleus is weaker than that of the phosphate groups of phosphorylated PVA, because the size and valence of sulfate group are smaller than those of the phosphate group. Therefore, the effect of sulfated PVA is weaker than that of phosphorylated PVA, even though the degree of esterification, charge density along the polymer chain, chain length of the polymer, and probably the thickness of the adsorption layer are larger in sulfated PVA than in phosphorylated PVA.

As for Na_2SO_4, desorption of inorganic sulfate ion from and, instead, selective binding of inorganic phosphate ion to the site on the nuclei easily occur through the competition between them by virtue of the absence of the polymer effect of adsorbed layer, mentioned above. On the other hand, it is rather difficult to explain the effect of Na_2Chs. It is an acidic mucopolysaccharide which has sulfate, carboxylate, and hydroxyl groups. The size of carboxylate group is smaller than that of sulfate group, which means that the affinity of carboxyl group for HAP is lower than that of sulfate group.[9] In addition, charge density along the polymer chain of Na_2Chs is lower than that of sulfated PVA. That is, one negative charge per 0.49 nm for Na_2Chs, as compared with 0.27 nm for sulfated PVA. Therefore, phosphate ion as one of the lattice ions can easily penetrates into the adsorbed layer of Na_2Chs. It gets to the site on the surface, and rejects the carboxylate and sulfate groups of Na_2Chs occupying the sites, resulting in the growth of the HAP crystal.

Effects of Phosphorylated Cellulose on the Formation of HAP

Duplicity of the Effects of Phosphorylated Cellulose. The induction time increased after attaining a minimum with a concentration of insoluble phosphorylated cellulose at given concentrations of Ca^{2+} and phosphate ion, as is shown in Figure 7. That is, small amount of phosphorylated cellulose accelerates the transformation, while much amount inhibits. Phosphorylated cellulose concentration at the minimum point increased, while the induction time decreased with the degree of supersaturation. In contrast with phosphorylated cellulose, cellulose (cotton fiber) was indifferent to the induction time, as shown by dotted

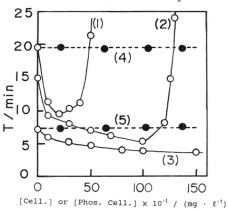

Figure 7. Effect of cellulose and phosphorylated cellulose on the induction time. Full ((1)-(3)) and broken ((4)-(5)) lines show the effect of phosphorylated cellulose and cellulose, respectively. Initial concentrations of Ca^{2+} and phosphate ion in mmol/l unit were 1.25 and 2.50 (curves (1) and (4)), 2.50 and 5.00 ((2)), and 5.00 and 10.00 ((3) and (5)). Temperature was 30°C.

[Cell.] or [Phos. Cell.] x 10⁻¹ / (mg · ℓ⁻¹)

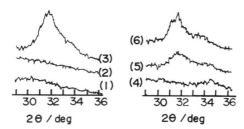

Figure 8. X-ray powder diffraction patterns in the absence (chart (1) - (3)) and presence (chart (4) - (6)) of phosphorylated cellulose. Phosphorylated cellulose was amorphous with respect to the X-ray.

lines in Figure 7. These facts suggest that phosphorylated cellulose plays dual significant roles; promoting and inhibiting the formation of HAP in specific manners through adsorption or consumption of Ca^{2+} by its ester phosphate groups and through offering the heterogeneous nucleation sites for the crystal growth by binding Ca^{2+}.

X-Ray Powder Diffractometry in the Presence of Phosphorylated Cellulose. In connection with the effect of phosphorylated cellulose on the induction time shown in Figure 7, crystallinity of the precipitate was studied by means of an X-ray powder diffractometry. The precipitate was prepared by mixing 2.50 mmol/l $CaCl_2$ with 5.00 mmol/l K_2HPO_4 in the presence or absence of 700 mg/l phosphorylated cellulose at 30°C. Figure 8 shows the diffraction patterns. The precipitates at 5 minutes after the mixing (patterns (1) and (4)) were amorphous, while those at 30 minutes (patterns (3) and (6)) crystalline irrespective of the presence or absence of phosphorylated cellulose. The precipitate after 10 minutes in the absence of phosphorylated cellulose (pattern (2)) was still amorphous, while that in the presence of phosphorylated cellulose (pattern (5)) already crystalline owing to the acceleration effect of phosphorylated cellulose on the crystallization, which is in accordance with the expectation from the data shown in Figure 7. Thus, it was confirmed again that insoluble phosphorylated cellulose contributes to promote the crystallization to HAP although soluble phosphorylated PVA inhibits it.

SEM Observation. Figure 9 is SEM photographs of the surfaces of cellulose fiber (A),partially hydrolyzed phosphorylated cellulose (B), and phosphorylated cellulose (C). They were smooth before the precipitate formation (see photographs (a) in Figure 9). The amount of the precipitate adhered on the surface of the phosphorylated cellulose increased with time after mixing Ca^{2+} with phosphate ion, while the surface of cellulose (A) was smooth during the examined time (photographs (b), (c), and (d) in (A)). Strictly speaking, the amount of the precipitate on the surface of phosphorylated cellulose (C) was more than that of partially hydrolyzed phosphorylated cellulose (B) at respective times shown in Figure 9. Unfortunately, it was technically difficult to prepare the samples for SEM before elapsing the induction time. However, these observations suggest again that the ester phosphate groups on the fibers play an important role in crystallization to HAP or formation of the nuclei and/or embryos for calcium phosphates.

Adsorption of the Calcium Ion by the Phosphorylated Cellulose. Figure 10 shows the adsorption amount of Ca^{2+} by phosphorylated cellulose in the absence of inorganic phosphate ion. It decreased with a concentration of NaCl. This means the binding is electrostatic and Ca^{2+} is competing with Na^+ for the adsorption sites of ester phosphate groups on phosphorylated cellulose. On the other hand, phosphate ion in the absence of Ca^{2+} was never adsorbed on phosphorylated cellulose.

In consideration of these facts, it could be assumed that the initial trigger for the formation of calcium phosphates on the surface of phosphorylated cellulose is capturing Ca^{2+}

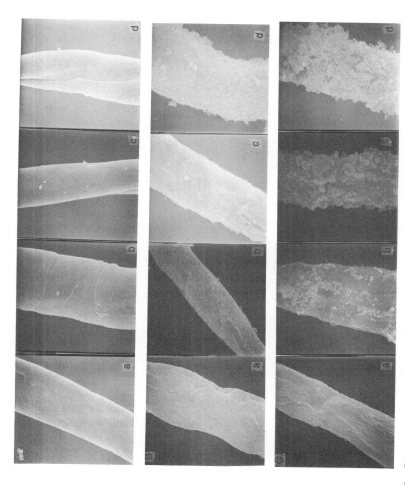

Figure 9. SEM photographs for fibers and precipitates. (A) cellulose fiber, (B) partially hydrolyzed phosphorylated cellulose, and (C) phosphorylated cellulose. Elapsed time/minutes = 0 (a), 14 (b), 18 (c), and 60 (d). The sample precipitate was prepared by mixing 1.25 mmol/l $CaCl_2$ with 2.50 mmol/l K_2HPO_4 in the presence of the fiber (250 mg/l).

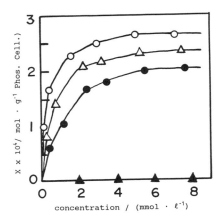

Figure 10. Adsorption isotherms for Ca^{2+} ($CaCl_2$) and phosphate ion (K_2HPO_4). Adsorption amount was determined at 0 (open circle), 50 (open triangle), and 100 mmol/l NaCl (closed circle). Phosphate ion was not adsorbed, as is shown by closed triangle.

by the ester phosphate group. When both Ca^{2+} and phosphate ion are added together into an aqueous system containing phosphorylated cellulose, Ca^{2+} is first of all bound by the ester phosphate group and forms a complex of Ca-Phos. Cell., onto which inorganic phosphate ion (Pi) is bound, resulting in a complex of Pi-Ca-Phos. Cell. This complex grows in its extent, and the CaPi clusters such as —Pi-Ca-Pi-Ca— are formed on the surface. This is the formation of crystal nucleus or embryo for calcium phosphates. In other words, the ester phosphate group of phosphorylated cellulose is an important core for the heterogeneous nucleation for calcium phosphates.

The clusters as the nuclei accelerate the formation of HAP. Therefore, the induction time decreases with the amount of phosphorylated cellulose at a given degree of supersaturation (see Figure 7). When the degree of supersaturation increases at a given amount of phosphorylated cellulose, the extent and rate of heterogeneous nucleation (and also homogeneous nucleation) increase, resulting in the decrease in the induction time.

On the other hand, when more amount of phosphorylated cellulose is added at a given concentration of Ca^{2+}, more amount of Ca^{2+} is bound along the fiber rather dispersedly/separately, compared to the manner in the presence of small amount of phosphorylated cellulose. This effect results also in more consumption of Ca^{2+}, more concurrent decreases in the degree of supersaturation and thermodynamic activity of Ca^{2+}, and smaller size of each cluster than those in the presence of small amount of phosphorylated cellulose. Therefore, the delay in the transformation was induced in the presence of much amount of phosphorylated cellulose (see Figure 7).

CONCLUSIONS

1. Water-soluble polymers (phosphorylated PVA and sulfated PVA) retarded the transformation and inhibited the crystallization of ACP to HAP.
2. This effect was explained in terms of adsorption of ester phosphate and sulfate groups onto the HAP nuclei/embryos through isomorphous substitution with inorganic phosphate ion on HAP and through electrostatic attractive force toward the surface Ca^{2+}. Significance of the adsorbed layer of the polymer was emphasized.
3. The effects of sulfated PVA on the retardation and inhibition were weaker than those of phosphorylated PVA, because the size and valence of sulfate group of sulfated PVA are smaller than those of phosphate group of phosphorylated PVA.

4. Insoluble phosphorylated cellulose was either a promotor or an inhibitor of the crystallization from ACP to HAP. This effect depended on the amount of added phosphorylated cellulose, the degree of supersaturation, and the extent of the clusters formed around phosphorylated cellulose fibers.

5. Soluble phosphorylated PVA could be regarded as a simplified model compound for the regulator phosphoproteins in animal body, while insoluble phosphorylated cellulose for the organic phosphorylated matrices where biological calcification occurs.

REFERENCES

1. Shimabayashi, S., and Tanizawa, Y., 1990, Formation of hydroxyapatite in the presence of phosphorylated polyvinylalcohol as a simplified model compound for mineralization regulator phosphoproteins, *Chem. Pharm. Bull.* 38: 1810-1814.
2. Shimabayashi, S., Kawamura, H., and Uno, T., 1993, Effect of sulfate group on the formation of hydroxyapatite in an aqueous phase, *Phosphorus Research Bulletin* 3: 1-6.
3. Shimabayashi, S., Hashimoto, N., and Uno, T., 1993, Effect of phosphorylated cellulose and bovine serum albumin on crystallization of hydroxyapatite, *Phosphorus Research Bulletin* 3: 7-12.
4. Gee, A., Domingues, L. P., and Deitz, V. R., 1954, Determination of inorganic constituents in sucrose solutions, *Anal. Chem.* 26: 1487-1491.
5. Moreno, E. C, Kresak, M, and Hay, D. I., 1984, Adsorption of molecules of biological interest onto HAP, *Calcif. Tissue Int.* 36: 48-59.
6. Aoba, T. and Moreno, E. C., 1985, Adsorption of phosphoserine onto HAP and its inhibitory activity on crystal growth, *J. Colloid Interface Sci.* 106: 110-121.
7. Shimabayashi, S., Moriwaki, A., and Nakagaki, M., 1985, Effect of condensed phosphate on the precipitate formation and dispersion of calcium phosphate in water, *Chem. Pharm. Bull.* 33: 4641-4648.
8. Shimabayashi, S., Tanizawa, Y., and Ishida, K., 1991, Effect of phosphorylated organic compound on the adsorption of bovine serum albumin by HAP, *Chem. Pharm. Bull.* 39: 2183-2188.
9. Shimabayashi, S., and Matsumoto, M., 1993, Non-stoichiometric dissolution of hydroxyapatite in the presence of simple salt, *J. Chem. Soc. Jap.* 1993: 1118-1122.

15

CRYSTAL GROWTH INHIBITION OF HYDROXYAPATITE BY POLYCARBOXYLATES

Role of Calcium and Polymer Molecular Weight

C. L. Howie-Meyers,[1] K. Yu,[1] D. Elliott,[1*] T. Vasudevan,[1] M. P. Aronson,[1] K. P. Ananthapadmanabhan,[1] and P. Somasundaran [2]

[1] Unilever Research United States
Edgewater, New Jersey 07020
[2] Columbia University
New York, New York 10023

ABSTRACT

Polycarboxylate polymers are well known as effective inhibitors of calcium phosphate crystal growth. Model crystal growth inhibition measurements show that polycarboxylates require the addition of calcium in the form of dissolved salt to increase their efficiency. Importantly, polycarboxylates exhibit maximum crystal growth inhibition activity in a narrow molecular weight range. The latter is around 1000 to 2000 for acrylate-maleate copolymers and around 5000 to 6000 for polyacrylates. The role of calcium and the dependence on polymer molecular weight have been studied by adsorption, calcium binding studies and electrokinetic measurements. The results obtained show that complete surface coverage is required for effective inhibition of crystal growth by polymers. Importantly, the molecular weight effect results from competition for polymer between the substrate and the dissolved calcium.

INTRODUCTION

Polymeric agents are often used to inhibit or modify the crystal growth of minerals such as calcium phosphate.[1-7] In such roles, polymers find use in water treatment as scale control agents or as anticalculus agents in oral cleaning compositions.[1-7] It is generally

---

* Currently of Calgon Corporation, Pittsburgh, PA 15230.

Mineral Scale Formation and Inhibition, Edited by Zahid Amjad
Plenum Press, New York, 1995

accepted that these agents operate by adsorption onto the mineral surface at growth sites, thereby preventing further crystal formation.

An agent's performance as a crystal growth modifier is determined by the degree of adsorption, its surface conformation, and binding strength. In each of these properties, molecular weight (MW) is a key parameter. Optimum agents must strike a balance between low MW for rapid and efficient adsorption and high MW to promote ion sequestration and stronger adsorption. Careful choice of MW range gives large differences in effectiveness for specific functions; for example, low MW materials (1000-2000) are reported to be good for scale inhibition, while higher MW fractions (4000-12,000) are better for antiprecipitation.[8] Products having combinations of polymers of varying MW (or broad MW distributions by design) have also been reported.[8,9] In these systems, synergistic activity is found because different MW materials perform a specific function. Thus, MW and MW distribution are important parameters which must be carefully controlled to achieve optimum performance.

Several studies of MW effect have been reported previously.[2-5,7,10] An early report on poly(sodium methacrylates) ranging from MW 5000 to 600,000 concluded that the precipitation of calcium carbonate was inhibited more efficiently as MW decreased.[10] Nestler[11] studied fractions of polyacrylic acid from 9,000 to 100,000 MW, and reported that adsorption capacity on calcium sulfate increased with increasing MW. The lower MW fractions were adsorbed in preference to the higher fractions, but surface coverage of the lower fractions was found to be less efficient than the higher MW fractions. Jones[12] studied the effect of MW of carboxymethyl celluloses (MW range 50,000 to 400,000) as corrosion inhibitors and found that effectiveness decreased with increasing MW. Other studies on the effect of MW have been reported.[3,13] Recently, Amjad has reported on a series of studies on the effect of MW on brushite[2] and calcium sulfate.[5] In both reports, polyacrylic acids of MW about 2000 were found to have optimum effectiveness as scale inhibitors.

Recently, we reported that low molecular weight acrylate/maleate cotelomers[7,14] were effective inhibitors of dental calculus. Interestingly, in these systems addition of calcium in the form of dissolved calcium salt was necessary to activate the polymers as crystal growth inhibitors. In addition, at a given calcium level, a MW dependency was observed for these materials as calcium phosphate crystal growth inhibitors, with optimum activity observed for materials of MW 1000-2000. To provide further understanding of the reasons for this behavior, we have compared crystal growth inhibition activity with polymer adsorption, zeta potential and Ca ion sequestration behavior for these acrylate/maleate copolymers (MW range 1200 to 50,000) and for a similar series of polyacrylic acids (MW range 1600 to 60,000). The results of this fundamental study are presented in this paper.

EXPERIMENTAL

Materials

Hydroxyapatite (HAP) obtained from Albright and Wilson (chemical formula $Ca_{10}(PO_4)_6(OH)_2$) was used for the experiments. The HAP sample had a Ca:P molar ratio of 1.61 (theoretical ratio 1.67) and a surface area of 26 m^2/g, as determined by the BET nitrogen adsorption technique. Butanetetracarboxylic acid (BTCA) was obtained from Chemie Linz and was used without further purification.

Two polymer series were studied, differing only in molecular weight. A series of copolymers of acrylic acid and maleic acid (molar ratio of acrylate to maleate 2:1) were tested. The cotelomers AM-C, AM-D, and AM-H were prepared as disclosed previously.[7,14] ^{14}C-labelled analogs of these cotelomers were also prepared. Sokalan CP7, an acrylate/maleate copolymer (ratio of acrylate to maleate 2:1), was obtained from BASF Corp. A second series was polyacrylic acids (PAA). Samples of PAA were obtained from BASF Corp. (Sokalan PA25PN) and from Rohm and Haas Co. (Acrysol A-1, Acrysol LMW-20N, and Acusol 410N).

Molecular Weight Determination

Molecular weights of the polymers were measured using aqueous gel permeation chromatography (GPC). The GPC analyses were performed using a Waters 150-C ALC/GPC having a series of columns of 120/250/500/1000 Å Ultra Hydrogel from Waters. Column temperature was 30°C, injection volume was 100 µl, and flow rate was 1.1 ml/min. Refractive index detection was used, and data collection was accomplished with Nelson GPC software Version 5.1 (P.E. Nelson) for absolute molecular weight determination. The mobile phase was 0.1M $NaNO_3$ at pH 7. Sample concentration was 4-5 mg/ml. Polyacrylate standards were used for calibration.

Adsorption Studies

Adsorption Procedure (UV Method). Samples of 0.1 gm of HAP in 5 ml of deionized water were pre-equilibrated by incubating on a Hematology/Chemistry Mixer (Fisher Scientific) overnight at 37°C. Five ml of polymer solution were added to these samples to give an initial concentration ranging from 0 to 1000 ppm of polymer in a 10 ml sample. These slurries were incubated as before for two hours then filtered using 0.22 mm cellulose acetate filters (Schleicher and Schuell). The remaining polymer in solution was determined by UV Spectroscopy (Perkin-Elmer 330 UV/VIS/NIR Spectrophotometer) at 210 nm. The amount adsorbed was determined by difference between the initial concentration and the final polymer concentration. The concentration of a polymer solution was obtained from the standard plot constructed for that particular polymer.

Adsorption Procedure (Radiolabelled Polymers). Samples of 0.1 gm of HAP in 5 ml of deionized water were pre-equilibrated by incubating on a Hematology/Chemistry Mixer (Fisher Scientific) overnight at 37°C. Five ml samples of polymer solutions were prepared by mixing volumes of 2.5% nonlabelled polymer and 5 ml of 2.5 wt% ^{14}C-polymer (420 mCi/ml) and water. The polymer treatment solutions were prepared so that the total polymer concentration ranged from 50-1000 ppm when added to the 5 ml of water used in the pre-equilibration. A 50 µl aliquot was taken from each sample to determine initial activity levels by scintillation counting (Beckman LS 5801 Scintillation Counter), using Scintiverse^{TM} BD scintillation cocktail (Fisher Scientific). These polymer solution were added to the pre-equilibrated samples and incubated as before for two hours. After incubation the samples were centrifuged at 3000 rpm (1287 x g) for 10 min, the supernatant was decanted and the HAP pellets were dissolved with 1N HCl. The activity of the supernatant and the dissolved pellet was measured as before by scintillation counting.

Adsorption Procedure with Added Calcium (Radiolabelled Polymers). Samples of 0.1 gm HAP in 4 ml of deionized water were pre-equilibrated on a Hematology/Chemistry Mixer overnight at 37°C. A 1.0% stock polymer solution (0.196 mCi/ml) containing non-labelled and ^{14}C-labelled polymer was prepared. Treatment solutions were made containing 1 ml of the polymer stock and $CaCl_2$ where the molar ratio of Ca:polymer repeat unit ranged from 0.1:1 to 2:1. A 30 µl aliquot of each polymer solution was removed for initial scintillation counting (Beckman LS 5801 Scintillation Counter), using Scintiverse scintillation cocktail (Fisher Scientific). The Ca/polymer solutions were added to the pre-equilibrated samples and incubated as before for two hours. After incubation the samples were centrifuged at 3000 rpm (1287 x g) for 10 min., the supernatant was decanted and the HAP or brushite was dissolved in 1N HCl. The activity of the supernatant and the dissolved pellet was measured as before by scintillation counting.

Crystal Growth Inhibition Assays

pH-Stat Assay. The pH-stat assay, which measures the ability of an agent to inhibit transformation of amorphous calcium phosphate to HAP, was conducted using a procedure described by Gaffar.[15] The assay was performed using a Radiometer TTT80 Titrator equipped with pH-meter, autoburette, and plotter. In the procedure, equimolar amounts of calcium chloride and sodium hydrogen phosphate solutions were mixed with known quantities of an agent and the onset of hydrolysis is measured via consumption of NaOH as determined by monitoring solution pH.

$$14 \; OH^- + 10 \; Ca^{++} + 6 \; NaH_2PO_4 \rightarrow Ca_{10}(PO_4)_6(OH)_2 + 6 \; Na^+ + 12 \; H_2O \qquad (1)$$

Delay time is the time which an agent delays the onset of HAP formation. It is calculated from the amount of NaOH consumed. Longer delay times denote better inhibition efficiency.

HAP Seeded Crystal Growth Inhibition Assay (SCGI). HAP crystal growth inhibition was measured by treating 12.5 mg of HAP, introduced as a 0.5 ml slurry of HAP and deionized water with 3.75 ml of an aqueous solution containing desired level of agent at pH 8.0 and 50 mM $CaCl_2$. After a treatment time of 5 minutes, the samples were centrifuged at 3000 rpm (1287 x g). The supernatant was discarded and the HAP pellet was washed twice with deionized water. The HAP pellet was then suspended in a calcifying solution containing 1.5 mM Ca^2+ and 4.5 mM PO_4^{3-} in 0.2 M barbitone. The samples were incubated on a Hematology/Chemistry Mixer for 60 minutes at 37°C. The slurries were filtered using 0.22 μm cellulose acetate filters. The remaining calcium in solution was analyzed by Atomic Absorption Spectroscopy. % Inhibition was determined using the following relationship:

$$\% \; inhib. = \frac{[Ca] \; depletion \; of \; control - [Ca] \; depletion \; of \; sample}{[Ca] \; depletion \; of \; control} \qquad (2)$$

Microbial Mineralization Assay. This in-house assay developed to mimic the presence of oral bacteria under calculus formation conditions was done as follows. Etched glass rods were placed in an aqueous solution containing *Streptococcus sobrinus 6715-14.* Microbes were allowed to grow onto the rods for about 2 days, and the rods were removed from the solution and placed in a treatment solution containing antitartar agents at levels of about 1.25 wt%. Treatment with a solution containing only deionized water was used as a control. After treatment for about 30 seconds, the rods were placed in a calcifying solution made up of calcium and phosphate at levels of 1.5 mM (from $CaCl_2$) and 5.0 mM (from KH_2PO_4), respectively. The standard calcifying solution was supplemented with 25% filtered human saliva; as a comparison, a calcifying solution with the same calcium and phosphate levels but with no saliva was used as well. The glass rods were mineralized in the calcifying solution for 4 days, after which the level of calculus formation was assessed via calcium (Atomic Adsorption Spectroscopy) and phosphorus analysis.[16] Results were reported as % reduction of calculus formation relative to the control.

Zeta Potential and Calcium Binding Measurements

Samples of HAP (typically 0.05 grams) were dispersed in 50 ml of a 2×10^{-3} M NaCl solution containing the desired concentration of polymer. The solution pH was preadjusted to 8. The solutions were stirred for 15 minutes and the final pH was recorded.

To conduct the experiments, about 30 ml of the suspension was transferred to the cell of the meter (Zeta Meter from Zeta Meter, Inc.). The electrodes were inserted and air bubbles were removed. Voltage and current were applied to the sample. The velocity of about 10-15 particles were measured, from which the average mobility was calculated. The zeta potential (in millivolts) was calculated from the average mobilities.

Calcium ion activity in solution was measured using a calcium sensitive electrode, Orion Model 93-20, under the conditions employed for zeta potential measurements. Since only the free calcium in solution will respond to the ion selective electrode, the difference in calcium activity measured in the presence and absence of polycarboxylate polymers was taken as the amount of calcium bound to the polymer.

RESULTS

Two types of polycarboxylates, polyacrylates and poly(acrylate-co-maleates), were selected with variations only in MW. The selected MW range covers from very low MW, which are typically useful as crystal growth modifiers, to higher MWs which are more effective as sequestrants or antinucleation agents. Table 1 summarizes the materials used in these studies.

Crystal Growth Inhibition by the Polymers

Role of Calcium. The HAP seeded crystal growth assay measures an agent's ability to inhibit the growth of HAP crystal on a HAP seed. The ability of AM-C polymer to inhibit the crystal growth of HAP is given in Figure 1 along with the data for sodium-pyrophosphate, a widely used low molecular weight crystal growth inhibitor. Interestingly, in the absence of added calcium, the AM-C polymer does not exhibit much activity. However, the addition of 1.5 mM $CaCl_2$ to the system improved the performance of the anionic polyelectrolyte. Similar enhancement by calcium chloride was also observed for the inhibition of HAP by polyacrylate polymers. For this reason, all the subsequent work on HAP inhibition was conducted in the presence of added calcium chloride. The exact level of calcium added is

Table 1. GPC determination of molecular weights of the polymers used in this study

Sample	Molecular weight		
	M_w	M_n	MWD
AM-C	1200	700	1.6
AM-D	2500	1200	2.1
AM-H	8400	2400	3.5
Sokalan CP7	48,500	9800	4.9
Acrysol LMW-20N	1600	1200	1.3
Sokalan PA25PN	5900	2100	2.7
Acrysol LMW-100N	11,800	5600	2.1
Acrysol A-1	60,000	—	—

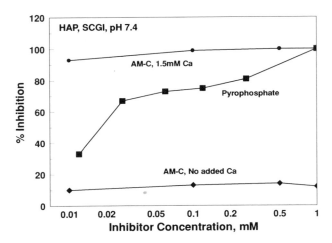

Figure 1. Seeded crystal growth inhibition of HAP by AM-C (acrylate-maleate cotelomer) in the presence and absence of added calcium as a function of reagent concentration. Results for sodium pyrophosphate is included for comparison.

specified in each of the figures. The reasons for the need for calcium itself is discussed in a later section.

Effect of Polymer Molecular Weight. The acrylate/maleate copolymers and the polyacrylic acids were compared for their ability to inhibit HAP crystal growth in seeded crystal growth inhibition assays. The results obtained are shown in Figures 2 and 3. Varying levels of calcium were added to the polymers to alter activity. With no calcium present in the treatment solutions, each of the polymers gives low inhibition and are about equal in effectiveness. With addition of calcium, activity levels change, and differences in MW begin to emerge. The results indicate that as calcium levels are increased, the activity of the polymers are affected, presumably due to differences in surface adsorption[14] and solution sequestration. The members of each series behave differently, however, based on molecular weight. In the acrylate/maleate series, sample AM-C shows optimum behavior, with maximum activity at molar ratios above 1:4 (Ca:polymer repeat unit). The lower MW analog BTCA and the highest MW sample (Sokalan CP7) show reduced activity. The CP7 sample actually shows a narrow window of strong inhibition activity between 1:4 and 4:1 Ca:repeat

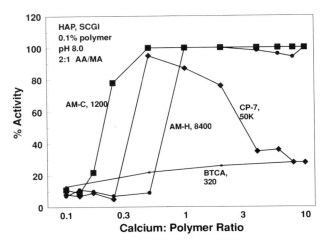

Figure 2. Seeded crystal growth inhibition of HAP as a function of concentration of added calcium in the presence of various acrylate-maleate copolymers, showing their molecular weight effect. Numbers next to the polymer indicate their molecular weight.

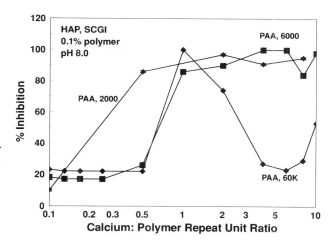

Figure 3. Seeded crystal growth inhibition of HAP as a function of concentration of added calcium chloride showing the effect of molecular weight of various polyacrylate polymers. Numbers next to the polymer indicate their molecular weight.

unit ratios, while the BTCA is ineffective at all ratios. AM-H has an activity profile intermediate to CP7 and AM-C, with reduced activity at higher calcium-containing ratios.

In the polyacrylate series, similar effects are observed, with the optimum MW range at about 5000-6000 (Sokalan PA25PN). The higher molecular weight Acrysol A-1 shows only a narrow window of efficient inhibition between 0.5:1 and 2:1 ratios.

Table 2 compares the inhibition activity of both series in the pH-stat assay, which measures the ability of an agent to inhibit the transformation of amorphous calcium phosphate to HAP. In both series the trend is increasing activity with decreasing MW. Each of the assays shows that distinct MW ranges give optimum inhibition. Thus, polymer MW is a key indicator of an agent's ultimate performance as an inhibitor of calcium phosphate crystallization. In summary, the optimum MW range for the acrylate/maleates is between 1000-2000, and for the polyacrylic acids slightly higher, at about 5000-6000.

The effect of various molecular weight polymers on microbial mineralization assay is given in Figure 4. Here the ability of polymer to inhibit the crystal growth in the presence of salivary components in a "plaque" matrix is evident. The results clearly show a molecular weight dependence with the AM-C polymer with a molecular weight of 1000-2000 exhibiting maximum activity.

Table 2. pH-Stat assay of Polyacrylic acids and Acrylate/Maleate copolymers of various molecular weights

	Sample, 30 ppm dosage	Delay time, minutes
Acrylate/maleate copolymers		
	MW 1200	23.2
	MW 2500	12.2
	MW 8000	3.4
	MW 50,000	0.0
Polyacrylic Acids		
	MW 1600	16.5
	MW 6000	10.7
	MW 10,000	0.0
	MW 60,000	0.0

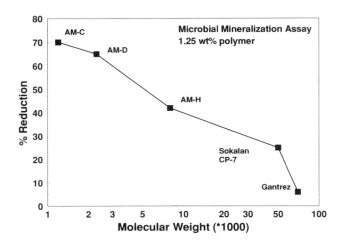

Figure 4. Percent reduction in microbial mineralization of calcium phosphate as a function of polymer molecular weight.

Polymer Adsorption and Zeta Potential Measurements

The degree of adsorption of an agent on the mineral surface is a strong indicator of its crystal growth inhibition activity. Adsorption isotherms of acrylate-maleate cotelomers, AM-C and AM-H are given in Figure 5. Interestingly, AM-C appears to adsorb more than AM-H at lower concentrations. However, at 1000 ppm, the extent of adsorption of both the polymers is essentially the same.

The effect of calcium on AM-C and AM-H adsorption at a constant concentration of 1000 ppm of the polymer is shown in Figure 6. The results indicate that the adsorption of AM-C is enhanced more than that of AM-H. At higher calcium levels precipitation was observed in the polymer-calcium chloride solutions and therefore adsorption beyond that region was not determined.

Zeta potential values of HAP in water, in the presence of AM-C are shown in Figure 7. In the absence of added calcium chloride (at 0 ppm $CaCl_2$), HAP exhibits a negative charge. With the addition of calcium, as expected, the zeta potential appears to increase and eventually become positive at about 2.5 ppm calcium. In the presence of both the polymer and calcium, HAP zeta potential becomes more negative than that in water and finally attains

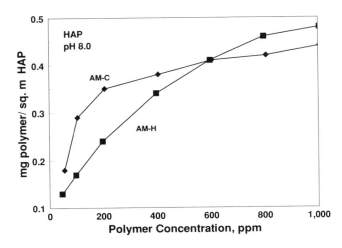

Figure 5. Adsorption of acrylate maleate copolymers AM-C (Mol. wt. 1200) and AM-H (mol. wt. 8400) on HAP as a function of polymer concentration.

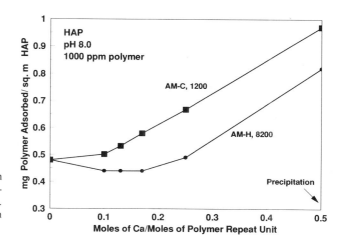

Figure 6. Effect of added calcium chloride on adsorption of acrylate-maleate copolymers AM-C (Mol.wt. 1200) and AM-H (Mol.wt. 8400) on HAP.

a constant value around -40 mV. This increase in the negative potential of HAP is attributed to the enhanced adsorption of the polymer in the presence of calcium. This is consistent with the earlier mentioned result that the polymer adsorption increases with increase in the added calcium concentration.

Zeta potential measurements of polymer-treated HAP surface are shown in Figures 7 and 8. In these figures, increased surface negative charge denotes greater adsorption of the polymer onto HAP surface. The AM-C and AM-H samples (MWs 1200 and 8000) appear to adsorb most, while the low MW BTCA adsorbs to the least extent. Higher MW samples appear to adsorb less efficiently. This behavior tracks the observed inhibition activity. Thus, there seems to be a correlation between activity and adsorption behavior.

Polymer Sequestration

Figure 9 gives a comparison of the calcium binding isotherms of acrylate/maleate copolymers. As shown, the extent of calcium binding appears to increase with increase in polymer molecular weight. In other words, the free calcium in solution decreases with increasing MW. This indicates that the mechanism by which inhibition of crystal growth

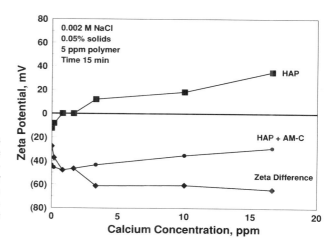

Figure 7. Zeta Potential of HAP as a function of added calcium chloride concentration in the presence and absence of acrylate-maleate copolymer AM-C. The curve designated as "Zeta Difference" refer to the difference in zeta potential of HAP after and before AM-C addition.

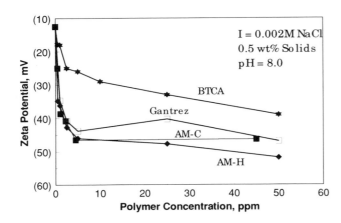

Figure 8. Zeta Potential of HAP in the presence of various acrylate-maleate copolymers as a function of polymer concentration.

occurs is not merely by surface adsorption, but possibly also by calcium sequestration. Note, however, that the effect of sequestration itself may be to reduce adsorption at the solid liquid interface, as the solution calcium, in principle, can compete with surface sites for the polymer.

DISCUSSION

The above discussed results show that effective inhibition of HAP crystal growth requires the addition of calcium in the form of soluble calcium chloride. Furthermore, the effectiveness of polycarboxylates to inhibit the crystal growth is dependent on polymer molecular weight with acrylate maleate polymers showing an optimum molecular weight around 1000-2000. The results are examined below in the light of the zeta potential and adsorption results.

As shown in Figure 7, HAP exhibits a negative zeta potential in water under the test pH conditions. This clearly implies that the surface of HAP has more negative sites than positive sites. Consequently, full coverage of the HAP surface by a polycarboxylate polymer cannot be expected under these conditions. Calculations using the average molecular weight

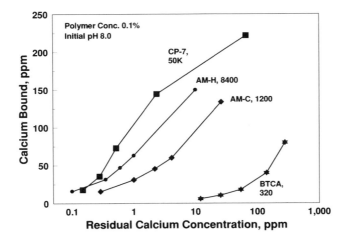

Figure 9. Calcium binding isotherms for various acrylate-maleate polymers showing the dependence of binding on polymer molecular weight.

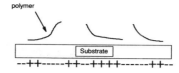

No Added Calcium
Incomplete
Coverage

Figure 10. A schematic diagram showing how calcium ions enhance adsorption of anionic polymers onto HAP.

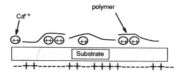

Calcium Bridging
Polymer. Adsorption
Enhanced. Complete
Coverage.

of the polymer, adsorption values in the absence of added calcium, and the assumption that polymer adsorbs with a flat configuration, shows that the extent of surface coverage by polymer is only about 40% of a monolayer. Addition of calcium, as supported by the zeta potential results, can make the surface more positive. Thus, in the presence of calcium, the adsorption of the polyanions should increase possibly by the bridging mechanism depicted in Figure 10. This is again consistent with the polymer adsorption results given in Figure 6.

A correlation between polymer adsorption and the crystal growth activity given in Figure 11 shows a good correlation between the two with 100% inhibition corresponding to monolayer coverage of the HAP surface by the polymer. This is indeed interesting because in most inorganic inhibitor systems coverage of active sites by the inhibitor is considered to be sufficient to inhibit the crystal growth. It can, however, be argued that as the active sites get inhibited by adsorption, other less active sites can become the "active sites" and therefore for good inhibition, complete surface coverage is desirable. At least, in the present case of inhibition by polymers, it appears that complete coverage is needed to achieve maximum crystal growth inhibition.

The polymer molecular weight dependence studies clearly showed that there exists an optimum molecular weight for inhibition of crystal growth by polymers. In the case of acrylate-maleate cotelomers, this optimum range appears to be around 600-2100. As mentioned earlier, other polyelectrolyte systems reported in the literature also exhibited activity maximum in certain optimum molecular weight range.[2,5] The origin of molecular weight

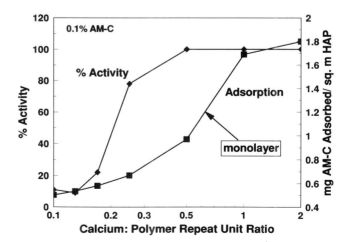

Figure 11. The correlation between adsorption of AM-C on HAP and the seeded crystal growth inhibition of HAP by AM-C as a function of added calcium chloride concentration.

dependence, on the other hand, is not clear from the reported studies. Some of the possible reasons for the molecular weight effect on crystal growth inhibition are examined below.

The crystal growth inhibition of HAP occurs as an agent blocks surface growth sites and prevents further precipitation or, as an agent sequesters ions in solution, competes with the surface for available calcium, and in effect lessens the thermodynamic driving forces for precipitation. In general, adsorption on seed crystals is essential for effective crystal growth inhibition. The correlation between AM-C adsorption on HAP and its crystal growth inhibition activity clearly shows that in the present case adsorption is critical to inhibiting HAP growth.

It is well known that polymer adsorption from poor solvents onto a relatively insoluble solids such as metal oxides and hydroxides increase with increase in polymer molecular weight.[17] This is because, in simplistic terms, the number of monomeric groups per molecule involved in the binding process will be higher for a high molecular weight polymer resulting in a higher enthalpic contribution for binding. Importantly, the entropic loss resulting from the binding will be less for a polymer molecule than for the same number of monomer molecules constituting the polymer. Thus, both entropically and enthalpically the adsorption of higher molecular weight polymer is favored more than that of lower molecular weight polymers.

While the above analysis can explain the low activity of very low molecular weight polymers such as BTCA, it can not explain the reasons for the reduction in activity above a certain molecular weight range. In this regard, it is important to recognize that crystal growth inhibition systems are different from relatively insoluble oxides in the sense that in general, these systems contain multivalent ions in solution. In the present case, in fact, without the addition of soluble calcium the polymers exhibited almost no activity. Therefore, adsorption results that should be correlated with inhibition activity should correspond to adsorption that occurred in the presence of added calcium ions. Addition of soluble calcium ions can be expected to have a significant effect on the adsorption behavior of the polycarboxylate polymers. Since adsorption is essentially a partitioning of the adsorbate between the substrate and the solution, factors that enhance the partitioning of the polymer into solution can be expected to lower its adsorption. In the present case, calcium ions can compete with the HAP surface for polycarboxylate polymers and in turn form solution complexes. An analysis of the adsorption, zeta potential and solution complexation results and their dependence on polymer molecular weight may provide a better insight into the mechanisms involved.

The low adsorption of BTCA (molecular weight 320) compared to AM-C(molecular weight 1200) on HAP is evident from the zeta potential results given in Figure 9 which shows that BTCA does not increase the negative charge of HAP compared to AM-C or other tested polycarboxylates. This is consistent with the observed inability of BTCA to inhibit crystal growth of HAP as well as the expected low adsorption of very low molecular weight polymers.

The polymer adsorption results obtained in the absence of added calcium shows that the lower molecular weight polymer AM-C adsorb more than the higher molecular weight AM-H in the low concentration range (See Figure 5). However, by about 1000 ppm, both the polymers exhibit about the same level of adsorption. Importantly, adsorption of AM-C and AM-H at a level of 1000 ppm as a function of added calcium (see Figure 6) indicate that the increase in the binding of the lower molecular weight polymer AM-C is significantly higher than that of the higher molecular weight AM-H polymer. Thus, the calcium addition appears to enhance the adsorption of the AM-C more than that of AM-H. The higher crystal growth activity of AM-C over AM-H is thus consistent with their measured adsorption.

The above adsorption results suggest that Ca ions enhanced the adsorption of low molecular weight polycarboxylates more than that of the high molecular weight ones. This may be because of several reasons. First of all, the tendencies of the lower and higher

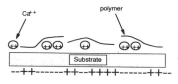

Lower Molecular Weight
More Adsorption, Less
Sequestration/Precipitation

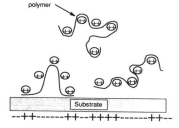

Figure 12. A schematic diagram showing the differences in the behavior of low and high molecular weight polycarboxylates adsorbing on HAP in the presence of dissolved calcium ions.

Higher Molecular Weight
Less Adsorption, More
Sequestration/Precipitation

molecular weight polycarboxylates to sequester calcium in solution are markedly different. This is evident from the calcium binding results given in Figure 9. It appears that, in the case of the high molecular weight polymer, sequestration and complex formation with soluble calcium is more favored than adsorption. The low molecular weight polymer, on the other hand tend to adsorb at the interface. Thus the delicate balance between adsorption and sequestration seem to control the molecular weight dependence of crystal growth inhibition. The lower molecular weight polymer may also have a kinetic advantage over their higher molecular weight counterparts during the adsorption process.

Yet another reason for the low inhibition efficiency of higher molecular weight polyelectrolytes might be due to their adsorbed layer conformation at the solid-liquid interface. Normally, polyelectrolytes adsorb on oppositely charged solids by a flat conformation(18). However, in the presence of multivalent ions, adsorption with loops and tails may be more favored since some of the charged groups on the polymer can interact with calcium ions as depicted in Figure 12. This may be energetically favorable since the loss of entropy for adsorption with loops and tails will be less than that for flat adsorption. If this occurs, incomplete surface coverage can lead to inefficient crystal growth inhibition. Note that such loopy structures are not possible with low molecular weight polymers because of the relatively low flexibility of the polymer molecule in that molecular range.

In systems with relatively high solubility, for example, brushite, a more soluble form of calcium phosphate than HAP, one can expect crystal growth inhibition to occur by the actives adsorbing at the solid-liquid interface as well by sequestration of precipitating ions in solution. In such cases, therefore, a combination of molecular weights may be beneficial.

CONCLUSIONS

Polycarboxylates are effective inhibitors of HAP crystal growth. In the present system, addition of calcium was necessary to potentiate the crystal growth inhibition activity of polycarboxylate polymers. Zeta potential and adsorption measurements showed that the role of calcium was to reverse the negative charge of HAP under the test conditions and in turn promote polymer adsorption at the interface.

The optimum MW range for the crystal growth inhibition of HAP appears to be about 1000-2000 for acrylate/maleate copolymers and slightly higher for polyacrylates. The reason for the existence of an optimum molecular weight range is attributed to the maximum adsorption on HAP of polymers in that molecular weight range. The lack of activity of very low molecular weight polymer (MW < 400) is attributed to its low adsorption at the solid-liquid interface resulting from the limited number of functional carboxylate groups. The presence of multivalent cations, calcium in the present case, results in competition between the surface and the solution calcium for polycarboxylate polymers. Interestingly, very high molecular weight polymers tend to form complexes in solution leading to precipitation rather than adsorbing at the HAP surface. Also, the possible non-flat conformation of high molecular weight polyelectrolytes adsorbing at the solid-liquid interface in the presence of multivalent cations in solution can lead to incomplete surface coverage. Thus, the delicate balance between adsorption, conformational effects and solution sequestration results in an optimum molecular weight range for maximum crystal growth activity by polycarboxylates.

REFERENCES

1. "Dispersants", Kirk-Othmer, 1982, Encyclopedia of Chemical Technology, Third Edition, v. 7: 833.
2. Amjad, Z., 1989, Constant composition study of dicalcium phosphate dihydrate crystal growth in the presence of (polyacrylic acids), Langmuir, 5:1222-1225.
3. Amjad, Z., 1989, Effect of precipitation inhibitors on calcium phosphate scale formation, Can. J. Chem., 67:850-856.
4.. Smith, B.R. and Alexander, A.E., 1970, The effect of additives on the process of crystallization: Further studies on calcium sulphate, J. Colloid Interface Sci., 34:81-90.
5. Amjad, Z., 1988, Calcium sulfate dihydrate (Gypsum) scale formation on heat exchanger surfaces: The influence of scale inhibitors, J.Colloid Interface Sci., 123:523-536.
6. Gaffar, A., Polefka, T., Afflitto, J., Esposito, A., and Smith, S., 1987, in vitro evaluations of pyrophosphate/copolymer/NaF as an anticalculus agent, Compend. continuing Education, 8(3), S242-250.
7. Elliott, D.L., Howie-Meyers, C.L., and Montague, P.G., Hypophosphite containing polymers as antitartar agents, US Patent 5,011,682.
8. Schiller, A.M., Goodman, R.M., and Neff, R.E., 1978, Preparation of anionic polymers for use as scale inhibitors and antiprecipitants, US Patent 4,072,607.
9. Song, D.S., Duffy, R.J., Witschonke, C.R., Schiller, A.M., and Higgins, A., 1977, Low molecular weight hydrolyzed polyacrylamide and use thereof as scale inhibitor in water systems, US Patent 4,001,161.
10. Williams, F.V. and Ruberwrin, R.A., 1957, Effect of polyelectrolytes on the precipitation of calcium carbonate, J. Amer. Chem. Soc., 79: 4898-4900.
11. Nestler, C.H., 1968, Adsorption and electrophoretic studies of poyacrylic acid on calcium sulfate, J. Coll. Inter. Sci., 26:10-18.
12. Jones, L.W., 1961, Development of a mineral scale inhibitor, Corrosion, 17:232-236.
13. Libutti, B. et al., 1984, The effects of antiscalants on fouling by cooling water, Mater. Perform., 23(11): 47-50.
14. Elliott, D.L., Howie-Meyers, C.L., and Kanapka, J.A., 1994, Polymeric anticalculus agents 1: In-vitro activity of phosphinate containing acrylate-maleate cotelomers, Submitted to Colloids and Surfaces.
15. Gaffar, A. and Moreno, E.C., 1985, Evaluation of 2-Phosno-butane 1,2,4, tricarboxylate as a crystal growth inhibitor in vitro and in vivo, J. Dent. Res., 64(1)-6-11.
16. Chen, P.S., Toribara. T.Y. and Warner, H., 1956, Microdetermination of Phosphorus, Anal. Chem. 8:1756-1758.
17. Scheutjens, J.M.H.M. and Fleer, G.J., 1985, Interaction between two adsorbed polymer layers, Macromolecules,18: 1882-1900.
18. Van der Schee, H.A. and Lyklema, J., A lattice theory of polyelectrolyte adsorption, 1984, J. Phys. Chem., 88: 6661-6667.

Mechanisms of Regulation of Crystal Growth in Selected Biological Systems

C.S. Sikes[1] and A. Wierzbicki[2]

[1] Department of Biological Sciences
University of South Alabama
Mobile, Alabama 36688
[2] Department of Chemistry
University of South Alabama
Mobile, Alabama 36688

INTRODUCTION

The design of inhibitors of crystallization is aided by an understanding of the binding of the inhibitors at the molecular level and the influence of the binding on crystal morphology. Small, anionic inhibitors such as phosphonates are effective in part because of their high affinity for cationic regions of surfaces of crystals. Polymeric, polyanionic inhibitors may have this affinity enhanced by spatial matching of anionic residues of the polymer with cationic binding sites of the mineral scale.

Another contribution to the mechanism of scale inhibition may be the specific orientation taken by an inhibitor as it binds to the crystal surface. For example, the interaction between both monomeric and polymeric carboxylate inhibitors and crystal surfaces may be favored by a perpendicular presentation of the antiscalant COO⁻ groups at the crystal surface in a manner similar to the orientation that would be taken by incoming lattice anions of the uninhibited crystal surface.[1,2,3,4,5,6]

The purpose of the present chapter is to examine the influence that the stereochemical orientation of an inhibitor relative to a crystal surface may have in determining the specific interaction and therefore the morphology of the resulting crystal. The beneficial effect of phosphorylating a carboxylate inhibitor is also demonstrated. The systems chosen for detailed analysis are the interaction of polyaspartate with calcite and the inhibition of calcium oxalate formation by citrate, a tricarboxylate, and phosphocitrate (Figure 1).

The (1 -1 0) planes of calcite have carbonate ions that are perpendicular to that crystal surface, which can offer an optimal opportunity for binding of a polyanion like polyaspartate. Calcium oxalate monohydrate (COM) has an especially interesting crystal lattice for this study in that there are two distinct surfaces, (-1 0 1) and (0 1 0) (Figure 2), in which the anionic oxalate groups are arranged perpendicularly to each other. This in turn creates a possibility of distinct interactions with the anionic inhibitors.

Mineral Scale Formation and Inhibition, Edited by Zahid Amjad
Plenum Press, New York, 1995

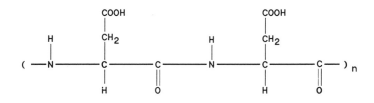

CITRIC ACID

PHOSPHOCITRIC ACID

POLYASPARTIC ACID

Figure 1. Citric, phosphocitric, and polyaspartic acids.

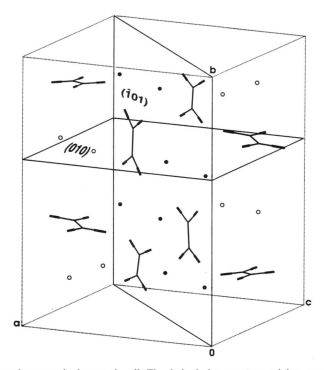

Figure 2. Calcium oxalate monohydrate, unit cell. The dark circles represent calcium atoms, the open ones represent water molecules. The two perpendicular orientations of oxalate ($^-$OOC-COO$^-$) in planes (-1 0 1) and (0 1 0) are shown.

Both calcite and COM crystals fulfill significant roles in nature[7,8,9,10,11] as well as presenting problems in a variety of settings ranging from industrial to medical.[12,13,14] For example, calcite is the most common component of both skeletal structures and industrial scale. Similarly, COM is a useful structural component and metabolite in organisms but it also is a common problem scale, particularly in the paper industry, and it creates problems when found as organ stones in the kidney and urinary tract.

METHODS

Preparation of Crystals of Calcite and COM

Calcite was spontaneously nucleated from supersaturated solution and collected as described previously.[9,15] In some cases, crystals were grown attached to glass discs. Inhibitors were added at doses that were known to reduce crystal formation but not stop it entirely. This produced crystals with altered morphology, which suggested specific locations and effects of adsorbates.

COM was prepared according to Sutor[16] as modified by Sallis and Lumley.[17] This involved diffusion of 0.1 M $CaCl_2$ and 0.1 M NaOx through paper wicks into a buffered solution of 0.2 M sodium acetate (pH 6.1) in the presence or absence of inhibitor. Growth of COM was initiated by immersing glass fibers into the reaction solution. After 20 h, the fibers were air-dried prior to microscopy.

Preparation of Adsorbates

Oyster shell protein, which has roughly half of its residues as aspartate and phosphoserine, was isolated by standard protocols.[18] The first peak from reversed phase liquid chromatography was the form of the protein used herein. A polyaspartate of 20 residues was prepared by solid-phase synthesis.[19,20] PC was prepared by phosphorylation of tribenzyl citrate followed by deprotection by hydrogenation.[21] The monosodium salt was crystallized from water and then subsequently brought to pH 7.0 with additional alkali. Sodium citrate was purchased from Sigma-Aldrich, Australia.

Microscopy

Scanning electron micrographs were produced by use of a Phillips SEM 505 and an AmRay 1000 A SEM. Atomic force micrographs were obtained by use of a Nanoscope II (Digital Instruments) operating in constant force mode with 100 m Si_3N_4 tips at forces ranging from 10^{-8} to 10^{-10} newtons.[9] Scanning forces and angles were varied to verify that images were free of artifactual influences of the tips.

Molecular Modeling of Inhibitor-Surface Interactions

Crystal models were prepared using unit cell parameters and fractional coordinates that are available in the literature.[9,10] The sizes of the crystal models were carefully chosen to assure that the boundary effect played no role in inhibitor-surface interaction.

Polyaspartate was positioned near the cleaved surface within the electrostatic range of interaction and energy optimization was applied. The minimization procedure takes into account the contributions of electrostatic forces, van der Waals forces, bonds, angles, and dihedral angles to the total energy of the system. Starting from the initial position of the inhibitor-crystal surface system, the total energy was minimized yielding the energy and geometry of the most favorable position of inhibitor on the crystal surface. The coordinates of lattice atoms were kept fixed and the inhibitor was allowed to translate, rotate, and adopt any conformation on the surface of calcite.

The binding energy was determined as the difference between the energy of the inhibitor-crystal system and the sum of the inhibitor and crystal energies when separated beyond the interaction distance. The effect of water on the system was analyzed by introducing an 8 Å thick hydration shell around the inhibitor and repeating the minimization procedure. The final configuration of the system was the same, since electrostatic interaction between the surface and the inhibitor molecule was the driving force in determining the final conformation of the system. The water molecules were accommodated around the inhibitor modifying the attraction of the molecule to the surface but having no major effect on the binding mechanism to the surface. For this reason, further consideration was given to the in-vacuo, polyaspartate-surface system only, using a constant dielectric model with dielectric constant equal to one.

Software programs used were CERIUS for modeling crystal surfaces, QUANTA for generating and positioning the adsorbates, and CHARMm 22 for energy minimization (all from Molecular Simulations, Inc., Burlington, MA).

Computational Structures of Inhibitors

Investigation of inhibitor - surface interaction by molecular modeling requires correct initial geometry and charge distribution for the inhibitor molecule and ions of the surface, both of which can be calculated efficiently using modern tools of computational chemistry. Where available, the predicted computational structure of the inhibitor should be compared to the x-ray molecular structure. For example, the predicted geometry of citrate[22] and phosphocitrate agreed very favorably with x-ray determined geometry of inhibitors.[10] The geometric structures of inhibitors and polyatomic surface anions were obtained using a series of geometry optimizations for a neutral inhibitor molecule applying both the *ab initio* Hartree-Fock method (Gaussian 92, Gaussian Inc., Pittsburgh, PA) using 6-31G** basis set, and the semiempirical modified neglect of differential overlap (MNDO, Spartan, Wavefunction, Inc., Irvine, CA). The Hartree-Fock method is based on the replacement of the full many-electron wave function by a single determinant of products of one-electron functions. The notation, 6-31G**, refers to a split-valence set with polarization and with the addition of p-functions to each hydrogen, which provides the best *ab initio* results for large second-row-element molecules.[23] In the MNDO method, certain integrals were taken as zero or derived from spectroscopic measurements.[24,25] Because deprotonation of some functional groups of the inhibitors would occur at the pH values of interest, the electrostatic charge distributions for the inhibitor ions were determined by single point calculations to fit charges from electrostatic potentials.[26,27] This method of charge determination has produced reliable atomic charges, in fact better than those derived from Mulliken population analysis.

RESULTS

SEM's of spontaneously nucleated, control, calcite crystals revealed well-formed rhombohedrons (Figure 3). Similar crystals grown in the presence of oyster shell protein were larger, less numerous, and had incomplete edges in the form of troughs, as well as some roughening of the cleavage surfaces (Figure 4). Crystals (not shown) grown in the presence of polyaspartate looked essentially the same as the protein-treated crystals.

Crystals grown on glass discs viewed by AFM often appeared pyramidal, sometimes exhibiting flat plateaus (Figure 5). The apparent angles of the sides of these crystals may

Figure 3. Scanning electron micrograph (SEM) of control calcite crystals, scale bar = 2 μm.

have been exaggerated due to the interaction of the inverted pyramidal AFM tip and the crystals as their angular surfaces are scanned under the tip.[28] However, the plateau regions were ideal surfaces for AFM.

The plateaus exhibited hexagonal arrays of atoms spaced 5 Å apart with each set of three closest atoms forming 60° angles, as occurs in the basal plane of calcium atoms of calcite (theoretically 4.99 Å, 60°; Figure 6). When Asp_{20} molecules were bound to this surface, AFM images revealed parallel arrays of the peptide (Figure 7). The length, width, and height of the imaged peptides were in good agreement with theoretical dimensions (~70 Å, 3-4 Å), particularly if the COO^- groups were drawn down onto an ionic surface rather than extended in a normal β-sheet conformation. When the oyster shell protein was bound

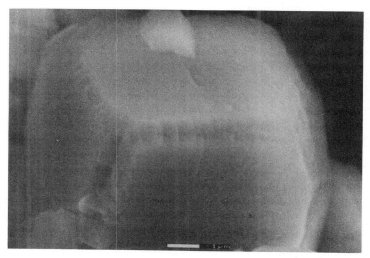

Figure 4. SEM of a calcite crystal grown in the presence of 0.1 mg oyster shell protein per ml, showing altered edges and corners, scale bar = 1 μm.

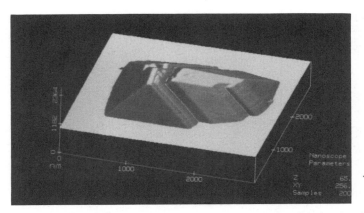

Figure 5. Atomic force micrograph (AFM) of calcite crystals grown on a glass disc showing the presence of a plateau region, the imaged area was 2.5 μ x 2.5 μ.

to the putative basal plane of calcium atoms, AFM images revealed a layer of globular molecules on this surface (Figure 8).

Oyster shell protein was also seen on surfaces within the troughs along the edges of spontaneously nucleated crystals, (Figure 9). The protein molecules lined up in regular fashion along terraces of 50 to 100 nm within the troughs, perhaps binding both the vertical

Figure 6. AFM's of the plateau region of Figure 5, showing atoms spaced 5 Å apart and forming angles of 60° among each set of 3 closest atoms. A. Magnification = 4.8 x 10^6. B. Magnification = 17.2 x 10^6.

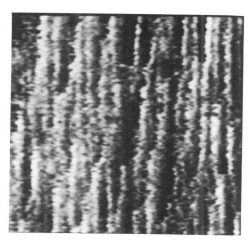

Figure 7. AFM of Asp_{20} molecules bound to the plateau region of Figure 5. The imaged area was 8 nm x 8 nm.

and horizontal surfaces of the terraces at once. The dimensions of the terraces seemed to be determined by the dimensions of the protein.

To assist in interpreting these images, a unit cell of calcite is shown (Figure 10), with critical planes such as (0 0 1) and (1 -1 0) designated. Computer modeling of the interactions of Asp_{15} with (1 -1 0) surfaces parallel to the c-axis (Figure 11) and perpendicular to the c-axis (Figure 12) are also shown. A schematic drawing of the possible interaction of the oyster shell protein with a calcite crystal is shown in Figure 13. Computer models of the atomic positions of the (1 -1 0) and (1 1 0) are given in Figure 14.

Figure 15 shows the computer-drawn COM crystal morphology depicting (-1 0 1) and (0 1 0). Figure 16 shows the computer model of COM used for the modeling. Although the planes (1 1 0) are considered to be higher probability growth faces than (1 2 0), as seen in Figure 17, it was the (1 2 0) apical planes that evidently occurred, based on the matching of the observed apical angle with the predicted apical angle of Figure 15. Control crystals of COM (Figure 17) also had clearly expressed (-1 0 1) and (0 1 0) faces. Some crystals showed evidence of twinning along (-1 0 1).

SEM's of COM crystals grown in the presence of citrate (Figure 18) and PC (Figure 19) revealed stabilized (-1 0 1) faces. An increase in non-specific morphological effects of PC was observed at higher doses (Figure 20). Molecular models of the adsorbates with both complete and incomplete (-1 0 1) surfaces are shown in Figures 21-24.

DISCUSSION

The unifying feature of the results was that the crystal surfaces that exhibited perpendicularity and favorable spacing of anionic lattice ions were the ones where the adsorbates exhibited stereospecific interactions. Details for each crystal are discussed.

Interaction of polyaspartate with (1 -1 0) faces of calcite

There are two energetically most favorable directions predicted for polyaspartate along the (1 -1 0) calcite face: parallel and perpendicular to the c-axis.

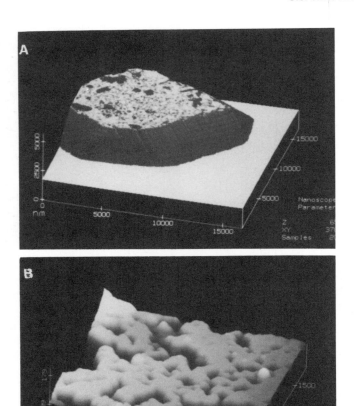

Figure 8. AFM's of oyster shell protein molecules bound to the plateau region of calcite. A. Imaged area = 16 μ x 16 μ. B. Imaged area = 1.8 μ x 1.8 μ.

Polyaspartate orientation parallel to the c-axis. Binding in this direction was predicted to occur very strongly yielding approximately -104 kcal/mole binding energy per residue (Table 1). As seen in Figure 11, the predicted binding was due to coordination of carboxylate groups of polyaspartate with calcium atoms parallel to (0 0 1). In the model, the carboxylate groups orient themselves to complete the coordination polyhedra of surface calcium ions. Carbonate ions parallel to (0 0 1) were mostly accommodated in neutral regions of peptide bonds of Asp_{15}. One carbonate ion coordinated with the terminal NH_3^+ group of the peptide. Three of the carboxylate groups per molecule of Asp_{15} were forced away from the calcite surface due to the translational mismatch between the surface and the chain as a result of protruding carbonate ions of the surface.

Polyaspartate Orientation Perpendicular to the c-Axis. Binding of Asp_{15} on (1 -1 0) in this direction was stronger yielding a binding energy of -117 kcal/mole per residue (Table 1). The proposed binding mechanism again involved coordination of carboxylate groups

Figure 9. AFM of oyster shell protein molecules bound to terraces of the trough regions at the edges of crystals as in Figure 4. Imaged area = 1 μ x 1 μ.

with rows of calcium atoms along (0 0 1). The overall fit of the polyaspartate molecule to the surface in the perpendicular direction (Figure 12) was better than in the parallel one. There was no chain strain in the perpendicular orientation due to translational mismatch between the surface lattice ions and the adsorbate: all 16 carboxylate groups per molecule of Asp_{15} interact with the surface.

It may be important to note, however, that the binding of individual carboxylate groups in the parallel rather than the perpendicular position was calculated to be stronger because the separation of adjacent rows of positive ions better matched the separation between carboxyl groups of polyaspartate in β-sheet conformation. But, due to the translational mismatch along the c-axis, approximately every fifth carboxylate was driven away from the surface, with a resulting decrease of the total binding energy. On the other hand, in

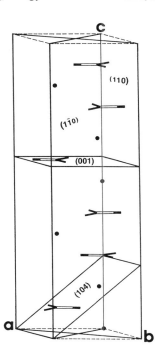

Figure 10. Calcite ($CaCO_3$) unit cell showing the basal plane (0 0 1), the cleavage plane (1 0 4) of the control crystal rhombohedrons, and the prism planes (1 1 0) and (1 -1 0).

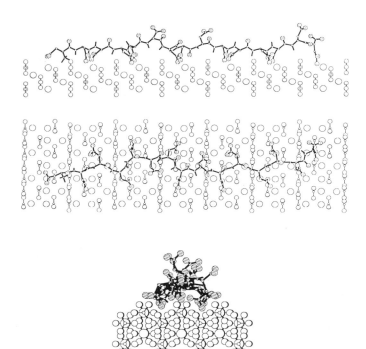

Figure 11. Computer models of Asp$_{15}$ bound to calcite, complete surface (1 -1 0), parallel to the c-axis. Shown are a full side view, top view, and axis side view.

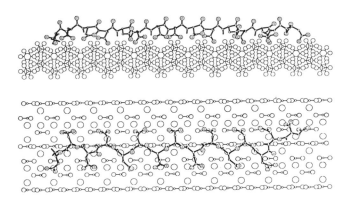

Figure 12. Computer models of Asp$_{15}$ bound to calcite, complete surface (1 -1 0), perpendicular to the c-axis. Shown are a full side view, top view, and axis side view.

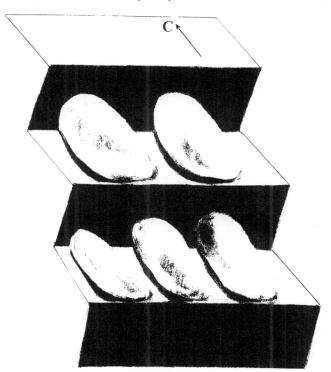

Figure 13. Schematic drawings of (1 -1 0) regions at the corners of a calcite rhombohedron (top) with protein molecules bound to terraces in these regions (bottom).

perpendicular binding, the 7 carboxylate groups along one side of the ß-sheet could easily bind to one row of calcium ions along (0 0 1). The carboxylate groups on the other side had weaker binding due to less favorable spacing to another surface row of calcium, 8.5 Å distant from the first.

Overall, the perpendicular binding of Asp_{15} on preexisting, complete (1 -1 0) surfaces was more advantageous energetically. Binding of a smaller molecule not subject to chain strain could differ from this. For example, the binding of Asp_5 was calculated as -794 kcal/mole in the parallel direction versus -773 kcal/mole in the perpendicular.

Substitution of Asp_n Carboxylate Groups Into Carbonate Lattice Positions. If the carbonate ions of the lattice that contributed to chain strain in the parallel position were replaced by carboxylate groups of Asp_{15}, the calculated binding energy in this position was greatly increased. Substitution of carboxylate for carbonate groups in the perpendicular orientation did not have a strong effect on binding (Table 1).

In a growing crystal, the surfaces would be transitional between solid and solution. Therefore, substitution of binding groups of adsorbates for lattice ions is likely and in fact has been postulated for binding of single aspartate and related molecules to (1 -1 0) surfaces of calcite, including orientations essentially parallel to the c-axis.[5,29]

Binding of Proteins to Calcite

Matrix proteins from biological minerals most likely also positioned along preferred vectors on the calcite surface, presumably with the Asp_n and other anionic domains participating in the binding. As seen in Figure 9, the oyster shell protein lined up consistently in a single orientation on calcite on a set of terraces in troughs along the edges of the crystal.

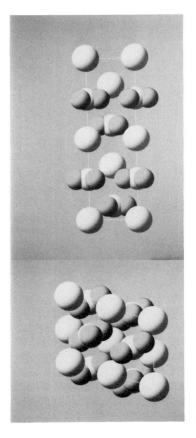

Figure 14. Computer models of (A) the (1 -1 0) and (B) the (110) surfaces of calcite.

Crystals grown in the presence of polyaspartate also exhibited this morphology. Based on the schematic model of Figure 13, the binding sites were identified as on (1 -1 0) terraces, parallel to the c-axis. Because the crystals were growing, the surfaces were presumably incomplete so that the parallel orientation was favored due to substitution of carboxylate groups of the proteins for carbonate ions of the lattice.

The binding orientation of the first adsorbate molecule apparently influenced the orientation of the rest of the adsorbates on parallel binding sites, as seen in the parallel arrays of peptides in Figure 7 on the basal plane. In this case, any binding direction along the a-axes would be equivalent because all of the calcium atoms are equidistant from each other and the underlying carbonate ions to which they are ionically bound. However, the peptides were always seen to be bound essentially parallel to each other. This result was predicted by the computer simulations of binding of Asp_{15} to (0 0 1) surfaces of calcite. The first molecule to bind influenced the alignment of the next because the C and N termini of one peptide would attract its counterpart of the other.

There also was some piling of molecules, but the binding was basically in a monolayer at the dose of peptide studied. Direct observation of monolayer binding and different classes of binding sites is of course useful in binding studies in general.

The binding of the oyster shell protein to the (1 -1 0) surfaces of completely separate terraces showed that the orientation was highly selective even for the initial adsorbate molecule in that binding on each terrace, which would be independent events, was the same. Confirmation of the identity of the surface will require atomic level resolution of the surface.

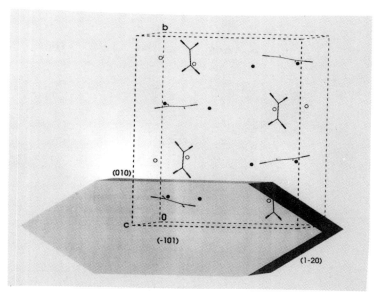

Figure 15. Computer model of the morphology of COM, showing the unit cell and the (-1 0 1), (0 1 0), and (1 2 0) surfaces.

Interaction of Polyaspartate with (1 1 0) and (1 0 4) Faces of Calcite

The (1 1 0) calcite surface has a different orientation of carbonate ions than does (1-10), with every other row of (0 0 1) carbonate ions of the (1 1 0) surface having two protruding oxygen atoms rather than one. Also, the atoms are more crowded (Figure 14).

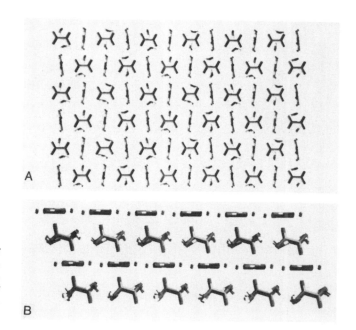

Figure 16. Computer models of (A) top view of the (-1 0 1) surface of COM and (B) the (0 1 0) surface, showing a side view of two (-1 0 1) planes that are perpendicular to (0 1 0).

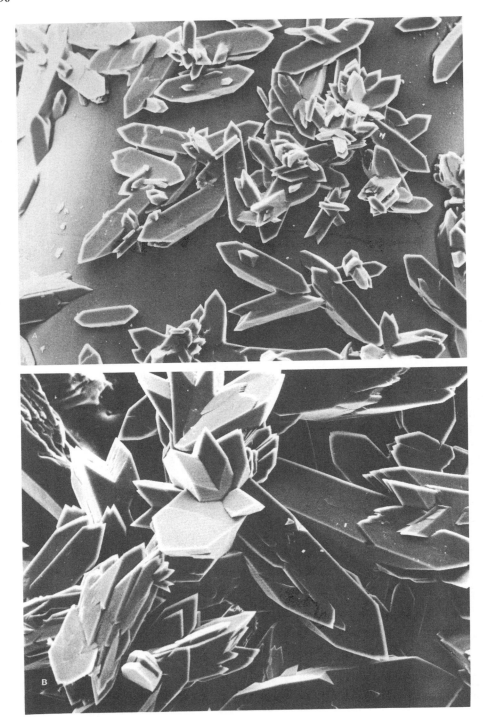

Figure 17. SEM's of COM crystals grown in the absence of inhibitors showing (A) clearly expressed (-1 0 1), (0 1 0), and (1 2 0) faces. In addition, in (B), twinning along (-1 0 1) is seen. The crystals were approximately 50 to 100 μ in length.

Table 1. Summary of molecular modeling of binding of some adsorbates to selected crystal surfaces. The "+" indicates that the adsorbate is predicted to have a specific direction on the surface and specific spatial matching of its functional groups with lattice positions of the surface. The "-" indicates that the binding is nondirectional and driven by nonspecific ionic interactions. If the crystal surfaces were modeled as perfect planes with all lattice positions occupied by lattice ions, they were termed "complete". If the surfaces had one or more lattice anions removed so that an anionic group or groups of the adsorbate could substitute into the lattice position, the surfaces were termed "incomplete". COM refers to calcium oxalate monohydrate

Molecule	Surface	Binding energies, kcal/mole		Stereospecific binding
		Complete	Incomplete	
ASP_{15}	Calcite			
	(1 -1 0) ‖ C	-1565	-4303	+,+
	(1 -1 0) ⊥ C	-1749	—	+
	(1 1 0)	-452	—	-
	(1 0 4)	-248	—	-
	(0 0 1)	-4893	—	-
Citrate	COM			
	(-1 0 1)	-546	-1071	+,+
	(0 1 0)	-108	-502	-,-
Phosphocitrate	(-1 0 1)	-717	-1424	+,+
	(0 1 0)	-142	-675	-,-
Antifreeze protein from winter flounder	Ice			
	(2 0 1)	-280	—	+

This would not permit strong, directional binding. Overall, the differences would reduce the fit of Asp_{15} such that the calculated binding energy to (1 1 0) was lowered to -30 kcal/mole per residue. The binding direction was nonspecific, as was binding to (1 0 4) cleavage faces of control crystals of calcite. This binding was reduced to -17 kcal/mole per residue (Table 1).

Such binding is likely to be reversible. In fact, repeated attempts failed to visualize polyaspartate molecules on (1 0 4) surfaces, even though binding studies with radiolabelled peptides suggested that they were there.[30,31] Such loosely bound adsorbates probably were not seen because they were dislodged by the AFM tip. For example, we have estimated that a force of the AFM probe of 10^{-9}N would be sufficient to dislodge an adsorbed species with a binding energy in the range of -700 kcal/mole.[9,32]

Interaction of Citrate and PC with the (-1 0 1) Face of COM

When COM crystals were grown with citrate and PC present, plane (-1 0 1) was more pronounced, with the length-to-width ratio of crystals significantly decreased (Figures 18,19). In the presence of even the low concentration of 0.05 mg/ml of PC, apical planes were not present at all and again (-1 0 1) faces were stabilized (Figure 19). This effect of PC has also been reported for nephrocalcin[33] a polyanionic protein of urine.

In some cases, the curvature of (-1 0 1) faces was enhanced. This indicated that there may have been some nonspecific interactions between PC and the crystal. At 0.25 mg/ml of

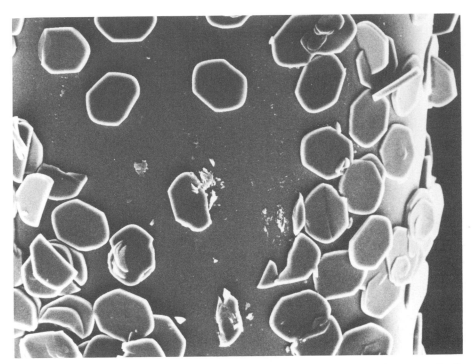

Figure 18. SEM of COM crystals grown in the presence of 1.3 mg citrate/ml. The crystals were smaller (length ~30 μ), with a significant decrease in the ratio of length to width.

PC, this effect was further enhanced, producing irregularly overgrown COM crystals (Figure 20). SEM images also revealed that PC-treated crystals were smaller.

According to the molecular model of binding to the complete (-1 0 1) surface of COM, all of the carboxylate groups of citrate interacted with the calcium ions of the surface, partially completing the calcium coordination polyhedra (Figure 21). The citrate molecule bound to a trapezoidal arrangement of calcium ions between oxalate groups. The carboxylate groups approached the surface perpendicularly, just like incoming oxalate ions would in control crystals. Since PC is bigger than citrate, it coordinated with six instead of four calcium ions (Figure 22). The binding energy was -717 kcal/mol as compared to -546 kcal/mole for citrate (Table 1). The binding-site recognition of PC matched a hexagonal arrangement of calcium ions.

As mentioned above for calcite, during COM crystal growth, inhibitor molecules may not only interact with complete crystal surfaces but also they are likely to substitute their own functional groups into lattice positions of the forming crystal. To investigate this, one oxalate group was removed from the (-1 0 1) surface and the energy minimization procedure was repeated for the inhibitor-surface system. The shape and the charge distribution of the inhibitor was even more important here since the inhibitor had to fit into the space normally occupied by the oxalate ion and to coordinate with neighboring calcium ions. The citrate molecule fit very well into the oxalate ion position. All three carboxylate groups were involved in binding, with the binding energy increased significantly to -1071 kcal/mol (Table 1). All six negatively charged oxygens of the carboxylate groups were less than 2.6 Å from the neighboring calcium ions (Figure 23).

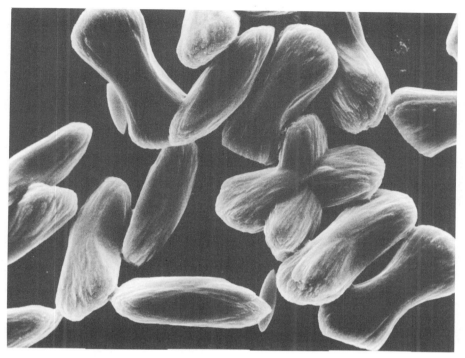

Figure 19. SEM of COM crystals grown in the presence of 0.05 mg phosphocitrate/ml. The crystals were 20 to 30 μ in length.

Binding was improved by replacing the hydroxyl group of citrate with the relatively small and negatively charged PO_4 group, resulting in better coordination with calcium ions (Figure 24). The optimal binding geometry of PC to (-1 0 1) incomplete planes of COM involved the coordination of the phosphate group with four calcium ions (Figure 22), with all of the oxygens of phosphate less than 2.7 Å away from the closest calcium ions. Moreover, the β carboxyl group was coordinated with three calcium ions and the binding energy was lowered to -1424 kcal/mol (Table 1). Analysis of PC binding to the (-1 0 1) surface indicated a high stereospecificity of inhibitor-substrate recognition and suggested why PC was a better inhibitor than citrate.

Interaction of Citrate and PC with (0 1 0) Face of COM

Citrate and PC binding to (0 1 0), another calcium rich plane of COM, was also compared. Again both complete and incomplete surfaces were considered (Table 1). Interactions of both citrate and PC with complete (0 1 0) surfaces were not stereospecific and the binding energies were much less favorable, -108 and -142 kcal/mols respectively. There was no specific inhibitor-surface recognition which instead was based on the overall attraction between rows of positively charged calcium ions and the negatively charged inhibitor molecule. Citrate could almost fit into the space created by oxalate ion removal. However, the calcium ion distribution did not match the distribution of the functional groups of citrate. Binding of PC to the incomplete (0 1 0) surface was also nonspecific and based predominantly on electrostatic interaction between the surface cations and the PO_4 group that was

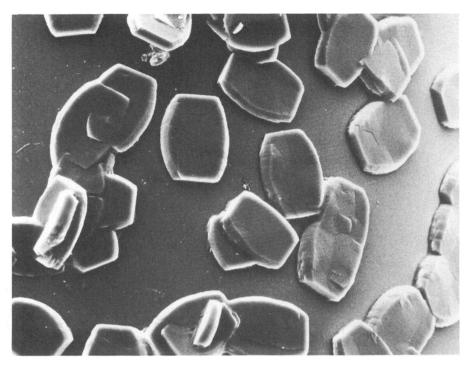

Figure 20. SEM of COM crystals grown in the presence of 0.25 mg phosphocitrate/ml. The crystals were ~20 μ in length.

drawn deep into the surface. Binding energies in this case were -675 kcal/mol and -502 kcal/mol for PC and citrate respectively (Table 1).

Polyanions, Polycations, Basal Planes

Commercial scale inhibitors are anionic rather than cationic, as anionic materials inhibit crystallization and cationic ones do not, at least not very well. For example, in the case of calcite formation, polyanions such as polyaspartate and polyacrylate are potent inhibitors but polycations like polylysine are not.[15,31,34] This observation is relevant to the occurrence of exposed basal (0 0 1) planes of calcite (Figs. 3-6). The nucleation on a surface of crystals on the basal plane with vertical growth along the c-axis has been observed previously for calcite[1,3] and in fact is a common occurrence in biomineralization.[35,36] Exposed basal planes of the biominerals calcite[37] and aragonite[38] as well as other minerals such as molybdenum sulfide[39] have been shown previously as well by AFM at the atomic level.

In calcite, the basal plane is the only surface that exhibits perfect hexagonal morphology. The basal plane may be either a homoplane of calcium or a homoplane of carbonate ions, both of which have hexagonal symmetry as observed in Figure 6. The observed atoms were interpreted as calcium cations because the polyanion, Asp_{20}, bound to them and they appeared to be single atoms rather than the 4-atom molecules of carbonate anions. It may be that no carbonate planes, to which a polycation would surely bind, were exposed under the conditions of crystal growth studied.

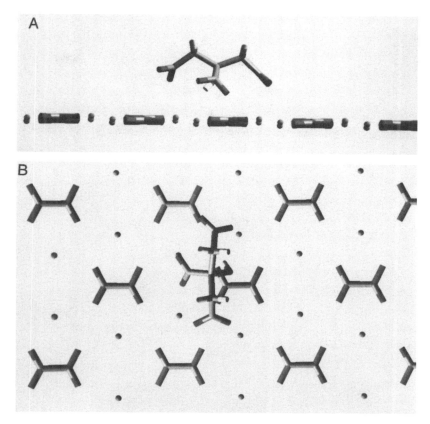

Figure 21. Computer models of citrate interacting with complete (-1 0 1) surface of COM: (A) Citrate above the surface (side view), (B) Citrate bound to the surface (top view).

Design of Scale Inhibitors

To be effective at low doses and to have stereochemically specific interactions with crystal surfaces, an inhibitor should have a combination of properties. Principal among these are strong-enough binding energies, which reflect the affinity of the inhibitor for the crystal, and proper spacing of the functional groups that interact with the lattice ions of the crystal surfaces. For many surfaces of common scales, it seems that anionic functional groups are more effective. These may be supplemented by some hydrophobic character of an adsorbate, which while not necessarily improving the affinity for the surface, may disrupt the approach of lattice ions to the surface, thereby hindering crystal growth.[40,41,42]

It is instructive to note that stereochemically active inhibitors may or may not have particularly strong binding energies. For example, both polyaspartate and phosphocitrate can interact with specific crystal surfaces with binding energies well into the (-)1000's of kcal/mole. On the other hand, proteins that stereochemically inhibit formation of ice crystals[43,44,45] have binding energies of about -300 kcal/mole (Table 1). This is more in the range of binding energies calculated for non-stereospecific interactions of polyaspartate, citrate, and PC with calcite or COM, which are accompanied by nonspecific crystal morphologies such as roughened, porous and often rounded cleavage surfaces. If the binding

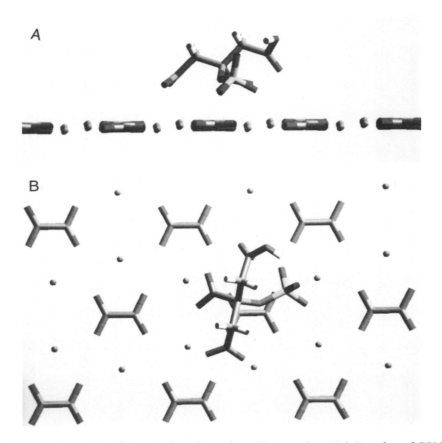

Figure 22. Computer models of phosphocitrate interacting with a complete (-1 0 1) surface of COM: (A) Phosphocitrate above the surface (side view), (B) phosphocitrate bound to the surface (top view).

energy is low (less negative) for a stereospecific inhibitor, the expression of a stereochemical effect on morphology of crystals will require higher doses.

This raises the question of cost of adsorbates that might be used for commercial applications. For many uses, this can be a determining factor in the choice and therefore the design of an inhibitor. Current uses, for example, dictate that an inhibitor must be available for around one dollar per pound for large scale applications such as detergent additives, and up to several dollars per pound for some common applications in water treatment. It's easy to see how this constraint can affect the options for designer antiscalants.

An inhibitor that lacks high affinity and stereospecific binding still may be an effective antiscalant. Lower-affinity, weaker binding would lead to nonspecific inhibition and produce greater morphological variability, including porous and other high-surface-area crystals, as mentioned above. This may be sufficient for many uses.

However, another factor to consider is that stereospecific inclusion of adsorbates into crystals may create composite materials of superior strength by blocking propagation of fractures along cleavage planes.[35,46,47,48,49] It may also be possible to regulate nucleation and growth of crystals by use of tailored adsorbates to produce specific shapes such as hexagonal tablets, which, for example, may pack into denser, stronger composites. Therefore, the design

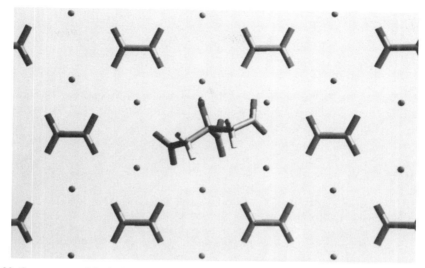

Figure 23. Computer model of citrate bound to an incomplete (-1 0 1) surface of COM (one oxalate ion removed from the surface).

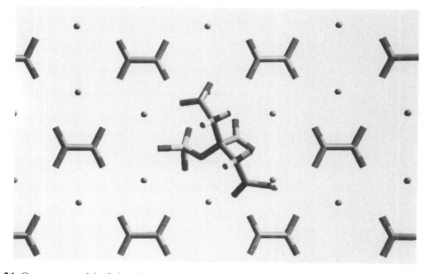

Figure 24. Computer model of phosphocitrate bound to an incomplete (-1 0 1) surface of COM (one oxalate ion removed from the surface).

of stereospecific inhibitors/nucleators of crystallization is likely to be very useful, particularly in special applications.

ACKNOWLEDGMENTS

We thank J. Sallis for preparing COM crystals and SEM's of them. This work was supported by grants from the National Science Foundation (grant EHR-9108761) and the Research Corporation (grant C-3662).

REFERENCES

1. Addadi, L., Weiner, S., 1985, Interactions between acidic proteins and crystals: stereochemical requirements in biomineralization, *Proc. Nat. Acad. Sci. USA* 82:4110-4114.
2. Addadi, L., and Weiner, S., 1986, Interactions between acidic macromolecules and structured crystal surfaces. Stereochemistry and biomineralization, *Mol. Cryst. Liq. Cryst.* 134:305-322.
3. Addadi, L., Moradian, J., Shay, E., Maroudas, H.G., and Weiner, S., 1987, A chemical model for the cooperation of sulfates and carboxylates in calcite crystal nucleation: relevance to biomineralization, *Proc. Nat. Acad. Sci. USA* 84:2732-2736.
4. Mann, S., Heywood, B.R., Rajam, S., and Birchall, J.D., 1988, Controlled crystallization of CaCO$_3$ under stearic acid monolayers, *Nature* 334:692-695.
5. Mann, S., Didymus, J.M., Sanderson, N.P., Heywood, B.R., and Samper, E.J.A., 1990, Morphological influence of functionalized and non-functionalized a, w-dicarboxylates on calcite crystallization, *J. Chem. Soc. Faraday Trans.* 86:1873-1880.
6. Mann, S., and Sparks, N.H.C., 1988, Single crystalline nature of coccolith elements of the marine alga *Emiliania huxleyi* as determined by electron diffraction and high-resolution transmission electron microscopy, *Proc. R. Soc. Lond. B.* 234:441-453.
7. Arnott, H.J., 1982, Three systems of biomineralization in plants with comments on the associated organic matrix, In: Nancollas, G.H. (ed), Biological mineralization and demineralization, Springer-Verlag, New York, 199-218.
8. Richardson, C.F., Johnsson, M., Bangash, F.K., Sharma, V.K., Sallis, J.D., and Nancollas, G.H., 1990, The effects of citrate and phosphocitrate on the kinetics of mineralization of calcium oxalate monohydrate, *Mat. Res. Soc. Symp. Proc.* 174:87-92.
9. Wierzbicki, A., Sikes, C.S., Madura, J.D., and Drake, B. 1994, Atomic force microscopy and molecular modeling of protein and peptide binding to calcite, *Calcified Tissue International* 54:133-141.
10. Wierzbicki, A., Sikes, C.S., Sallis, J.D., Madura, J.D., Stevens, E.D., Martin, K.L., 1995, Scanning Electron Microscopy and Molecular Modeling of Inhibition of Calcium Oxalate Monohydrate Crystal Growth by Citrate and Phosphocitrate, *Calcified Tissue International,* 56:297-304.
11. Sikes, C.S., Wierzbicki, A., and Fabry, V., 1994, From atomic to global scales in biomineralization, *Bulletin de l'Institut Oceanographique*, Monaco, n. spécial 14:1-14.
12. Sikes, C.S., Mueller, E.M., Madura, J.D., Drake, B., and Little, B.J., 1993, Polyamino acids as antiscalants, corrosion inhibitors, and dispersants: atomic force microscopy and mechanisms of action, *Corrosion 93*, paper 465:1-21.
13. Sikes, C.S., and Wierzbicki, A., 1994, Mechanistic studies of polyamino acids as antiscalants, *Corrosion 94*, paper 193:1-19.
14. Sikes, C.S., Wierzbicki, A., and Sallis, J.D., 1995b, Molecular mechanisms of control of formation of calcium oxalate, *Corrosion 95*, paper 473:1-15.
15. Wheeler, A.P., and Sikes, C.S., 1989, Matrix-crystal interactions in CaCO$_3$ biomineralization, In: Mann, S., Webb, J., and Williams, R.J.P. (eds), Biomineralization: chemical and biochemical perspectives, VCH, Weinheim, p. 95.
16. Sutor, D.J., 1969, Growth studies of calcium oxalate in the presence of various ions and compounds, *Brit. J. Urol.* 41:171-178.
17. Sallis, J.D., and Lumley, M.F., 1979, On the possible role of glycosaminoglycans as natural inhibitors of calcium oxalate stones, *Invest. Urol.* 16:296-299.

18. Rusenko, K.W., Donachy, J.E., and Wheeler, A.P., 1991, Purification and characterization of a shell matrix phosphoprotein from the American Oyster, In: Sikes, C.S., and Wheeler, A.P. (eds), Surface reactive peptides and polymers: discovery and commercialization, ACS Books, Washington, p. 107.

19. Sikes, C.S., 1991, Inhibition of mineral deposition by phosphorylated and related polyanionic peptides, U.S. Patent 5,051,401.

20. Sikes, C.S., Yeung, M.L., and Wheeler, A.P., 1991, Inhibition of calcium carbonate and phosphate crystallization by peptides enriched in aspartic acid and phosphoserine, In: Sikes, C.S., and Wheeler, A.P. (eds), Surface reactive peptides and polymers: discovery and commercialization, ACS Books, Washington, p. 50.

21. Pankowski, A.H., Meehan, J.D., and Sallis, J.D., 1994, Synthesis via a cyclic dioxatrichloro-phosphorane of 1,3-dibenzyl-2-phosphonoxy citrate, Tetrahedron Letters 35:927-930.

22. Glusker, J.P., 1980, Citrate conformation and chelation: enzymatic implications, *Acc. Chem. Res.* 13:345-352.

23. Hehre, W.J., Radom, L., Schleyer, V.R., and Pople, J.A., 1986, *Ab initio* molecular orbital theory, Wiley and Sons, New York.

24. Dewar, M.J.S., Thiel, W., 1977, Ground states of molecules. 38. The MNDO method. Approximations and Parameters, *JACS* 99:4899-4906.

25. Hirst, D.M., 1990, A computational approach to chemistry, Blackwell Scientific, Boston.

26. Chirlian, L.E., and Francl, M.M., 1987, Atomic charges derived from electrostatic potentials: a detailed study, *J. Computational Chem.* 8:894-905.

27. Breneman, C.M., and Wiberg, K.B., 1990, Determining atom-centered monopoles from molecular electrostatic potentials: the need for high sampling density in formamide conformational analysis, *J. Computational Chem.* 11:361-373.

28. Thundat, T., Zheng, X.Y., Sharp, S.L., Allison, D.P., Warmack, R.J., Joy, D.C., and Ferrell, T.L., 1992, Calibration of atomic force microscope tips using biomolecules, *Scanning Microscopy* 6:903-910.

29. Didymus, J.M., Oliver, P., Mann, S., DeVries, A.L., Hauschka, P.V., and Westbroek, P., 1993, Influence of low-molecular-weight and macromolecular organic additives on the morphology of calcium carbonate, *Journal of Chemical Society Faraday Trans.* 89(15):2891-2900.

30. Wheeler, A.P., Low, K.C., and Sikes, C.S., 1991, CaCO$_3$ crystal-binding properties of peptides and their influence on crystal growth, In: Sikes, C.S., and Wheeler, A.P. (eds), Surface reactive peptides and polymers: discovery and commercialization, ACS Books, Washington, p. 72.

31. Mueller, E.M., Sikes, C.S., 1993, Adsorption and modification of calcium salt crystal growth by anionic peptides and spermine, *Calcif. Tiss. Int.* 52:34-41.

32. Frommer, J., 1992, Scanning tunneling microscopy and atomic force microscopy in organic chemistry, *Angew. Chem. Int. Ed. Engl.* 31:1298-1328.

33. Deganello, S., 1991, Interaction between nephrocalcin and calcium oxalate monohydrate: a structural study, *Calcified Tissue International* 48:421-428.

34. Sikes, C.S., and Wheeler, A.P., 1985, Inhibition of inorganic or biological CaCO$_3$ deposition by polyamino acid derivatives, U.S. Patent 4,534,881.

35. Addadi, L., and Weiner, S., 1992, Control and design principles in biological mineralization, *Angew. Chem. Int. Ed. Engl.* 31:153-169.

36. Lowenstam, H., and Weiner, S., 1989, On biomineralization, Oxford Press, New York.

37. Drake, B., Hellman, R., Sikes, C.S., and Occelli, M.L., 1992, Atomic scale imaging of albite feldspar, calcium carbonate, rectorite, and bentonite using atomic force microscopy, *SPIE Proceedings* 1639:151-159.

38. Friedbacher, G., Hansma, P.K., Ramli, E., and Stucky, G.D., 1991, Imaging powders with the atomic force microscope: from biominerals to commercial materials, *Science* 253:1261-1263.

39. Kim, Y., and Lieber, C.M., 1992, Machining oxide thin films with an atomic force microscope: pattern and object formation on the nanometer scale, *Science*, 257:375-377.

40. Sikes, C.S., and Wheeler, A.P., 1988, Regulators of biomineralization. *CHEMTECH* 18:620-626.

41. Sikes, C.S. and Wheeler, A.P., 1989, Inhibition of mineral deposition by polyanionic/hydrophobic peptides and derivatives thereof having a clustered block copolymer structure, U.S. Patent 4,868,287. 20 pgs.

42. Garris, J., Sikes, C.S., 1993, Use of polyamino acid analogs of biomineral proteins in dispersion of inorganic particulates important to water treatment. *Colloids and Surfaces* 80:103-112.

43. Knight, C.A., Cheng, C.C., and DeVries, A.L., 1991, Adsorption of a-helical antifreeze peptides on specific ice crystal surface planes, *Biophysical Journal* 59:409-418.

44. Wen, D., and Laursen, R.A., 1992, A model for binding of an antifreeze polypeptide to ice, *Biophys. J.* 63:1659-1662.

45. Madura, J.D., Wierzbicki, A., Harrington, J.P., Maughon, R.H., Raymond, J.A., and Sikes, C.S., 1994, Interactions of the D- and L- forms of winter flounder antifreeze peptide with the (2 0 1) planes of ice, *JACS* 116:17-418.
46. Berman, A., Addadi, L., and Weiner, S., 1988, Interactions of sea-urchin skeleton macromolecules with growing calcite crystals - a study of intracrystalline proteins. *Nature* 331:546-548.
47. Berman, A., Addadi, L., Kvick, A., Leiserowitz, L., Nelson, M., and Weiner, S., 1990, Intercalation of sea urchin proteins in calcite: study of a crystalline composite material, *Science* 250:664-667.
48. Berman, A., Hanson, J., Leiserowitz, L., Koetzle, T.F., Weiner, S., and Addadi, L., 1993, Biological control of crystal texture: a widespread strategy for adapting crystal properties to function, *Science* 259:776-779.
49. Mann, S., 1993, Biomineralization: the hard part of bioinorganic chemistry!, *J. Chem. Soc. Dalton Trans.* 1993:1-9.

EVALUATION OF PHOSPHONO-, HYDROXYPHOSPHONO-, α- HYDROXY CARBOXYLIC AND POLYCARBOXYLIC ACIDS AS CALCIUM PHOSPHATE DIHYDRATE CRYSTAL GROWTH INHIBITORS

Zahid Amjad

Advanced Technology Group
The BFGoodrich Company
9921 Brecksville Road, Brecksville, Ohio 44141

INTRODUCTION

It is generally agreed that during the precipitation process at ambient temperature, different calcium phosphate phases, arranged in order of increasing solubility, hydroxyapatite ($Ca_5(PO_4)_3OH$, HAP); tricalcium phosphate ($Ca_3(PO_4)_2$, TCP); octacalcium phosphate ($Ca_8H_2(PO_4)6 \bullet 5H_2O$, OCP) and dicalcium phosphate dihydrate ($CaHPO_4 \bullet 2H_2O$, DCPD) may be formed depending upon the pH, ionic strength and level of supersaturation. Results of previous studies on the precipitation of calcium phosphates at physiological conditions indicate that the kinetically favored precursor phases such as DCPD, OCP, and TCP are formed prior to the formation of the thermodynamically stable HAP phase.[1-4] The crystal growth and inhibition of calcium phosphates has recently received considerable attention due to their importance in biological calcification process such as the formation of teeth and bone, the initiation of renal calculi, and in industrial water systems where deposition of calcium phosphates on heat exchanger surfaces results in decreased system efficiency.

The inhibition of calcium phosphates is normally accomplished by using the so-called "crystal growth inhibitors". Several studies have reported that phosphorous containing compounds such as pyrophosphate, tripolyphosphate, hexametaphosphate, and phosphonate inhibit crystal growth of DCPD and HAP by blocking active growth sites via adsorption.[5,6,7,8] This surface adsorption mechanism has been reported to reduce crystal growth of other sparingly soluble salts such as CaF_2, $CaSO_4 \bullet 2H_2O$, and $CaCO_3$.[9,10,11]

Recently, several studies have been reported on the influence of di- , tri- , and polycarboxylic acids on the crystal growth of calcium phosphates, calcium carbonate, and barium sulfate. Results of these studies indicate that, compared to citric, tricarballylic and benzene hexacarboxylic acids, poly (acrylic acid) greatly inhibits the rate of crystallization of DCPD, HAP, CaCO$_3$, and CaF$_2$.[12-15] A surface adsorption mechanism involving a simple Langmuir adsorption model has been proposed to account for the crystal growth inhibition of sparingly soluble salts in the presence of inhibitors. Nancollas et al.[16] investigated the influence of tricarboxylic acids on calcium phosphate precipitation. Comparison of the inhibitory effect of citric, isocitric, and tricarballylic acid suggests that the hydroxyl group in the molecular backbone plays a key role in the effectiveness of these tricarboxylate ions as inhibitors.

Polymeric inhibitors such as poly (acrylic acid), poly (maleic acid) and acrylic acid - based copolymers have been investigated as possible alternatives to phosphorus containing compounds.[17-20] It has been shown that polymers which have an effect on inhibiting the precipitation of sparingly soluble salts are highly substituted with carboxyl groups. In addition, it has been shown that one of the factors determining the effectiveness of a polymer is its molecular weight (M W). Amjad[21] in studies on the influence of poly (acrylic acid), PAA, of varying M W in controlling calcium sulfate scale formation on heat exchanger surfaces and on the precipitation of calcium phosphate and calcium fluoride from aqueous

Table 1. Structure of inhibitors evaluated

Inhibitor	Structure	Acronym	Mol. Wt.
hydroxyphosphono acetic acid	OH $\mid$ HC—COOH $\mid$ PO$_3$H$_2$	HPA	156
2-phosphono butane 1,2,4-tricarboxylic acid	PO$_3$H$_2$ $\mid$ CH$_2$ — C — CH$_2$ — CH$_2$ $\mid\quad\mid\qquad\quad\mid$ COOH COOH COOH	PBTC	270
glycolic acid	OH $\mid$ H$_2$C — COOH	GLA	76
malonic acid	COOH $\mid$ H$_2$C $\mid$ COOH	MNA	104
malic acid	OH $\mid$ HC— COOH $\mid$ H$_2$C— COOH	MLA	134
mandelic acid	OH $\mid$ ⬡—CH — COOH	MDA	152
poly (acrylic acid)	—(CH$_2$ — CH)$_n$— $\mid$ COOH	PAA	1500

solutions reported an optimum effectiveness at ~ 2000 M W. On the basis of the results for the similar M W, Smith and Alexander[22] reported that PAA was more effective than poly (methacrylic acid), in controlling the precipitation of the calcium sulfate from aqueous solutions.

In this paper we present results on the effect of a variety of inhibitors at constant supersaturation upon the kinetics of crystal growth of DCPD. The conditions were chosen so that no other calcium phosphate could be formed. The influence of inhibitors on the kinetics of crystal growth process was monitored by a constant composition technique in which the solution composition is kept constant, by maintaining the species concentration in solutions. Under these conditions, the rate of crystal growth of DCPD in the presence of inhibitors could be accurately measured. The inhibitors evaluated include: hydroxyphosphono acetic acid, 2- phosphonobutane 1,2,4- tricarboxylic acid, α- hydoxy carboxylic acids (i.e., glycolic, malic, malonic, and mandelic) and poly (acrylic acid), MW 1500. The structures of these inhibitors are shown in Table 1.

EXPERIMENTAL

Reagent grade chemicals and carbon dioxide free distilled water were used in the preparation of solutions. Potassium hydroxide solutions were prepared from dilut-it reagents which were standardized by titration with potassium hydrogen phthalate using phenolphthalein as indicator. Phosphate solutions were standardized spectrophotometrically at 420 nm as the molybdovanadium phosphate complex as well as by titration against standard potassium hydroxide solutions. Calcium chloride solutions were analyzed by atomic absorption spectroscopy and by ion exchange. DCPD crystals were prepared and characterized as described previously.[5] The specific surface as determined by B.E.T. method was found to be 36 m^2/g.

Crystal growth experiments were made in a double-walled Pyrex cell at 37°C using the constant composition technique. The stable supersaturated solutions of calcium phosphate with a molar ratio of total calcium, T_{Ca}, to total phosphate, T_p = 1.00, were prepared by adjusting the pH of the premixed solution of calcium chloride and potassium hydrogen phosphate to a value of 6.00 by slow addition of 0.10 M potassium hydroxide. Hydrogen ion measurements were made with a glass / Ag - AgCl electrode pair equilibrated at 37°C. The electrode pair was standardized before and after each experiment using NBS standard buffer solutions. The solutions were continuously stirred (~ 300 rpm) while nitrogen gas, presaturated with water at 37°C, was bubbled through the solution to exclude carbon dioxide.

Following the introduction of DCPD seed crystals into the supersaturated solutions, crystallization began immediately without any induction period. The crystal growth reaction was monitored by the addition of titrant solutions from mechanically coupled automatic burets mounted on a modified pH stat (pH meter, model 632; dosimat, model 655; impulsomat, model 614; Brinkmann Instruments, Westbury, N.Y.). The titrant solutions in the burets consisted of calcium chloride, potassium phosphate, potassium hydroxide, and inhibitor. The molar concentration ratio of the titrants corresponded to the stoichiometry of DCPD. Potassium chloride was added to the calcium phosphate supersaturated solutions in order to maintain the ionic strength to ± 1%. The constancy of solution composition was verified by analyzing the filtered samples which were withdrawn at various time intervals, for total calcium and total phosphate by a combined spectrophotometric method.[5] The rates of crystallization were determined from the rates of addition of titrant, and corrected for surface area changes.[5]

Table 2. Crystallization of DCPD on DCPD seed
crystals in the presence of inhibitors[a]

Exp.	Inhibitor	Conc. (10^6 M)	10^4 rate, mole DCPD m^{-2} min^{-1}
1	none	0.0	1.52
2	none	0.0	1.61
3	none	0.0	1.01 [b]
4	none	0.0	2.82 [c]
5	HPA	0.05	1.07
6	HPA	0.080	0.863
7	HPA	0.125	0.705
8	HPA	0.250	0.475
9	HPA	0.250	0.792 [d]
10	HPA	0.500	0.238
11	HPA	2.50	0.125
12	PBTC	0.050	0.880
13	PBTC	0.125	0.611
14	PBTC	0.250	0.325
15	PBTC	1.00	0.082
16	PA	0.250	0.052
17	HEDP	0.250	0.295
18	GLA	125.0	1.14
19	GLA	250.0	0.938
20	GLA	333.0	0.763
21	GLA	500.0	0.588
22	MLA	125.0	0.82
23	MNA	125.0	0.92
24	MDA	125.0	1.04
25	MA	0.125	0.44
26	PAA	0.125	<< 0.1

DCPD seed (mg): [a]13, [b]8.5, [c]23.5, [d]22

RESULTS AND DISCUSSION

The experimental conditions used in this study are summarized in Table 2. Typical plots of moles of DCPD grown on DCPD seed crystals as a function of time, are illustrated in Figure 1. It can be seen that crystal growth began immediately upon the addition of seed crystals. Changes in surface area were calculated from these data, assuming the crystallites to be perfect spheres and the ordinate values, normalized to initial seed surface area, are also shown as corrected curves in Figure 1. The slopes of the lines are used to calculate the growth rates expressed as moles of DCPD grown per square meter of surface in Table 2. Experiments performed both in the presence and absence of inhibitors (exp. 2, 4, 8, 9 Table 2) at different seed concentrations showed that crystallization took place on the seed crystals exclusively.

The influence of HPA (hydroxyphosphono acetic acid) on the crystal growth of DCPD from calcium phosphate supersaturated solutions was studied by a series of experiments summarized in Table 2. The analytical data of a typical crystal growth experiment in the presence of HPA (8.0×10^{-8} M, exp. 6) are summarized in Table 3. It can be seen that the stoichiometry of the precipitating phase (Table 3) was constant with a calcium to phosphate molar ratio of 1.00 ± 0.02 for more than 100 minutes of reaction. The amount of newly grown DCPD phase was more than 300 % of the original seed materials. The results of experiments to test the effect of HPA concentration on DCPD crystal growth are shown in Figure 2 and rate data summarized in Table 2.

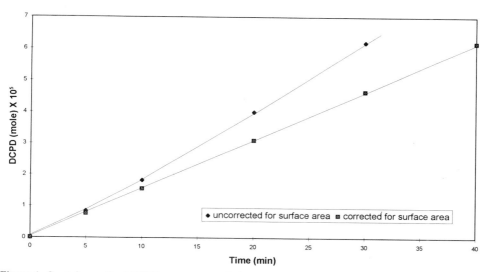

Figure 1. Crystal growth of DCPD at constant solution composition. Plots of DCPD (mole) crystallized as function of time. Dotted line correspond to experiment 2, uncorrected for surface area changes.

As illustrated in Figure 2, the crystallization rates of DCPD are highly sensitive to the concentrations of HPA in solution. Interestingly, a HPA concentration as low as 0.050×10^{-6} M significantly reduces the crystal growth rate of DCPD. Figure 2 further shows that, at 0.125×10^{-6} M concentration, the growth rate is reduced by ~ 50 % and at 2.50×10^{-6} M HPA concentration, the growth reaction is completely inhibited.

The results of the experiments to evaluate the performance of PBTC (2- phosphono butane 1,2,4- tricarboxylic acid) as a DCPD crystal growth inhibitor are summarized in Table 2. Plots of moles of newly grown DCPD as a function of time for crystal growth experiments in the presence of varying concentrations of PBTC are shown in Figure 3. It can been be seen that the presence of low concentrations of PBTC markedly reduces the rate of DCPD crystallization. Figure 4 illustrates the plots of newly grown DCPD on DCPD seed crystals as function of time in the presence of 2.50×10^{-7} M concentration of each

Table 3. Crystallization of DCPD on DCPD seed crystals in the presence of hydroxyphosphonic acid[a]

Time (min)	TCa (mM)	TP (mM)	Extent of crystallization (as % original seed)
0	4.160	4.160	0
15	4.181	4.172	24
30	4.166	4.186	55
40	4.171	4.162	80
60	4.162	4.186	140
80	4.176	4.158	220
105	4.164	4.185	340

[a]Hydroxyphosphonic acid = 8.00×10^{-8} M, DCPD seed = 13 mg/ 150 ml

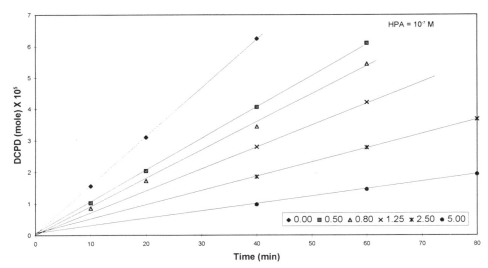

Figure 2. Crystal growth of DCPD at constant supersaturation. Plots of DCPD (mole) crystallized as a function of time in the presence of varying concentrations of HPA.

phosphonates (i.e., HPA, PBTC, and HEDP, hydroxy ethylidine 1,1, diphosphonic acid) and PA, phytic acid, present initially in the supersaturated solutions.

The kinetic data summarized in Table 2 (exp. 8, 14, 16, 17) clearly show that PBTC exhibits stronger inhibitory activity than HPA in reducing the crystal growth of DCPD. The overall order, among phosphorus containing inhibitors, in terms of decreasing effectiveness is: PA > HEDP > PBTC > HPA.

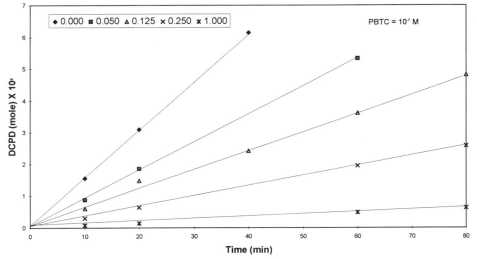

Figure 3. Crystal growth of DCPD at constant supersaturation. Plots of DCPD crystallized as a function of time in the presence of varying concentrations of PBTC.

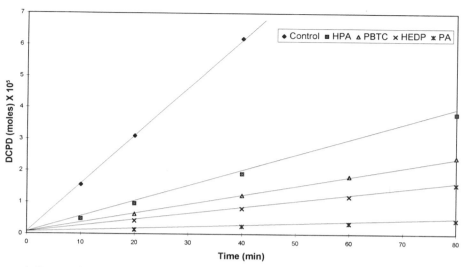

Figure 4. Comparative effect of phosphorus containing inhibitors on crystal growth of DCPD at constant supersaturation. Plots of DCPD (moles) crystallized as a function of time in the presence of 2.500×10^{-7} M phosphonate and phytate.

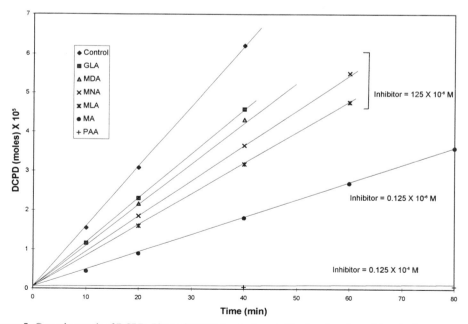

Figure 5. Crystal growth of DCPD. Plots of DCPD (moles) crystallized as function of time in the presence of glycolic acid.

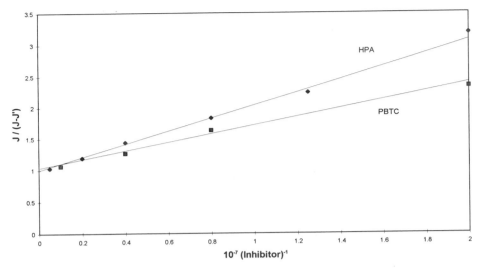

Figure 6. Comparative effect of inhibitors on crystal growth of DCPD. Plots of DCPD (moles) crystallized as a function of time in the presence of polycarboxylic acids.

The effect of α- hydroxy carboxylic acids (i.e., glycolic acid, GLA; malic acid, MLA; malonic acid, MNA; mandelic acid, MDA, etc.) and low MW PAA as crystal growth inhibitors are summarized in Table 2. Figure 5 details the profiles of moles of DCPD grown per unit area of DCPD seed crystals as a function of time in the presence of varying concentrations of glycolic acid. Figure 6 shows a comparison of various carboxylic acids at 2.50×10^{-4} M of each of these inhibitors. As can be seen among the two dicarboxylic acids

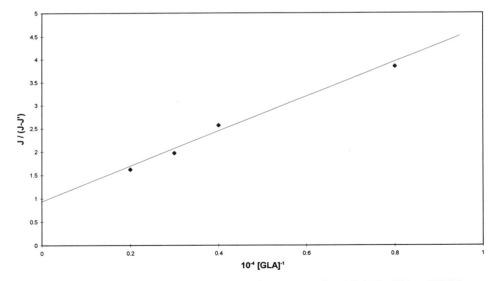

Figure 7. Langmuir adsorption isotherm. Plots of J / (J - J′) against 1/ inh) for HPA and PBTC.

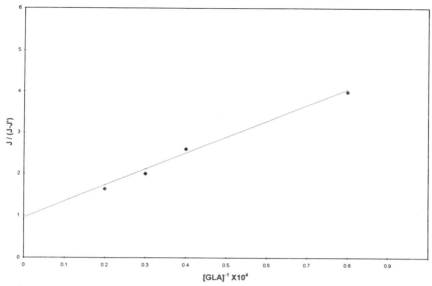

Figure 8. Langmuir adsorption isotherm. The influence of glycolic acid on the crystal growth of DCPD.

(i.e., MLA and MNA) the acid containing a hydroxyl group (i.e., MLA) is more effective as DCPD crystal growth inhibitor.

It is noteworthy that PAA which does not contain any hydroxyl group shows the best performance as a DCPD crystal growth inhibitor. On the basis of kinetic data the order in terms of decreasing effectiveness is as follows: PAA >> MLA > MNA > GLA > MDA > control (no inhibitor).

The crystal growth inhibition experiments using DCPD seed crystals were all made at constant supersaturation with respect to DCPD. Therefore, the reduction in the rate of crystallization of DCPD in the presence of inhibitors could be attributed to inhibitor adsorption at active growth sites. Since the rate of crystal growth in the presence of inhibitors appeared to depend on the surface coverage, and assuming that adsorption equilibrium are fast, especially when compared to the crystal growth process, the relationship between the rates of crystal growth in the absence, J, and in the presence, J', of inhibitor is given by:

$$J / (J - J') = 1 + k / (Inh)$$ (1)

where k is the "affinity constant" and (Inh) is the concentration of inhibitor. Plots of J / (J – J') against 1/(Inh) for phosphonates and glycolic acid are shown in Figures 7 and 8, respectively. The excellent linearity suggests that the inhibitory effect is due to adsorption at active growth sites. The best - fit linear relation and the intercept (within experimental uncertainty) of unity strongly suggests that the mechanism of inhibition is the same as proposed for the Langmuir adsorption isotherm, namely, the formation of a monolayer of inhibitor ion at growth sites on the crystal face that blocks further growth.

The values of the "affinity constant", k, x 10^5 as calculated from Figure 7 for HPA and PBTC are 50 and 80, respectively, compared to 312 and 170 reported for AMP (amino tris methylene phosphonic acid) and HEDP. Also, summarized in Table 4 are the k values for various polyphosphates. As shown in Table 4 both HPA and PBTC are weaker inhibitors compared to hexametaphosphate and phytic acid. Among the phosphonates, AMP and HEDP which have 3 functional groups (phosphono and hydroxy + phosphono) are approximately

Table 4. Affinity constants for various inhibitors
of DCPD crystal growth at 37°C

Inhibitor	k , 10^5	log K_{ML}
Pyrophosphate, PYP	21.4[a]	3.2
Hexametaphosphate, HMP	210[a]	7.6
Phytic acid, PA	410[a]	3.2
Mellitic acid, MA	11[b]	3.5
Hydroxyethylidine 1,1- diphosphonic acid, HEDP	170[b]	7.1
Amino tris (methylene phosphonic acid), AMP	312	6.9
Hydroxyphosphono acetic acid, HPA	50[c]	3.7
2-Phosphonobutane 1,2,4 - tri carboxylic acid, PBTC	80[c]	4.9
Glycolic acid, GLA	0.0267[c]	1.6
Malic acid, MLA	0.071[c]	2.7
Mandelic acid, MDA	0.0250[c]	1.3
Malonic acid, MNA	0.055[c]	2.5

[a]Reference 5, [b]Reference 13, [c]This work.

4 to 6 times stronger inhibitors than HPA and PBTC. The comparison of k values for these inhibitors suggest that inhibitor power increases with the number of phosphonate groups.

The differences in the ability of the various phosphonates (viz., AMP, HEDP, HPA, PBTC, etc.) to retard or inhibit crystallization of DCPD could be rationalized in terms of their abilities to form a complex with the calcium ion. Table 4 summarizes the comparative data on calcium - phosphonate complexation constant, K_{ML} and the affinity constant for various phosphonates. It can been seen that on the basis of affinity constant values, the order of effectiveness for the phosphonates is AMP >> HEDP >> PBTC > HPA., whereas the values of the complexation constants follows the order HEDP > AMP > PBTC > HPA. On the basis of data presented in Table 4, it seems evident that it is not simply the abilities of these phosphonates to complex calcium ion, but also the affinity of their adsorption to the crystal surface that influences their ability to inhibit crystal growth. Among the phosphono carboxylic acid inhibitors i.e., HPA and PBTC, the ranking in terms of their effectiveness as DCPD crystal growth inhibitor is consistent with their ability to form ca-phosphonate complex. To determine the effect on inhibitor power of the proximity of relevant groups, the potencies of HPA and PBTC were compared with glycolic and malic acids. The k values for these inhibitors (Table 4) show that inhibitor power increases with the incorporation of phosphonic acid group.

Table 4 summarizes the comparative data on affinity constants and K_{ML} for various carboxylic acids. It can be seen that, on the basis of affinity constant values, the order of effectiveness for α-hydroxycarboxylic, dicarboxylic, benzene hexacarboxylic, and polyacrylic acids is PAA >> MA >> MLA > MNA > GLA > MDA, whereas the values of the complexation constants follow the order MA > MLA > MNA > GLA > MDA. It is noteworthy that the inhibitory activity among hydroxy acids and dicarboxylic acids is consistent with the values of Ca-carboxylate complex constants. Regarding the performance of polycarboxylic acids (benzene hexacarboxylic acids and polyacrylic acid) compared to other carboxy containing acids, it appears that the most effective inhibitors for DCPD crystallization are those which are highly charged.

REFERENCES

1. Nancollas, G.H., Amjad, Z., Koutsoukos, P.G.,1979, Calcium phosphates —— Speciation, Solubility and Kinetic Considerations, ACS Symposium Series 93, 475.

2. Amjad, Z.,1990, Influence of polyelectrolytes on the precipitation of amorphous calcium phosphate, *Colloids and Surfaces*, 48, 95.

3. Brecevic, L., Sendijarevic, A., and Furedi-Milhofer, H.,1984, Precipitation of calcium phosphates from electrolyte solutions. VII. The influence of di- and tricarboxylic acids, *Colloids and Surfaces*, 11, 55.

4. Amjad, Z., 1988, Effect of precipitation inhibitors on calcium phosphate scale formation, *Can. J. Chem.*, 67, 850.

5. Amjad, Z., 1988, Constant composition study of crystal growth of dicalcium phosphate dihydrate. The influence of polyphosphates, phosphonates, and phytate, *Can. J. Chem.*, 66, 2281.

6. Meyer, J. L. and Nancollas, G. H., 1973, The influence of multidentate organic phosphonates on crystal growth of hydroxyapatite, *Cal. Tiss. Res.*, 13, 295.

7. Koutsoukos, P. G., Amjad, Z., and Nancollas, G. H.,1981. The influence of phytate and phosphonate on the Crystal growth of fluoroapatite and hydroxyapatite, *J. Colloid Interface Sci.*, 83, 599.

8. Amjad, Z., 1987. The influence of polyphosphates, phosphonates, and poly (acrylic acid) on the crystal growth of hydroxyapatite, *Langmuir*, 3, 1063.

9. Tomson, M. B. Nancollas, G.H., and Kazmierczak, T., 1976, Controlled composition seeded growth kinetics of calcium carbonate, Paper No. SPE 6591, Society of Petroleum Engineers of AIME, Dallas, TX.

10. Shyu, L. J.,1982., Ph.D. Thesis, State University of New York at Buffalo, N.Y.

11. Amjad, Z., 1985, Applications of antiscalants to control calcium sulfate scaling in reverse osmosis systems, *Desalination*, 54, 263.

12. Leung, W. H. and Nancollas, G.H., 1978, A kinetic study of the seeded growth of barium sulfate in the presence of additives, *J. Inorg. Nucl. Chem.*, 40, 1871.

13. Amjad, Z., 1987, The inhibition of dicalcium phosphate dihydrate crystal growth by polycarboxylic acids, *J. Colloid Interface Sci.*, 117, 98.

14. Amjad, Z. 1993, Performance of inhibitors in calcium fluoride crystal growth inhibition, *Langmuir*, 9, 597.

15. Amjad, Z. 1987, Kinetic study of the seeded growth of calcium carbonate in the presence of benzene polycarboxylic acids, *Langmuir*, 3, 324.

16. Nancollas, G. H. and Tomson, M. B. 1976, Precipitation of calcium phosphates in the presence of polycarboxylic acids, *Faraday Discuss. Chem. Soc.*, 61, 2976.

17. Zuhl, R.W., Amjad, Z. and Masler, W.F., 1987, *J. Cooling Tower Inst.*, 8, 41.

18. Oath, H. and Kawamura, T., Furukawa, Y. ,1989, Zinc and phosphate scale inhibition by a newly developed polymer. Paper No. 432, Corrosion / 89.

19. Van der Leeden, M. C., Van Rosmalen, G. M., 1987, Aspects of additives in precipitation processes: Performance of polycarboxylates in gypsum growth prevention, *Desalination*, 66, 185.

20. Wilkes, J. F., 1993, A historical perspective of scale and deposit control , Paper No. 458, Corrosion / 93.

21. Amjad, Z. 1987, Calcium sulfate dihydrate(gypsum) scale formation on heat exchanger surfaces. The influence of scale inhibitors, *J. Colloid Interface Sci.*, 123, 523.

22. Smith, B. R. and Alexander, A. E., 1970, The effect of additives on the process of crystallization II. further studies on calcium sulfate, *J. Colloid Interface Sci.*, 34, 81.

18

DICALCIUM PHOSPHATE DIHYDRATE CRYSTAL GROWTH IN THE pH REGION OF 4.5 TO 6.2

Kostas D. Daskalakis[1] and George H. Nancollas[2]

[1] National Status and Trends
NOAA/NOS/ORCA-21
1305 East West Hwy., Silver Spring, Maryland 20910
[2] Department of Chemistry
State University of New York at Buffalo
Buffalo, New York 14260-3000

ABSTRACT

The Dual Constant Composition (DCC) method was used to study the crystal growth rates of dicalcium phosphate dihydrate (DCPD) in the pH range of 4.5 to 6.2 at 37.0°C with ionic strength, I = 0.15 mol l^{-1}. DCPD crystal growth was retarded at pH values of 5.5 or lower, when Reagent grade chemicals were used in place of Ultrapure reagents, presumably due to chemical impurities. At constant relative supersaturation with respect to DCPD, σ_{DCPD} = 0.294, growth was pH dependent following the order 4.5 >> 5.0 > 6.2 ≈ 5.3 > 5.5. A minimum rate of 4.6 x 10^{-5} mol m^{-2} min^{-1} was found at pH 5.5, in agreement with previous DCPD growth and dissolution experiments. This pH value was near the point of zero charge (PZC) of DCPD, suggesting the participation of surface complexation in controlling the growth rate. The crystal growth rate could be expressed as a first order dependence on the pH with slopes of -0.61 and 0.16 on the acidic and basic sides of the PZC, respectively. Mostly rectangularly shaped crystals with well defined edges were grown at a pH of 6.2, in contrast to lower pH values of 5.0 and 4.5 where the grown phase had more irregular morphology.

INTRODUCTION

Dicalcium phosphate dihydrate CaHPO$_4$•2H$_2$O (DCPD) has been proposed as a precursor to the formation of hydroxyapatite (HAP) in hard tissue including teeth and bones. It has been identified in embryonic chick bone, young dental calculus, caries, plaque and renal stones.[1,2] It is stable in aqueous suspensions at pH values less than 6.4[3,4], and it is probably the first crystalline phosphate phase to precipitate at high solution supersaturations,

Mineral Scale Formation and Inhibition, Edited by Zahid Amjad
Plenum Press, New York, 1995

transforming to octacalcium phosphate (OCP) at higher pH values.[3] Constant composition (CC) studies of the kinetics of DCPD crystal growth and transformation[5,6] concluded that in the pH range of 4.6 to 6.0 DCPD growth followed a surface controlled mechanism and there were indications of a growth rate minimum at pH 5.5. Zhang and Nancollas[7] studied the dissolution of DCPD and concluded that the rate was controlled by desorption and volume diffusion processes depending on the undersaturation, with a minimum rate of reaction at pH of 5.0-5.6. In another study of DCPD dissolution[8] the use of Ultrapure reagents demonstrated that the initial reaction rates were slower when Reagent grade chemicals were employed to prepare the solutions.

EXPERIMENTAL

The DCC method[3] was used in the present work to investigate the growth of DCPD in the pH range of 4.5 to 6.2 at constant supersaturation. Growth experiments were made in

Table 1. Conditions used in DCC experiments

Run Number	pH	Ca=PO$_4$ 10^{-3} molL^{-1}	DCPD mg	Run Number	pH	Ca=PO$_4$ 10^{-3} molL^{-1}	DCPD mg
1	5.50	10.00	10.1	33	5	15.68	8.5
2	5.50	9.00	10.2	34	4.5	28.7	8.9
3	5.50	9.00	9.9	35	4.5	28.7	8.3
4	5.50	9.00	10.4	36	6.2	4.47	9.8
5	5.50	9.00	9.9	37	4.5	28.7	12.8
6	5.50	9.00	10.6	38	5	15.68	6.7
7	5.50	9.00	10.7	39	5	15.68	9.8
8	5.50	9.00	10.5	40	5	15.68	5.2
9	5.50	8.00	20.0	41	5	15.68	6.0
10	5.50	9.00	10.8	42	6.2	4.47	9.1
11	5.50	9.00	#*	43	6.2	4.47	3.6
12	5.50	9.00	5.7	44	6.2	4.47	5.0
13	5.50	9.00	5.9	45	4.5	28.7	4.7
14	5.50	9.00	6.8	46	4.5	28.7	3.9
15	5.50	9.00	6.8	47	4.5	28.7	3.5
16	6.20	4.30	10.7	48	4.3	36.3	3.4
17	6.20	4.30	7.2	49	4.3	36.3	4.8
18	6.20	4.47	6.8	50	4.3	36.3	4.3
19	6.20	4.47	8.1	51	4.3	36.3	4.2
20	6.20	4.47	8.1	52	5	15.68	5.0
21	6.20	4.47	9.9	53	5	15.68	5.0
22	6.20	4.47	8.6	54	5.5	9	4.9
23	6.20	4.47	10.6	55	5.5	9	4.7
24	6.20	4.47	10.2	56	5.5	9	7.1
25	6.20	4.47	8.6	57	5.5	9	5.6
26	5.00	15.68	9.0	58	5	15.68	6.0
27	5.00	15.68	20.4	59	5	15.68	6.1
28	5.00	15.68	14.8	60	4.5	27.9	6.0
29	5.00	15.68	18.4	61	4.5	27.9	6.1
30	5.00	15.68	10.0	62	4.5	27.9	10.9
31	5.00	15.68	7.4*	63	4.5	27.9	6.1*
32	5.00	15.68	7.9*				

#Slurry suspension used.
*More DCPD crystals were added during the experiment.

solutions prepared with either Reagent or Ultrapure chemicals, in order to determine possible effects of trace impurities on the crystal growth kinetics.

Reagents (NaCl, CaCl$_2$, KH$_2$PO4) of analytical grade (Baker Analyzed) were used for experiments 1-38 (see Table 1), and Ultrapure NaCl, KH$_2$PO$_4$ (Aldrich), and CaCl$_2$ (Alfa) were used for the remainder. Analytical concentrate KOH (Dilute-It, J.T. Baker Chemical Co.) was used under N$_2$ to exclude CO$_2$ for all experiments. Stock solutions were prepared using deionized triple-distilled water (D-water) and filtered twice through 0.2 μm Millipore membrane filters. DCPD seed material was prepared from KH$_2$PO$_4$ and Ca(NO$_3$)$_2$ as described previously.[3] The seed molar calcium/phosphate ratio was 1.00 ± 0.02 and the solid was stored dry. The specific surface area, 0.69 ± 0.06 m^2 g^{-1}, was measured by BET nitrogen adsorption (30% nitrogen 70% helium mixture, Quantasorb II, Quantachrome).

Supersaturated solutions (100 ml) were prepared in water jacketed Pyrex vessels under nitrogen gas, at 37.0 ± 0.1°C and I = 0.15 mol l^{-1} adjusted with sodium chloride, by the slow mixing of stock solutions, and the pH was adjusted by the addition of 1.00×10^{-2} mol l^{-1} KOH. Following the introduction of DCPD seed crystals the ionic concentrations were maintained constant by the addition of two titrants (KOH + KH$_2$PO$_4$) and CaCl$_2$ controlled by glass and calcium electrodes (Orion Research), respectively. A Brønsted Ag/AgCl reference electrode with a 0.15 mol l^{-1} KCl salt bridge was used as reference. The cells were stirred by means of an overhead stirrer at 380 rpm, to avoid damage to the crystals. The stirring rate was sufficient to keep all crystals suspended in solution. The titrant effective concentration (C$_{eff}$) with respect to DCPD was usually 6.00×10^{-3} mol l^{-1} Experiments 1-22 (Table 1) were made using DCC instrumentation[4], consisting of a pair of 605 pH meters, 614 Impulsomats, 665 Dosimats (Metrohm) and dual-channel BD41 chart recorders (Kipp & Zonen). Experiments 23-63 were performed using a computerized version that included a PC computer controlling a dual-channel Orion 720A pH meter (Orion, CT) through the RS-232 serial port. Two stepper motor assemblies acting as titrant delivery systems were controlled by the computer through a 24 channel digital I/O card (National Instruments, Austin, Texas). The minimum volume delivery (one step of the stepper motor) for the 10 ml burets was 3 μl. The Graphical User Interface (GUI) and software controlling the DCC apparatus was developed under LabWindows (National Instruments, Austin, Texas). During the experiment each cycle consisted of pH and calcium electrode potential measurements and addition of titrants when required. Each of these cycles was completed in less than 12 seconds. The limiting factor in the frequency of data retrieval was the communications with the pH meter at 1200 baud. The data were processed and displayed in the computer monitor in two graphs per channel, i.e. electrode potential and titrant volume against time. Data were stored automatically in files for further analysis. In general, the sampling rate was very high for most experiments and only one tenth of the data points were saved and approximately 50 points were used for the rate calculations.

The constancy of the concentrations in the supersaturated solutions was verified in the DCC experiments, by removing aliquots periodically, filtering through 0.22 μm membrane filters (Millipore), and analyzing for total calcium by atomic absorption and for phosphate spectrophotometrically at 420nm as the vanadomolybdate complex.[9] The solid phases were examined using powder X-ray diffraction (XRD), and electron microscopy (Hitachi S800 Field Emission Microscope (FEM) operated at 25 kV with PGT IMIX X-ray microanalyser). Water content was measured by thermogravimetric analysis (TGA) (DuPont Instruments 591 thermogravimetric analyzer). The particle size distributions of the seed and grown materials were determined with a Malvern 3600-E particle sizer; electrophoretic mobility measurements were made in solutions saturated with respect to DCPD at 37°C at a total ionic strength adjusted to 0.15 mol l^{-1} with NaCl (Malvern Zetasizer II C).

Solution relative supersaturation with respect to DCPD (σ_{DCPD}) defined as:

Table 2. Concentrations and Calculated Supersaturation
(σ_{DCPD}) in DCC Experiment

pH	$Ca = PO_4$ 10^{-3} $molL^{-1}$	σ_{DCPD}
4.50	28.70×10^{-3}	0.331
4.50	27.90×10^{-3}	0.295
5.00	15.68×10^{-3}	0.294
5.30	11.20×10^{-3}	0.294
5.50	9.00×10^{-3}	0.294
6.20	4.47×10^{-3}	0.294

$$\sigma_{DCPD} = 1 - \left[\frac{(Ca^{2+})\,(HPO_4^{2-})}{K_{sp}} \right]^{1/2} \tag{1}$$

was calculated using ionic species concentrations. These were derived from the experimental total concentrations using the computer speciation program AESPEC, based on expressions for ion-pair formation and solubility product constants, mass balance and electroneutrality. Experimentally determined values[10] of 1.35×10^6, 591, 2.7 and 2.20×10^{-7} were used for the association constants of $CaPO_4^-$, $CaHPO_4$, $CaH_2PO_4^+$ and DCPD solubility product (K_{sp}), respectively. Changes in surface area, S, of the crystals during the reactions were calculated using equation 2:

$$S_t = S_o \left[\frac{m_t}{m_o} \right]^p \tag{2}$$

where m_o and m_t are the masses of crystallites initially and at time t, respectively. The value of 2/3 was usually assumed for p in 3-dimensional isotropic crystal growth.

The growth rate (R) was calculated using equation 3.

$$R = \frac{dn}{dt} \frac{1}{S_t} \tag{3}$$

where n is the DCPD moles grown. The value of n was obtained from the volume of the titrant solution added, V, and the effective concentration of the titrant, C_{eff}, using equation 4

$$n = V\,C_{eff} \tag{4}$$

The total concentrations for all experiments with the calculated σ_{DCPD} values are presented in Table 2.

RESULTS

Influence of Chemicals on the Growth Rate

Typical DCPD growth rate curves for experiments at a pH value of 6.2 , using Reagent grade chemicals, presented in Figure 1, illustrate the excellent reproducibility of the experimental results. The observed initial apparent rate decrease probably reflects the initial adaptation of DCPD crystal surfaces due to ion exchange and the reduction of the dislocation densities.[8,11,12] Dislocation reduction would be expected to be more pronounced at the lower

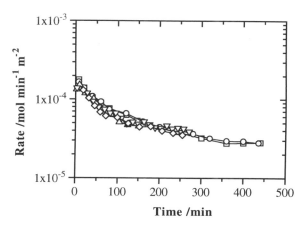

Figure 1. DCPD growth rates as a function of time at pH 6.20 using Reagent grade chemicals, and various amounts of seed material: Exp. 19, 8.12 mg (square); Exp. 20, 8.1 mg (circle); Exp. 21, 9.9 mg (up triangle); Exp. 23, 10.6 mg (down triangle); and Exp. 24, 10.2 mg (diamond). Growth rate appears to be independent of seed mass.

driving forces with relative supersaturation $\sigma_{DCPD} = 0.295$.[8] At higher σ_{DCPD} values growth is expected to be affected less by dislocation density reduction.

In Figure 2, DCPD growth data at pH 5.0 using Reagent grade substances are plotted as a function of time for experiments with varying amount of DCPD seed (open symbols). As in Figure 1 there was a reduction in the initial rate, the decrease being inversely proportional to the seed mass. However, in contrast to the results at pH 6.2, initial growth rates at pH 5.0 were greater for higher DCPD seed concentration. In Figure 2 experiments 28 (14.8 mg DCPD seed) and 29 (18.4 mg) reached minima of 5×10^{-6} and 1.6×10^{-5} mol min^{-1} m^{-2}, respectively, followed by an increase and finally the rate reached a plateau at approximately 5×10^{-5} and 8×10^{-5} mol min^{-1} m^{-2}. In experiment 30 (10.0 mg DCPD) the

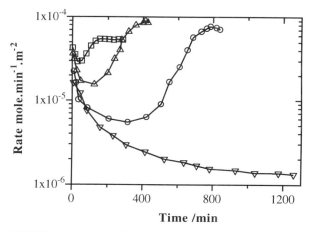

Figure 2. Influence of DCPD seed concentration on DCC growth rates at pH 5.00 using Reagent grade substances. Exp. 30, 10.0 mg (inverted triangle); Exp. 28, 14.8 mg (circle); Exp. 29, 18.37 mg (triangle); and Exp. 24, 20.4 mg seed (square).

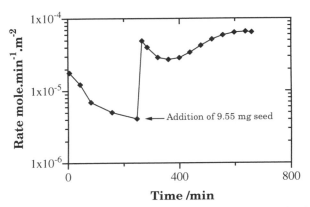

Figure 3. Influence of newly added seed in a "Dynamic test". Experiment 32 initiated with 7.9 mg DCPD crystals. After 270 min. of reaction 9.6 mg DCPD were added, dramatically increasing the rate.

rate dropped to approximately 1.3×10^{-6} mol min^{-1} m^{-2} and did not increase during the duration of the run.

In experiment 32 (Figure 3) the initial rate with 7.9 mg seed was similar to that of experiment 30. However, a rapid rate increase was induced by the addition of 9.6 mg DCPD seed after 270 minutes of reaction. These "Dynamic test" results were similar for DCPD growth at pH values of 5.5 and 4.5. The data suggest that growth was inhibited by impurities contained in the solutions, and that growth reached a constant rate only after new growth sites were formed. In experiment 32, the inhibitors were reduced through adsorption on the surface of the crystallites introduced initially, allowing growth to commence on the new crystals introduced at the latter time.

To minimize the influence of chemical impurities, Ultrapure chemicals were used, starting with experiment 38, and the growth rates were compared with those obtained in the Reagent grade experiments. The calculated rates of DCPD crystal growth experiments using Ultrapure and Reagent grade substances are shown in Figures 4 and 5 as a function of time

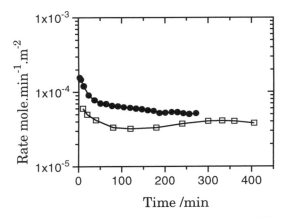

Figure 4. DCPD growth rates at pH 5.50 using similar masses 7.1 mg, of seed and Ultrapure chemicals (closed circles), and 6.8 mg and Reagent grade (open squares) chemicals. Initial rates are greater for Ultrapure solutions, but the rates become similar as the growth progresses.

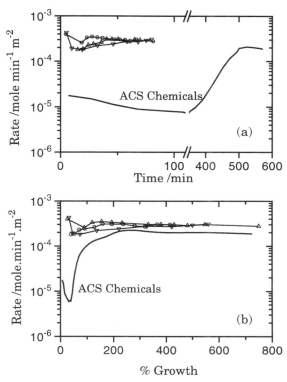

Figure 5. DCPD growth at pH 4.50 and σ_{DCPD} = 0.330 using Ultrapure (open symbols) and reagent grade (solid line) substances as a function of a) time and b) % overgrowth. Note the break in the time axis, and the rise of the rate for reagent grade after approximately 350 minutes. Rates were approximately equal after 200% overgrowth.

for pH 5.5 (σ_{DCPD} = 0.294) and 4.5 (σ_{DCPD} = 0.330), respectively. The initial rates for experiments at pH 5.5 were almost twice as fast in Ultrapure chemicals, but the difference decreased as the reaction proceeded. At pH values of 4.5 (Figure 5a), rates calculated from growth experiments using Ultrapure electrolyte solutions were greater by a factor of more than 10 and only after a considerable length of time did the Reagent grade growth rate (experiment 34) achieve a comparable value. In Figure 5b where DCPD growth rates at pH 4.5 are plotted as a function of % overgrowth, the minimum for experiment 34 occurred at 35% overgrowth and the rate reached a constant value at 200% overgrowth.

pH Dependence of Growth Rate

In Figure 6 plots of average DCPD growth rate, for experiments using Ultrapure substances, at various pH values are presented as a function of time. It can be seen that the initial rates were greater for experiments made at the higher pH values, probably reflecting ion exchange processes at the crystal surface. However, as the reaction proceeded the growth rates eventually reached a constant value. Interestingly, at pH values of 6.2 and 5.5 the growth rate decreased monotonically, while at pH 5.0 and 4.5 there was a minimum after which the rate increased, the increase being greater for experiments at pH 4.5. The final growth rates

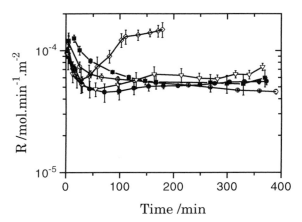

Figure 6. DCPD growth rates with the calculated standard error as a function of time at $\sigma_{DCPD} = 0.294$ for pH values of 4.50 (diamonds), 5.00 (inverted triangles), 5.30 (closed circles), 5.50 (open circles), and 6.20 (squares).

plotted as a function of pH in Figure 7, decreased in the order $4.5 \gg 5.0 > 6.2 \approx 5.3 > 5.5$ revealing a minimum at pH 5.5 (Figure 7).

X-ray powder diffraction analysis showed that the seed crystals consisted of DCPD with a small amount ($\approx 5\%$) of monetite ($CaHPO_4$), presumably introduced by partial dehydration of the seed during storage and XRD analysis. Solid material collected during the DCC experiment were also DCPD with traces of monetite, from the seed crystals. The grown phases were analyzed by thermogravimetric analysis, and the water content was found to be $20.1\pm0.5\%$ in good agreement to the theoretically calculated value of 20.9% for DCPD. The traces of monetite (<5% based on the H_2O content) that were discovered in the crystallites were from the original seed material, while the new grown phase was DCPD. Particle size distribution demonstrated that there was a general increase in the average

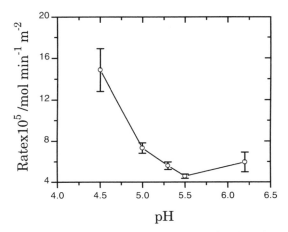

Figure 7. Plot of DCPD growth rate at constant supersaturation as a function of pH. Rate is much greater at pH 4.5, and a minimum is exhibited at pH 5.50.

Figure 8. Field emission micrograph of DCPD grown at pH a) 6.2 and b) 5.00 reveal differences in the crystal habit.

spherical diameter of the grown phosphate phases, at all pH values used in these experiments. This suggests that growth occurred on the seed crystals, without secondary nucleation.

Field emission micrographs of DCPD crystals grown at a pH of 6.2 and 5.0 are shown in Figure 8a and 8b, respectively. Mostly trapezium shaped crystals with well defined edges were grown at a pH of 6.2. In contrast, crystals grown at a pH of 5.0 or lower, had a different symmetry with one faster growing axis. This suggests that even under relatively low driving forces the shape of the crystals is kinetically controlled, some crystal axes growing faster than others. The crystal structure of DCPD ($CaHPO_4 \bullet 2H_2O$) is very similar to gypsum ($CaSO_4 \bullet 2H_2O$) which has a needle-like shape believed to be due to elongation of (010) plates.[13] According to Heijnen and Hartman[13], the (010) phase in gypsum can grow at $\frac{1}{2}d_{011}$ steps, which is faster than the growth rate of (111) at the same solution supersaturation. Actually it is known that $CaHPO_4 \bullet 2H_2O$ is normally needle-like or prismatic to tabular (010).[13] It could, therefore, be argued that it is the (010) phase in DCPD that grows faster under these conditions, resulting in needle-like crystals. Interestingly, precipitated DCPD crystals stored in aqueous solution at pH 5.5 for approximately 2 years, showed a needle-like morphology and grew extremely slowly in DCC runs (experiment 11).

Electrophoretic mobility measurements of DCPD in NaCl solution plotted in Figure 9 as a function of pH show that the isoelectric point (IEP) occurred at approximately pH 5.3, which was similar to the IEP in 0.1 mol l^{-1} KNO_3.[6] Because of the use of supporting electrolyte in these experiments, the surface charge may have been affected. However, since both NaCl and KNO_3 supporting electrolytes give similar results, only small changes in the surface charge are expected. Thus, it is expected that the point of zero charge (PZC) of DCPD should be near to the IEP, i.e. 5.3.

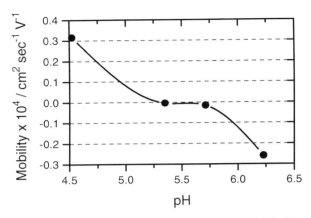

Figure 9. Electrophoretic mobility measurements of DCPD crystals in 0.15M NaCl solution, plotted against pH. The isoelectric is approximately at 5.5.

DISCUSSION

In the present work it has been shown that initial DCPD growth rates in Reagent grade electrolyte solutions are inhibited. The magnitude of the inhibition decreased with increasing DCPD seed concentration. A smaller rate dependence on seed mass for HAP dissolution was also attributed to the influence of impurities from the solution.[14,15] Increased rates with the use of Ultrapure chemicals have been observed for DCPD, OCP and HAP dissolution experiments.[8,14] It has been argued that the adsorption of these impurities is very fast[14] and that further rate decreases were due to reduction of active sites.[8] Reduction of active growth sites may also explain the decrease in the initial rate of the growth experiments using Reagent grade chemicals presented here, but not the dependence of the rate on pH. Moreover, the rate increased up to 50-fold at later stages of the experiment (Figures 2 and 5). The rate decreases were probably caused by impurities in the reagent grade chemicals poisoning the seed surfaces. The inhibitory influence of impurities diminished as the extent of reaction increased probably due to their incorporation into the crystals. This is also supported by the "Dynamic test" in Figure 3 and similar results for HAP dissolution.[14] Incorporation of foreign ions in crystals is well documented. For example, it has been shown that strontium ions reduce the rate of DCPD growth and are incorporated into the grown phase.[16]

The pH dependence of the inhibitory effects is interesting. It was determined that the IEP of DCPD was approximately 5.3 (Figure 9), and it is expected that the point of zero charge (PZC) would be close to the IEP. The inhibition induced by impurities is much greater at pH values below the PZC, when the surface had a net positive charge. Thus it appears that the trace impurities were negatively charged species such as pyrophosphate, carbonate, arsenate and silicate ions present in the solutions. Previous work in silica gel[17], has demonstrated that DCPD does not grow in the presence of carbonate and the growth was hindered in pyrophosphate containing solutions. These results emphasize the need to take into account possible poisoning of the crystal surfaces by trace impurities. Dissolution rates are probably not affected by inhibitors to the same degree.

The dependence of DCPD growth on pH followed the order 4.5 >> 5.0 > 6.2 ≈ 5.3 > 5.5 (Figure 8), exhibiting a minimum near the PZC of DCPD. A minimum at the pH range of 5.0 to 5.6 was also observed for the dissolution reaction.[7] The minimum pH value depended on the driving force and was approximately 5.3 for $\sigma_{DCPD} = -0.3$, and 5.6 for $\sigma_{DCPD} = -0.1$.[7] Thus growth and dissolution rates of DCPD exhibit similar pH dependence. It has been observed[17] that DCPD growth in various types of gels was much slower in the pH range of 4-5 as compared to pH 6. However, the rate dependence observed by LeGeros et. al.[17] may have resulted from factors associated with the gel matrix or inhibition at acidic pH values, as has been observed in the present work.

The pH dependence of the dissolution rate has been well documented for silicates[18,19,20] where it has been argued that H^+ and OH^- catalyze the reaction depending on pH, and that surface speciation is the controlling factor in their dissolution rate. According to Murphy and Helgeson[20] "... surface charge, rather than pH, reflects the concentration of the adsorbed OH^-... The influence of surface charge on the rate of hydrolysis is indicated in part by the coincidence of the zero point of charge".

The existence of a minimum in the DCPD growth near the PZC suggests that the surface charge influences the rate. In solutions more acidic than the PZC, the rate increases dramatically being catalyzed by H^+ that protonate surface phosphates to form $H_2PO_4^-$. In these regions, the growth rate can be expressed as a function of pH using equation 5[21]:

$$R \propto k\, pH^n \tag{5}$$

The values of n are -0.61 and 0.16 for $4.5 \leq pH \leq 5.3$ and $5.5 \leq pH \leq 6.2$, respectively. The negative slope on the acidic side of the PZC suggests that increased positive charge on the crystal surface due to H^+ incorporation (decreasing pH) accelerates the growth rate. At pH values more basic than PZC we observe a much lower rate dependence on the activity of H^+ presumably due to very weak Ca^{2+}-OH^- complexes. Because DCPD converts to OCP above pH 6.4, the range of OH^- activity was limited to less than $10^{-7.8}$ mol l^{-1} in our experiments.

CONCLUSIONS

Solution impurities affect the growth of DCPD crystals at low supersaturation. The pH dependence of the rate in solutions containing Ultrapure chemicals exhibits a minimum near the PZC in agreement with previous DCPD and silicate dissolution studies. The growth rates are proportional to the activity of proton ions according to the relationships: $pH^{-0.61}$ and $pH^{0.16}$ for $4.5 \leq pH \leq 5.3$ and $5.5 \leq pH \leq 6.2$, respectively. Changes in the crystal habit are observed for growth at pH values more acidic than the PZC.

ACKNOWLEDGMENTS

We thank Dr. Arman Ebrahimpour for running the X-ray powder diffraction analysis, and Dr. Jingwu Zhang for providing the DCPD seed crystals and for useful discussion. The work was supported by a grant from the National Institute of Heath (#DE03223).

REFERENCES

1. Rogers A.L. and Spector M., 1986, Pancreatic Calculi Containing Brushite: Ultrastructure and Pathogenesis, *Calcified Tissue Int.*, 39, 342.
2. Roufosse A.H., Landis W.J., Sabine W.K., and Glimcher M.J., 1979, Identification of Brushite in Newly Deposited Bone Mineral from Embryonic Chicks, *J. Ultrastructure Res.*, 68, 235-255.

3. Ebrahimpour A., Zhang J., and Nancollas G.H., 1991, Dual Constant Composition Method and its Application to Studies of Phase Transformation and Crystallization of Mixed Phases, *J. Crystal Growth*,113, 83-91.
4. Perez L., Shyu L.J., and Nancollas G.H., 1989, The Phase Transformation of Dicalcium Phosphate Dihydrate into Octacalcium Phosphate in Aqueous Suspensions, *Colloids and Surfaces*, 38, 295-304.
5. Hohl H., Koutsoukos P.G., and Nancollas G.H., 1982, The Crystallization of Hydroxyapatite and Dicalcium Phosphate Dihydrate; Representation of Growth Curves, *J. Crystal Growth*, 57, 325.
6. Salimi M.H., 1985, The Kinetics of Growth of Calcium Phosphate, Ph.D. Dissertation, Department of Chemistry, SUNY at Buffalo.
7. Zhang J. and Nancollas G.H., 1994, Unexpected pH Dependence of Dissolution Kinetics of Dicalcium Phosphate Dihydrate, *J. Phys. Chem.*, 98, 1689.
8. Zhang J. and Nancollas G.H., 1992, Dissolution Deceleration of Calcium Phoshate Crystals at Constant Undersaturation, *J. Crystal Growth*,123, 59-68.
9. Tomson M.B., Barone J.B., and Nancollas G.H., 1977, Atomic Absorption Newsletter 16, 117.
10. Zhang J. and Nancollas G.H., 1992, Interpretation of Dissolution kinetics of Dicalcium Phosphate Dihydrate, *J. Crystal Growth.*, 125, 251-269.
11. Kirtisinghe D., Morris P.J., and Strickland-Constable R.F, 1968, Retardation of the Rate of Growth of Salol Crystals in Capillary Tubes, *J. Crystal Growth.*, 3/4, 771-775.
12. Nancollas G.H., LoRe M., Perez L., Richardson C., and Zawacki S.J., 1989, Mineral Phases of Calcium Phosphate, The Anatomical Record, 224, 234-241.
13. Heijnen W.M.M. and Hartman P., 1991, Stuctural Morphology of Gypsum, $CaSO_4.2H_2O$, Brushite, $CaHPO_4.2H_2O$, and Pharmacolite(CaHAsO_4.2H_2O), *J. Crystal Growth*, 108, 290-300.
14. Budz J.A. and Nancollas G.H., 1988, The Mechanism of Dissolution of Hydroxyapatite and Carbonate Apatite in Acidic Solutions, *J. Crystal Growth*, 91, 490-496.
15. Daskalakis K.D., Fuierer T.A., Tan J., and Nancollas G.H., 1995, The Influence of b-Lactoglobulin on the Growth and Dissolution Kinetics of Hydroxyapatite, *Colloids and Surfaces A*, 96, 135-141.
16. LeGeros R.Z., Lee D., Quirolgico G., Shirra W.P. and Reich L., 1983, In Vitro Formation of Dicalcium Phosphate Dihydrate, $CaHPO_4 \bullet 2H_2O$, *Scanning Elec. Microsc.*, 407-418.
17. LeGeros R.Z. and LeGeros J.P., 1972, Brushite Crystals Grown by Diffusion in Silica Gel and in Solution, *J. Crystal Growth*, 13/14, 476-480.
18. Blum A. and Lasaga A., 1988, Role of Surface Speciation in the Low-Temperature Dissolution of Minerals, *Nature*, 331, 431-433.
19. Carroll-Webb S.A. and Walther J.V., 1988, A Surface Complex Reaction Model for the pH-Depcence of Corundum and Kaolinite Dissolution Rates, *Geochim. Cosmochim. Acta.*, 52, 2609-2623.
20. Murphy W.M. and Helgeson H.C., 1987, Thermodynamic and Kinetic Constants on Reaction Rates Among Minerals and Aqueous Solutions. III. Activated Complexes and the pH-Dependence of the Rates of Feldspar, Pyroxene, Wollostonite, and Olivine Hydrolysis, *Geochim. Cosmochim. Acta.*, 51, 3137-3153.
21. Chou L. and Wallast R., 1984, Study of the Weathering of Albite at Room Temperature and Pressure with a Fluidized Bed Reactor, *Geochim. Cosmochim. Acta.*, 48, 2205-2217.

19

METASTABLE EQUILIBRIUM SOLUBILITY DISTRIBUTION AND DISSOLUTION KINETICS OF CARBONATE APATITE POWDERS

J. L. Fox, Z. Wang, J. Hsu, A. Baig, S. Colby, G. L. Powell, M. Otsuka, and W. I. Higuchi

Department of Pharmaceutics and Pharmaceutical Chemistry
University of Utah
Salt Lake City, Utah

ABSTRACT

Carbonated apatites (CAP's) have been shown to exhibit a phenomenon that we refer to as metastable equilibrium solubility (MES). The MES behavior occurs when the kinetics of crystal growth and dissolution are such that dissolution of the crystalline material can relatively rapidly lead to a solution composition such that no further dissolution occurs, even though the degree of saturation of the solution is well below that required for growth. The solution composition at this point is indicative of the MES, and is most conveniently expressed in terms of a solution ion activity product (IAP).

For CAP's, an ion activity product of the form $K_{HAP} = (a_{Ca})^{10} (a_{PO4})^6 (a_{OH})^2$ has been shown to describe experimental data over a wide range of conditions, even though the stoichiometry implied by this expression is not that of CAP's. This is consistent with the hypothesis that both dissolution kinetics and solution equilibration behavior observed within the time frame of these experimental studies are governed by a surface complex with the stoichiometry of hydroxyapatite (HAP). A further complicating factor in CAP dissolution and equilibration behavior is that a given CAP sample does not exhibit a single value for the MES, but rather a distribution of MES's. Novel experimental techniques have been developed to quantitate these MES distributions and to apply the MES hypothesis to the dissolution kinetics of CAP's.

Each of several synthetic CAP's studied has been found to exhibit MES behavior, as has human enamel. Furthermore, it appears that systematic relationships exist between the values of the means of the MES distributions in a series of CAP samples and physical parameters such as crystallinity and carbonate content.

Mineral Scale Formation and Inhibition, Edited by Zahid Amjad
Plenum Press, New York, 1995

OVERVIEW

The quantitation of the dissolution kinetics and solubility behavior of human enamel and synthetic apatites has been a focus of work in our laboratory for some 3 decades[1-21] The MES concept has evolved over the past 2 to 3 years and is at a comparatively early stage of development. This chapter is therefore a snapshot of the state of our thinking up to the present time, and is necessarily incomplete.

The organization of this chapter is as follows. First, we will briefly review the previous state of affairs with respect to our understanding of CAP behavior to illustrate the need for the development of new concepts. Then the MES concept is described, including some speculation as to the underlying physical causes for this behavior. Following this, we present the experimental approach to characterizing the MES distribution of a CAP preparation and then show how the MES approach can be used to study dissolution kinetics. We then explore the issue of how generalizable the MES concept is, including both a range of experimental conditions for dissolution studies and the effect of heat treatment on the MES behavior of CAP's. Then we discuss how the MES might be quantitatively related to measurable physical parameters of CAP's, such as carbonate content and parameters related to crystallinity. We conclude by suggesting directions that future exploration of the MES concept might take.

PROBLEMS WITH CONVENTIONAL APPROACHES

It has long been known that dental caries is the result of dissolution of enamel by organic acids produced by oral bacteria as an end product of carbohydrate metabolism. An understanding of the solubility properties of enamel in oral fluids is therefore a prerequisite to a quantitative understanding of the process of dental caries formation. An obvious first step in this understanding would be the determination of the ion activity product of enamel. Attempts at this determination have shown that measured IAP's vary significantly[22-25] and it has been suggested that this variation in IAP is caused by variations in the impurities and defects present in enamel crystals.[22,23]

The usual strategy for avoiding these variations has been to use synthetic apatites as model systems for the study of enamel demineralization. HAP has been widely used for this purpose, but has not completely eliminated the question of variation among samples. Even today, there does not exist a universally accepted value for the solubility product of HAP and some have even questioned the general applicability of the solubility product principle to HAP.[26,27,28]

Several investigators[29-32] have noted that the solubility products of HAP preparations vary with the initial composition of the solution and with the slurry density (solid to solution ratio) employed in the determination, contrary to expectations based on thermodynamic principles. Furthermore, the dissolution of HAP may be non-stoichiometric and this has led investigators to postulate that HAP solubility might be governed by precipitated surface phases or surface complexes with stoichiometries different from that of HAP such as $CaHPO_4$[32] or $Ca_2(HPO_4)(OH)_2$.[30]

As human enamel contains on the order of 2-4% carbonate, one might expect that CAP's might be more appropriate model substances for enamel than is HAP. Consequently, the dissolution behavior of CAP's has been studied by a number of investigators and has been found to be related to the carbonate content and/or the crystallinity of the CAP preparation.[33,44] It has been shown that dissolution of enamel more closely parallels that of CAP than that of HAP[44] and that the responses of CAP and enamel to thermal treatment are similar and occur at much lower temperatures than are required for a similar response from HAP.[45] Interestingly, CAP's share the anomalous HAP behavior of the experimentally observed solubility depending on slurry density.[46]

The studies cited above are but a representative sample of the many studies done on dissolution and solubility properties of enamel, HAP and CAP's. Although they represent a great deal of work, they also show that there are two major loose ends that are not addressed by conventional thermodynamic approaches: the apparent solubility depends on both the stoichiometry of the material and its defect properties; and the apparent solubilities of both HAP's and CAP's also depend on the initial solution composition and/or solid to solution ratios employed in equilibration experiments.

Clearly a need exists for new approaches to the enamel / HAP / CAP dissolution behavior problem. As we shall see, the Metastable Equilibrium Soulubility (MES) distribution and other concepts described in this chapter provide both a means for dealing with the loose ends noted above and a framework for approaching a number of other issues.

The Metastable Equilibrium Solubility (MES) Concept

Over the past 2-3 years it has become clear that the key to understanding the behavior of CAP's is the concept of MES and the recognition that a given CAP preparation exhibits a distribution of MES's, with the mean MES value depending upon the carbonate content and/or the crystallinity.

The MES concept is based on the acknowledgement that within the time scale of interest (seconds to hours or days), CAP's do not reach thermodynamic equilibrium with their surrounding solution. Experimentally, this is manifested by dissolution kinetics that allow dissolution of a CAP crystallite to occur readily until a metastable equilibrium state is reached (in a few hours) after which further dissolution occurs exceedingly slowly. Just as the thermodynamic solubility (and corresponding solubility product, K_{sp}) of a solid can be calculated from the composition of a solution in equilibrium with the solid, so can the MES solubility product be calculated from the composition of the solution when the metastable equilibrium state has been reached. The concept of MES as the dissolution driving force has been extremely useful; we have shown that this driving force can be used to accurately describe dissolution kinetics over the entire time scale from about 1 second until metastable equilibrium is reached at about 10^5 seconds.

Because this metastable equilibrium is not a true equilibrium, but rather a state after which further dissolution occurs only very slowly, we should not expect growth or precipitation to occur at solution compositions with ion activity products (IAP's) that slightly exceed the MES IAP. Experimentally, we have shown that no growth is discernable for solutions that exceed the MES IAP by up to 5 orders of magnitude (e.g., 5 units in the pIAP). Such a behavior, where there is a window of saturation levels where neither growth nor dissolution occurs at an appreciable rate can be rationalized by a variety of possible atomic level models and is not an uncommon behavior for inorganic solids with very low solubilities.

The second key aspect of CAP dissolution behavior that has confounded attainment of a quantitative understanding in the past is that a given CAP preparation behaves as though it were a mixture of CAP's with a continuum of values of the MES IAP. Although we certainly did not expect this *a priori*, it seems quite reasonable in retrospect. Consider, for example, a CAP preparation containing 2% carbonate. This amounts to one carbonate ion for each three unit cells (based on HAP stoichiometry of $Ca_{10}(PO_4)_6(OH)_2$). If one imagines these carbonate ions being randomly distributed throughout the HAP lattice, one would expect to find many unit cells with no carbonate ion, some with one carbonate, and even a few with two or more carbonates. On the basis of charge alone, we would expect a carbonate for phosphate substitution to lead to a local increase in free energy, even after allowing for other substitutions to maintain overall charge balance. Of course, the distortion of the lattice resulting from the difference in geometry between carbonate and phosphate may also extend to neighboring unit cells and either increase or decrease the likelihood of subtitutions occuring in them. As we step back and observe the lattice, we expect to see this more or less random distribution of carbonates leading to localized domains, perhaps on the order of several unit cells, that have a distribution of carbonate content and a consequent distribution of energies.

Thus it seems entirely reasonable that a crystal lattice with an incorporated foreign ion such as carbonate would have such localized domains and, consequently, a distribution of solubilities. In fact, it would seem likely that such a distribution would be a common occurrence, rather than an exception and that as a crystal equilibrates, one would observe some sort of average behavior and a well defined (for a given degree of substitution) solubility. However, if these domains individually exhibited MES behavior (that is, little or no growth at modest saturaturations above the MES), then the overall crystal lattice would exhibit the behavior of a solid with a distribution of MES's.

It should be noted that the heterogeneous distribution of carbonate described above is but one of several possible explanations for the MES distribution behavior observed with CAP's. Any other property that could be heterogeneously distributed and affect the dissolution behavior at a domain level could lead to the MES distribution behavior. For example, defects could also be distributed in the manner described above for carbonate. In fact, one would expect a carbonate for phosphate substitution to lead to a local higher energy (than the non-substituted case) defect. The defect explanation is thus more general than the carbonate explanation. The defect explanation could be applicable to HAP if it were found to exhibit MES behavior, whereas the carbonate explanation obviously could not.

One other explanation that should be considered is the application of the Gibbs-Kelvin effect to a distribution of crystallite sizes. Crystallite size will, of course, affect dissolution rate via the specific surface area. However, the effect being considered here is not one of rate but the MES at which no further dissolution is observed and would be more closely related to the free energy of the domain in question than to the quantity of the domain exposed to solution.

The other topic to be considered is the suggestion that the behavior of a CAP might be governed by a surface complex with a stoichiometry different than that of the CAP itself, as has been suggested by others for HAP[30,32] and by us for HAP in the presence of solution fluoride[3] The fact that there are no reported solubility products for CAP's is an indication that the determination of the IAP values governing the behavior of CAP's may be a non-trivial issue. We shall later see that this is the case.

Preparation of CAP's

The CAP's used in the work described here were prepared following a modification of the procedure described by LeGeros et al.[35] in which anhydrous $CaHPO_4$ was hydrolyzed at 95°C in a solution containing $NaHCO_3$. All samples exhibited typical X-ray diffraction and IR absorption spectra. Ca/P ratios were measured by chemical analysis and carbonate content was determined from the IR spectra as described by Featherstone et al.[47], and for some samples was checked by the Conway microdiffusion method.[48]

Experimental Determination of MES Distributions

The design of an experimental procedure to characterize the MES distribution of a CAP preparation follows in a straightforward manner from the preceeding description of the MES phenomenon.

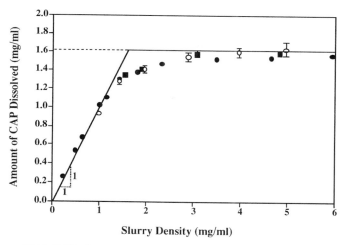

Figure 1. Amount of CAP (6.5%) dissolved in 0.1M acetate buffer (pH = 4.5, ionic strength = 0.5M) at 24-hour equilibrium vs. slurry density (the initial amount of CAP powder to solution ratio). Different symbols represent determinations by different investigators with the same sample.

A preliminary slurry density study can be done to demonstrate that the material possesses an MES distribution, as opposed to being characterized by a single MES value. Figure 1 shows the amount of CAP (carbonate content about 6.5%) dissolved in acetate buffer (0.1 mol/l, pH = 4.5, ionic strength = 0.5 mol/l) after 24 hours as a function of slurry density. For a pure material, all points for slurry densities less than the solubility should lie on the diagonal line (slope = 1) since the material should dissolve completely. For slurry densities greater than the solubility, all points should lie on a horizontal line, since an amount just equal to the solubility should dissolve, but no more. As is apparent, the observed points differ significantly from this expected behavior.

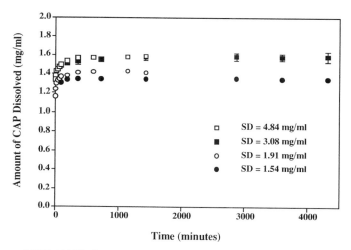

Figure 2. Amount of CAP (6.5%) dissolved in 0.1M acetate buffer (pH = 4.5, ionic strength = 0.5M) as a function of time and slurry density (SD).

To test whether the observed MES has been reached, the amount dissolved in these slurry density experiments can be monitored as a function of time. The result for several representative slurry densities is shown in Figure 2.

This figure shows that a plateau is reached at about 12 hours that persists with very little change for the 72 hours that the solutions were monitored. Hence we conclude that the 24 hour point is a fair representation of the plateau value for a given slurry density.

With evidence in hand that the CAP sample does possess a distribution of MES's, we can now design an experiment to measure this distribution. The means of doing this is suggested by studying the data shown in Figure 1.

Each of these points represents an observed solution composition, from which an observed IAP value can be calculated. From the initial slurry density and the amount of solid not dissolved, one can also calculate the fraction of the CAP that had an apparent solubility less than this observed IAP. If this exercise is carried out for each point, a dependence could be constructed for the fraction of solid dissolved as a function of solution IAP.

A much cleaner method of measuring fraction dissolved as a function of solution IAP is to actually measure the fraction of solid material dissolved in each of a series of experiments in solutions with various pre-calculated IAP's. The solution speciation calculations used to design these solutions include accounting for soluble complexes of calcium with carbonate, bicarbonate, various phosphate species and with acetate. A sodium acetate soluble complex is also included. Activity coefficients are calculated by an extended Debye-Huckel approach described by Bockris and Reddy[49] in which the reduced activity of water due to being tied up by hydration with ions is included. This correction is necessary for the high ionic strengths employed here and generally leads to agreement within 0.01 pH unit between calculated and observed pH values. All speciation calculations used in this work were carried out with EQUIL, a commercially available computer program.[50] The use of a relatively large solution volume guarantees that the composition of the solution does not change appreciably during the course of the equilibration and also necessitates the calcula-

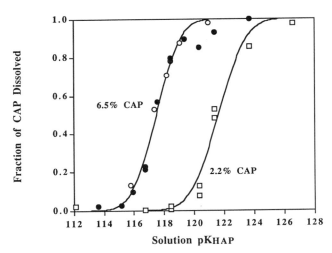

Figure 3. Cumulative solubility distribution for 2.2% CAP (□) (duplicate runs) and 6.5% CAP (○) and (●) after 24-hour equilibrium at room temperature. The abscissa is expressed as the solution ion activity product based on HAP stoichiometry. Solutions with the Ca/P ratio of 1.0 were used in these experiments except for (open circle) with solution Ca/P of 1.8. Solid lines represent best fit to the assumption of a normal distribution for MES. Reprinted from Hsu et al.[20] With permission of the publisher.

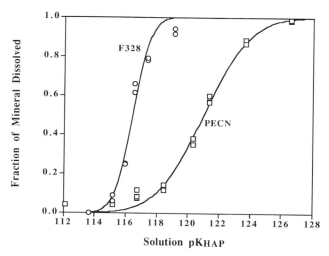

Figure 4. Cumulative solubility distribution for sample F328 (open circle) and sample PECN (open square) after 24 hours of equilibrium. The abscissa is expressed as the solution ion activity product based on HAP stoichiometry. Solid lines represent best fit to the assumption of a normal distribution for MES. Reprinted from Hsu et al.[20] With permission of the publisher.

tion of amount dissolved via gravimetric determination of the amount of undissolved crystals recovered from the experiment. Complete experimental details are available elsewhere.[20]

The cumulative solubility distributions for CAP's prepared as described above and containing 2.2% and 6.5% carbonate are shown in Figure 3. Note that the units on the abcissa are pK_{IAP} units (labeled as pK_{HAP} since the IAP used was that of HAP).

This means that larger numbers represent solutions with lower concentrations of calcium and phosphate. If these preparations had a single MES value rather than MES distributions, the plots would consist of several points with zero dissolved (for abcissa values corresponding to solutions supersaturated with respect to the solid) followed by several points with 100% dissolved (for abcissa values corresponding to solutions undersaturated with respect to the solid). By inspection we see that both these CAP samples exhibit a distribution of MES's and that the 2.2% CAP has a lower MES than the 6.5% CAP. This is consistent with the observation of numerous investigators that dissolution tendency increases with increased carbonate content.[33-36,38-44] Figure 4 shows a similar distribution for a CAP sample (F328) prepared by Dr. John Featherstone of The University of California at San Francisco.

This sample was prepared at a controlled constant temperature and pH and contained 6.4% carbonate. The figure also shows data obtained from a powdered enamel sample (PECN) provided by Dr. Conrad Naleway of the American Dental Association. This sample also exhibits a distribution of MES, noticeably broader than those from the synthetic samples.

The smooth curves in these plots represent the best fit of the cumulative fraction dissolved versus pK_{HAP} data to an error function. Differentiating this function gives the normal distributions of fraction dissolved per pK_{HAP} unit shown in Figure 5.

In this plot it is clear that the sample prepared by precipitation (F328) has a narrower MES distribution than the two samples prepared by hydrolysis (2.2% and 6.5%). Samples prepared in our laboratory by the precipitation technique also have narrower distributions than those prepared by hydrolysis. The human enamel sample (PECN) has the broadest distribution of all. This is not surprising since this enamel sample was a pooled sample.

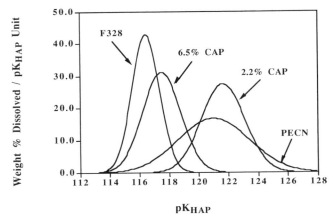

Figure 5. MES distribution for 2.2% and 6.5% CAP and for samples F328 and PECN after 24 hours of equilibration (in terms of the ion activity product expressed with respect to HAP stoichiometry). These distributions were based on the assumption of a normal distribution of MES's for each sample. Reprinted from Hsu et al.[20] With permission of the publisher.

Unfortunately, the technique for determining MES precludes the use of samples of a size representative of a selected presumably homogeneous region of a single tooth.

In each of these MES determinations we have chosen to express the solution composition in terms of the IAP for HAP, rather than for CAP. With only the data shown here, the choice of how to express the solution is arbitrary; we could just as well have chosen to express it in terms of an IAP corresponding to $CaHPO_4$ or $Ca_2(HPO_4)(OH)_2$ as has been suggested by others for HAP.[30,32] We chose not to use the obvious choice of a CAP-based IAP corresponding to the stoichiometry of the solid because solution carbonate does not have any measurable effect on the MES under the conditions employed in these studies. This is

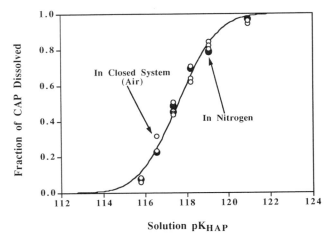

Figure 6. Effect of gas-phase environment on the determination of the MES distribution for 6.5% CAP after 24 hours of equilibrium at room temperature: (open circle), equilibrium in a closed system (air); (closed circle), equilibrium in a nitrogen chamber. Reprinted from Hsu et al.[20] With permission of the publisher.

shown by the data in Figure 6, in which MES determinations are compared for equilibration in a closed system and equilibration in a nitrogen chamber in which CO_2 is removed by continuous flushing.

If there were a carbonate common ion effect, as one might expect, the experiments conducted under nitrogen purging would have resulted in complete dissolution of the CAP samples. Instead, we see virtually no difference between the nitrogen system and the closed system in which carbon dioxide is retained in the head space in the vials in which the equilibrations are carried out. Although the calculated CO_2 partial pressure in these vials (≈ 0.004 atm) is about one order of magnitude greater than atmospheric CO_2 partial pressure[51], it could be argued that the corresponding solution total carbonate concentrations ($\approx 10^{-4}$ mol/l) are too small to rule out the possibility of a carbonate common ion effect. However, previous unpublished work conducted in our laboratory under 1 atm of CO_2 partial pressure was also unable to detect a carbonate common ion effect at pH 4.5.

The lack of a detectable carbonate common ion effect does not imply a violation of thermodynamics. Indeed, we would fully expect there to be a carbonate common ion effect if a true equilibrium were established. However, the metastable equilibrium reached in these experiments is not a true equilibrium, but rather marks a practical point demarcating the rapid dissolution kinetics observed in solutions far from equilibrium from the very much slower kinetics of the approach to true equilibrium.

At any point in the dissolution process, we might expect the rate of transfer of ions from the crystal to the solution phase to be a function of the composition of the surface of the crystal. We might also expect that if the process of ion exchange is more rapid than the net rate of dissolution, then the surface composition might be influenced by the composition of the ambient solution. In particular, we might expect that a carbonate ion on the surface might be released into solution relatively rapidly, as least in comparison to a phosphate ion. This expectation follows from the fact that carbonate would not be expected to fit as well into the structure of the apatitic lattice as does phosphate. At the same time, we have estimated the maximum carbonate concentration to be about 0.1 mmol/l, whereas the concentrations of phosphate employed in these studies is on the order of several mmol/l. Thus we expect the crystalline site vacated by the loss of carbonate to be much more likely to be filled by incorporation of a solution phosphate than by the re-incorporation of carbonate. The net result would be a carbonate-free surface complex, assumed for now to possess the stoichiometry of HAP. If the process of ion exchange is rapid relative to the net rate of dissolution, then the rate at which such a surface complex is lost into the solution might be expected to be the determinant of the dissolution rate.

The above discussion of ion exchange might suggest that growth of a phase with HAP stoichiometry might be expected to occur when the solution composition is sufficiently supersaturated with respect to the MES of a given site. Although such growth will clearly occur if the degree of saturation is high enough, the use of the MES model assumes that this degree of saturation is not reached in the course of these studies. This assumption has been tested in two ways. First, the carbonate content of residues after 4 days equilibration and 8 days equilibration have been compared and found to be not different from each other.[20] Had precipitation of a phase with HAP stoichiometry occurred, the fraction of carbonate in the residue would have decreased. Secondly, residues from equilibration studies have been re-equilibrated in solutions more saturated than the initial equilibration solution and no detectable growth of the solid phase has been found.[20] For example, the residue from a 4 day equilibration of 2.2% CAP in a buffer with $pK_{HAP} = 121.7$ was placed in a buffer with $pK_{HAP} = 118.5$ and re-equilibrated for another 4 days. No increase in the amount of solid phase was observed, in spite of the fact that the latter solution had over 50% more calcium and phosphate than did the original equilibration solution.

Applications of the MES Concept to Dissolution Kinetics

Thus far we have seen that the MES distribution concept works well in describing solution equilibration data taken at the end of 24 hours of exposure. It has already been acknowledged that the state of the system is not a true thermodynamic equilibrium at this point, but rather the end point of an initial phase where dissolution kinetics is relatively rapid. If the MES concept, with surface complexes with HAP stoichiometry, is the determinant of the dissolution driving force, as has been asserted, then these same concepts should be extensible to a model describing dissolution kinetics from time zero up to the attainment of the metastable endpoint of these experiments.

We have utilized two very different experimental approaches to this problem: dissolution studies using compressed pellets of CAP and studies using suspensions of CAP crystals. The pellet system has the advantage that the magnitude of the diffusional barrier to dissolution can be accurately calculated; the disadvantage is that there is reduced sensitivity of the model to subtleties of the kinetics of dissolution, since these subtleties are masked by the diffusion of ions out of the porous CAP pellet. Suspensions of CAP crystals, on the other hand, do not suffer this disadvantage because the diffusion of ions away from the crystal surface is fast relative to the dissolution step. The problem with this approach is that dissolution occurs so rapidly that details of the kinetics at early experimental times are not accessible by conventional experimental techniques. Regardless of which of these experimental approaches is taken, a quantitative physical model is required for critical data interpretation.

MES Distribution Based Models for CAP Dissolution Kinetics

Virtually all models for dissolution are based on a dissolution driving force equal to the solubility of the substance. In the simplest such model, the dissolution rate is directly proportional to the solubility of the substance in the ambient solution at the surface of the crystal. If the dissolution step is slow relative to the diffusion of solute away from the crystal surface, the diffusion step is ignored and the dissolution is said to be surface controlled. On the other hand, if the dissolution step is very fast, the process is said to be diffusion controlled, the solution at the crystal surface is then assumed to be saturated, and the observed rate of dissolution is given by Fick's first law applied to the aqueous diffusion layer. Although these two extremes are the cases most commonly treated by investigators, dissolution kinetics of enamel and apatites most often fall into an intermediate regime.[7,9,11,12,21,52]

Whether dissolution is surface reaction controlled, diffusion controlled, or in the intermediate regime, a key part of any dissolution model is the calculation of the solubility of the solute in the solution in which dissolution is occurring. Thus the solubility will depend on solution composition variables such as pH, buffer capacity, concentrations of calcium and phosphate, etc. The relevant solubility-like quantity to be used should be based on the observed MES and not the true thermodynamic solubility, at least for these studies in the kinetic regime where the solution is approaching the MES rather than approaching a true thermodynamic solubility. This approach, involving what we have referred to as an "effective solubility", has been extensively used in our laboratory.[1,2,6,7,10-12,52]

The obvious difference between a dissolution model for a substance with a distribution of MES's and a single effective solubility is that this distribution must be accounted for. The continuous distributions displayed in Figure 2 are discretized into a histogram composed of a finite number of elements ($\approx$ 6 - 10). For each element, we then have a relative amount and a calculated molar solubility in the dissolution medium. In the case of CAP, this solubility is calculated as the amount of CAP that would have to be dissolved in the dissolution medium to reach an IAP equal to that of the fraction of CAP being considered. We thus construct a table of molar MES's for the CAP preparation for each dissolution medium to be studied.[21,46,53]

In all the foregoing discussions, we have assumed that the MES was governed by a surface complex with the stoichiometry of HAP. This seems intuitively plausible, based on the previous discussion invoking a possible ion exchange mechanism. However, we would also like to accommodate alternative models involving surface complexes with other stoichiometries such as that of $CaHPO_4$ or octacalcium phosphate (OCP, $Ca_4H(PO_4)_3$). This is accomplished by expressing the solution concentrations in the MES determination in terms of the relevant IAP for the proposed complex. For the fraction with this IAP, the molar MES to be used in the model is the amount of CAP that would have to be dissolved in the dissolution medium to achieve a solution with the IAP based on the proposed complex. This recalculation of the apparent solubility tables allows us to do model calculations to test each of the several proposed surface complex models.

Crystallite Suspension Dissolution Kinetics Studies

A kinetic model for dissolution kinetics of crystallite suspensions has been developed by Wang.[53] A discretized MES distribution was constructed as described above and a table of molar MES's in various dissolution media prepared for each fraction of the distribution. The model assumes that: the accessible surface of each MES fraction of the distribution is proportional to the percentage of that fraction present at a given time; the total surface area of the CAP sample is proportional to the mass raised to the 0.66 power; and dissolution rate

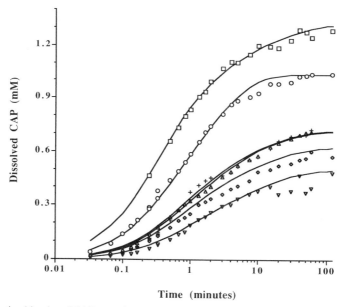

Figure 7. Dissolution kinetics of CAP crystal suspensions plotted vs. the logarithm of time. Dissolution media: open square, sink buffer (2 mg/ml); open circle, sink buffer (1 mg/ml); open triangle, acetate buffer with 30 mmol/l PO_4; plus; with 15 mmol/l Ca; open diamond, with 60 mmol/l PO_4; and open inverted triangle, with 30 mmol/l Ca. Except for the open squares, the initial slurry density of each is 1 mg/ml. Solid lines represent model simulations for dissolution in each medium.

of a given fraction is proportional to the product of its molar MES in the dissolution medium, its surface area at a given time, and a first order kinetic rate constant. A differential equation is written for the change in the mass of each fraction of the MES distribution and the resulting set of differential equations is solved numerically.

A novel experimental technique involving chemical quenching of the dissolution process has also been developed.[53] A slurry of CAP powder (50 mg / 25 ml) is sonicated and an aliquot of 0.5 ml added to a 6 ml plastic vial. Then 0.5 ml of double strength dissolution medium is added to start the dissolution. Thus the solution at the beginning of dissolution is the prescribed dissolution medium and the slurry density is 1 mg/ml. At a preset time the dissolution reaction is quenched by the addition of 1.0 ml of a solution containing 0.12 mol/l of $NaHCO_3$ and 0.3 mmol/l of sodium pyrophosphate. The quenched suspension is filtered immediately and the supernatant collected for chemical analysis. Time points can be taken as early as 2 seconds after the beginning of dissolution and, because a different aliquot is used for each time point, as many as are needed can be collected. The quenching solution has been shown to immediately stop dissolution even for solutions that are very undersaturated (i.e., at early times). Moreover, even for solutions that are near the plateau reached after longer times, there is no precipitation observed upon addition of the quenching solution. This effectiveness is a result of the quenching solution being designed to bring the solution IAP into the region where neither growth nor dissolution occurs at an appreciable rate, with some assistance from the dissolution and growth inhibitor, pyrophosphate.

The agreement between experimental data and model calculations is shown in Figure 7. Note that the time axis is logarithmic so that the entire range of data points from 2 seconds up to 2 hours is visible.

There are several points regarding this figure that are worth comment. First the data points range from 2 seconds to 2 hours - a span of three and one-half orders of magnitude. The MES distribution used as the basis for the model calculations was based on an experiment on the order of days, so it is fair to say that the MES distribution is being used to describe data ranging over at least 5 orders of magnitude.

All these model curves are calculated with a single set of physical constants. The only fitted parameters were the set of the rate constants for the MES fractions. The rate constants were fit to the data for dissolution into sink buffer for a suspension with a slurry density of 1 mg/ml. The other curves were calculated with no fitted parameters. It should be pointed out that a single common value of the rate constant for all the MES fractions did not work as well.[53] The rate constants that were used were such that the least soluble MES fraction had a rate constant about 200 times greater than that for the most soluble fraction. A constant multiplicative relationship between rate constants for successive fractions was assumed, so that the the set of kinetic rate constants correspond to fitting two parameters, rather than a single rate constant. It could be argued that this combination of first order rate law for dissolution and rate constant variation as a function of MES might be equivalent to a rate law with a different functional form.

This is an argument that has not yet been critically tested, but such a test would surely entail dissolution studies into a range of solution compositions such as various levels of calcium and phosphate common ion as has been done. The calculated dissolution curves for 4 different common ion solutions (2 levels each of calcium and phosphate) in Figure 7 show excellent quantitative agreement with the experimentally observed points over the entire time range measured.

Finally, the model accurately and quantitatively predicts the experimentally observed slurry density effect. That is, the plateau reached in the 2 mg/ml slurry density experiment is significantly higher than that reached in the 1 mg/ml slurry density experiment. This slurry density effect is *prima facie* evidence for a distribution of MES's, or for at least more than one MES. If the CAP only had a single MES, then the curves would reach exactly the same plateau.

A number of simulations have been done with various perturbations of the dissolution kinetics model to test its ruggedness.[53] These include showing that: a distribution of MES's is required (as noted above); a distribution of kinetic rate constants as described above is required; and that the ability of the model to describe the data is critically dependent upon the use of the experimentally measured MES distribution. Changing the mean MES value by one unit in either direction leads to a noticeably poorer match of model calculations with experimental data. In each of these cases of perturbing the model, the kinetic rate constants were used as adjustable parameters to fit the data for the dissolution in sink buffer of the 1 mg/ml suspension. All other simulated curves were then calculated as was done for the preparation of Figure 7.

Perhaps the most interesting perturbation of the model was in the assumption of the stoichiometric form for the dissolution rate determining surface complex. The entire data analysis was repeated using $CaHPO_4$ stoichiometry and also using the stoichiometry of octacalcium phosphate (OCP). Both these alternative stoichiometries were found markedly inferior to the assumption of HAP stoichiometry for the surface complex.[53] Thus the dissolution kinetics model, when used in conjunction with crystallite suspension dissolution experiments under the range of solution conditions used here, provides the ability to clearly discriminate between the various proposed surface complex models.

Compressed CAP Pellet Dissolution Kinetics Studies

Hodes[54] introduced the use of the rotating disk system for studying the diffusion controlled dissolution kinetics of $CaHPO_4$ pellets in 1972. Wu et al.[52] extended this model and applied it to HAP dissolution kinetics. The important extensions were to recognize that HAP dissolution in this system was intermediate between diffusion controlled and control by the kinetics of ion disengagement from crystal surfaces and to account for the fact that

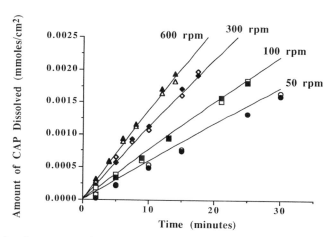

Figure 8. Comparison between experimental and model calculated dissolution kinetics for 2.2% CAP pellets in acetate buffer at rotation speeds of 50-600 rpm. Experimental results: (closed and open circle), 50 rpm; (closed and open square), 100 rpm; (closed and open diamond), 300 rpm; (closed and open triangle), 600 rpm. Closed symbols based on calcium analysis, open symbols based on phosphate analysis, plus indicates duplicate runs based on calcium analysis. Solid lines are model calculations. Reprinted from Hsu et al.[20] With permission of the publisher.

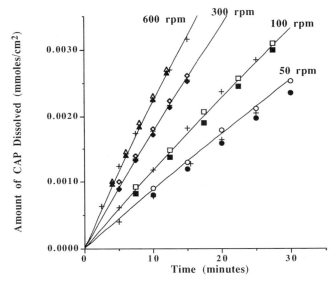

Figure 9. Comparison between experimental and model calculated dissolution kinetics for 6.5% CAP pellets in acetate buffer at rotation speeds of 50-600 rpm. Experimental results: (closed and open circle), 50 rpm; (closed and open square), 100 rpm; (closed and open diamond), 300 rpm; (closed and open triangle), 600 rpm. Closed symbols based on calcium analysis, open symbols based on phosphate analysis, plus indicates duplicate runs based on calcium analysis. Solid lines are model calculations. Reprinted from Hsu et al.[20] With permission of the publisher.

HAP pellets were porous and that aqueous solutions could readily penetrate the spaces between crystallites. The combination of porosity and relatively slow surface dissolution kinetics led to dissolution occuring from a broad zone extending from the pellet surface inward, rather than just from the pellet surface. Wu's model assumed that porosity was unchanging with time and the model was therefore applicable only to initial dissolution rate studies. Wu also showed how dissolution rate data as a function of the angular velocity of the rotating disk could be used to deduce the dissolution driving force and first order kinetic rate constant. Patel extended this model for HAP dissolution kinetics to the non-steady state case[11] and showed that it could accurately predict dissolution kinetics of dental enamel in several different weak acid buffer systems under a variety of experimental conditions. [12]

Finally, Hsu extended the non-steady state model to accommodate the measured MES distributions of CAP's[21,46], in much the same manner as has been described above for building the MES distribution into the CAP suspension dissolution model. This model provides excellent agreement with observed dissolution kinetics data for CAP's as is shown in Figures 8 and 9.

Note that the model accurately simulates the dependence on rotation speed, which is a weaker dependence than the direct proportionality to square root of rotation speed that would be expected for purely diffusion controlled dissolution. This work also included common ion studies, for which the model also replicated the experimental data very well. More importantly, the common ion data was used to test alternate models for the surface complex stoichiometry ($CaHPO_4$, OCP and CAP) and all were shown to be inferior to the HAP surface complex model, although the ability to discriminate was weaker than that of the suspension dissolution system described previously.

The compressed pellet model was also able to describe the dissolution kinetics adequately with a single value of the rate constant for all the fractions of the MES distribution[21] as opposed to the set of rate constants found necessary for the suspension dissolution kinetics studies.[53] This does not necessarily represent a conflict between the two approaches, but serves to underscore the greater sensitivity of the suspension technique to details of the dissolution kinetics. This is to be expected, since the suspension does not suffer from having these details masked by a diffusion step that is a major determinant of the observed dissolution rate in the compressed pellet system.

The compressed pellet model provides an ample description of the interplay of diffusion and surface reaction acting to determine the observed rate of dissolution of an enamel block or compressed pellet of synthetic apatite. This model does not stand on its own, however, but utilizes MES distribution data determined in independent experiments. Moreover, the details of the dissolution kinetics can be much more sensitively determined with the suspension dissolution kinetics technique. Thus these three types of studies complement each other and can be most effectively used together in a comprehensive approach to understanding the dissolution kinetics of CAP's.

The Generalizability of the MES Concept

To date, MES determination experiments have been carried out in our laboratory for about 20 different synthetic CAP preparations, as well as for human enamel. In each case, MES distributions similar to those shown in Figure 6 have been found. One particularly interesting finding is that human enamel subjected to heat treatment exhibits a similar MES distribution to untreated enamel, but with a new MES value corresponding to a low apparent solubility.[19] These distributions are shown in Figure 10.

Note that as heat treatment temperature is increased, the MES distribution is shifted toward lower solubilities (higher pK_{HAP} values). The temperature dependence of this shift in MES has a very close parallel in the decreasing dissolution rates for CAP samples treated at the same temperatures[19] as shown in Figure 11.

This similarity in sensitivity of human enamel and CAP to heat treatment, in contrast to the complete lack of responsiveness of HAP shown in Figure 11, is the basis for the

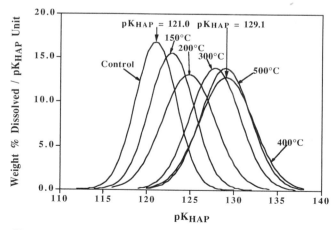

Figure 10. Effect of heat treatment on MES distribution of human enamel powders. Reprinted from Hsu et al.[20] With permission of the publisher.

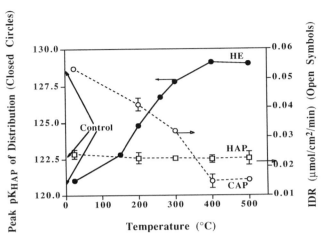

Figure 11. Mean of the MES distribution for human enamel (closed circles) and initial dissolution rates (IDR's) for HAP (open squares) and CAP (open circles) as functions of the temperature of heat treatment. Error bars represent standard deviations of duplicate runs. Reprinted from Hsu et al.[20] With permission of the publisher.

assertion that CAP is a better model for enamel than is HAP, in regards the responsiveness to heat treatment (and by implication, laser treatments designed to achieve the same temperatures at or near the surface of enamel[45,55]).

No published information yet exists on how the interaction of solution fluoride with CAP's can be interpreted by using the MES distribution model for CAP's. One might expect that a surface complex with FAP stoichiometry $(Ca_{10}(PO_4)_6F_2)$ might govern dissolution as was found by Mir using a simple HAP model for tooth enamel.[3] Such a model could be readily extended to the MES distribution, with the MES of each fraction governed by an FAP surface complex. Preliminary data indicates that such a model can, in fact, account for the influence of solution fluoride on the dissolution kinetics of CAP suspensions.[56]

Similarly, no published data is yet available for the ruggedness of the MES distribution model over ranges of pH and solution Ca/P ratios. Preliminary data[57] indicate that the MES distribution expressed as cumulative amount dissolved vs. solution pK_{HAP} determined at pH 4.5 is superimposable with the distribution determined at pH 5.5 and 6.5, even though the solution concentrations of calcium and phosphate are about 25 times less at pH 6.5 than at pH 4.5 for the same solution pK_{HAP} values.

The issue of Ca/P ratio has not yet been directly explored. However, the common ion conditions in the suspension dissolution studies do encompass the entire range of Ca/P ratios in the course of going from time zero (Ca/P = zero for phosphate common ion and Ca/P = infinity for calcium common ion) to the end of the experiment (Ca/P ratios nearer stoichiometric for either starting condition). This at least indirectly implies that the MES distribution also holds independently of the solution Ca/P ratio.

Systematic Variations of MES with Physical Parameters

It is well known that dissolution behavior of CAP's is related to the carbonate content and/or the crystallinity of the CAP's.[33,44] We would therefore expect that the mean and/or breadth of the MES distribution would also be a function of carbonate content and/or crystallinity of the CAP. To test this expectation, a series of 9 CAP samples has been prepared by hydrolysis of CaHPO4 as described by LeGeros et al.[35] Three samples with carbonate contents ranging from about 1% to about 6% were prepared at 50°C, 70°C and 95°C. As expected, both increasing carbonate content and decreasing synthesis temperature are correlated with an increase in the peak width of the 002 reflection of the powder X-ray diffraction pattern. This peak broadening has been interpreted to indicate a decrease in crystallinity. Interestingly, the mean of the MES distribution has been found to be a single-valued function of this peak width, irrespective of the carbonate content of the samples.[58,59] It has not yet been shown whether this peak broadening is mainly a function of structural disorder or crystallite size or both.

Further Applications of the MES Concept

The metastable equilibrium solubility concept provides a framework for understanding the entire range of dissolution kinetics and solution equilibration behavior over time scales ranging from seconds up to days. This time scale falls short of that required to reach a true thermodynamic equilibrium. It does, however, encompass the entire time scale of interest in the formation of dental caries. While clinical dental caries may be formed over a period of months or years, this really represents the cumulative effect of many cycles of exposure to demineralizing and remineralizing conditions, with the duration of each cycle on the order of hours at most.

The MES concept and its relationship to dissolution kinetics has been firmly established at pH 4.5.[20,21,46,53] Limited data obtained to date suggests that the MES distribution measured at pH 4.5 will also hold at pH 5.5 and 6.5, but no data is yet available on the applicability of the MES model to dissolution kinetics at these pH's. The situation with respect to solution fluoride is similar, with the MES being determined by a surface complex with FAP stoichiometry. A thorough test of the MES concept as a function of pH and solution fluoride is needed to assure the general applicability of the model over the range of solution conditions encountered in demineralization and remineralization.

The mean of the MES has been found to correlate with the crystallinity of CAP samples prepared by hydrolysis, irrespective of the carbonate content. The general applicability of this observation should be tested with samples prepared by precipitation or other techniques as well as with samples containing other ions such as fluoride, strontium or magnesium.

The MES concept has been shown to hold up for human enamel samples exposed to heat treatment at temperatures up to 500°C, with the mean of the MES distribution being shifted toward a lower solubility with increasing temperature of treatment. One would expect that laser irradiation treatments resulting in similar temperatures might induce similar shifts in the MES distribution. The establishment of a firm relationship between laser irradiation and the resulting MES distribution could provide a framework for the quantitative understanding of the effect of laser irradiation on the susceptibility of enamel to demineralization.

Finally, the MES concept, along with the surface complex model, has provided a means of describing the behavior of CAP's (and, by implication, dental enamel) over a wide range of conditions. A convincing molecular level explanation for the origins of this behavior not yet been put forth. To do so on the basis of first principles will be a worthy challenge for the future.

ACKNOWLEDGMENT

The authors wish to acknowledge support from NIDR Research Grant DE-06569 and an Undergraduate Fellowship from the Pharmaceutical Manufacturers Association.

REFERENCES

1. W. I. Higuchi, J. A. Gray, J. J. Hefferren and P. R. Patel, 1965, "Mechanisms of Enamel Dissolution in Acid Buffers." *J. Dent. Res.*, **44**(2): 330-341.
2.. W. I. Higuchi, N. A. Mir, P. R. Patel, J. W. Becker and J. J. Hefferren, 1969, "Quantitation of Enamel Demineralization Mechanisms: III, A Critical Examination of the Hydroxyapatite Model." *J. Dent. Res.*, **48**(3): 396-409.
3. N. A. Mir, W. I. Higuchi and J. J. Hefferren, 1969, "The Mechanism of Action of Solution Fluoride Upon the Demineralization Rate of Enamel." *Arch. Oral. Biol.*, **14**: 901-920.
4. M. G. Dedhiya, F. Young and W. I. Higuchi, 1973, "Mechanism for the Retardation of the Acid Dissolution Rate of Hydroxyapatite by Strontium." *J. Dent. Res.*, **52**(5): 1097-1109.
5. M. G. Dedhiya, F. Young and W. I. Higuchi, 1974, "Mechanism of Hydroxyapatite Dissolution. The Synergistic Effects of Solution Fluoride, Strontium, and Phosphate." *J. Phys. Chem.*, **78**(13): 1273-1279.
6. M. B. Fawzi, J. L. Fox and W. I. Higuchi, 1978, "A Possible Second Site for Hydroxyapatite Dissolution in Acidic Media." *J. Colloid Interface Sci.*, **67**(2): 304-311.
7. J. L. Fox, W. I. Higuchi, M. B. Fawzi and M. S. Wu, 1978, "A New Two-Site Model for Hydroxyapatite Dissolution in Acidic Media." *J. Colloid Interface Sci.*, **67**(2): 312-330.
8. E. N. Griffith, A. V. Katdare, J. L. Fox and W. I. Higuchi, 1978, "Transmission Electron Microscopic Confirmation of the Morphological Predictions of the Two-Site Model for Hydroxyapatite Dissolution." *J. Colloid Interface Sci.*, **67**(2): 331-335.
9. W. I. Higuchi, E. Y. Cesar, P. W. Cho and J. L. Fox, 1984, "Powder Suspension Method for Critically Re-examining the Two-Site Model for Hydroxyapatite Dissolution Kinetics." *J. Pharm. Sci.*, **73**: 146-153.
10. W. I. Higuchi, P. W. Cho, J. L. Fox and K. Yamamoto, 1986, "Unifying Criteria for Dissolution Kinetics of Various Hydroxyapatite Preparations." *J. Colloid Interf. Sci.*, **110**: 453-458.
11. M. V. Patel, J. L. Fox and W. I. Higuchi, 1987, "Physical Model for the Non-Steady State Dissolution of Dental Enamel." *J. Dent. Res.*, **66**: 1418-1424.
12. M. V. Patel, J. L. Fox and W. I. Higuchi, 1987, "Effect of Acid Type on the Kinetics and Mechanism of Dental Enamel Demineralization." *J. Dent. Res.*, **66**: 1425.
13. J. S. Chu, J. L. Fox and W. I. Higuchi, 1989, "Quantitative Study of Fluoride Transport during Subsurface Dissolution of Dental Enamel." *J. Dent. Res.*, **68**(1): 32-41.
14. J. S. Chu, J. L. Fox, W. I. Higuchi and W. P. Nash, 1989, "Electron Probe Microanalysis of Subsurface Demineralization and Remineralization of Dental Enamel." *J. Dent. Res.*, **68**(1): 26-31.
15. J. L. Fox, S. C. Dave and W. I. Higuchi, 1989, "Growth Kinetics of Fluorapatite Deposition on Synthetic Hydroxyapatite Seed Crystals." *J. Colloid Interf. Sci.*, **130**(1): 236-253.
16. Otsuka, M., J. Wong, W. I. Higuchi, and J. L. Fox, 1990, "Effects of Laser Irradiation on the Dissolution Kinetics of Hydroxyapatite Preparations." *J. Pharm. Sci.*, : 79:510-515.
17. J. L. Fox, D. Yu, M. Otsuka, W. I. Higuchi, J. Wong and G. L. Powell, 1992, "The Combined Effects of Laser Irradiation and Chemical Inhibitors on the Dissolution of Dental Enamel." *Caries Res.*, **26**: 333-339.
18. J. L. Fox, D. Yu, M. Otsuka, W. I. Higuchi, J. Wong and G. L. Powell, 1992, "Initial Dissolution Rate Studies on Dental Enamel after CO_2 Laser Irradiation." *J. Dent. Res.*, **71**: 1389.
19. J. Hsu, J. L. Fox, W. I. Higuchi, M. Otsuka, D. Yu and G. L. Powell, 1994, "Heat-treatment-induced Reduction in the Apparent Solubility of Human Dental Enamel." *J. Dent. Res.*, **73**(12): 1848-1853.
20. J. Hsu, J. L. Fox, W. I. Higuchi, G. L. Powell, M. Otsuka, A. Baig and R. Z. LeGeros, 1994, "Metastable Equilibrium Solubility Behavior of Carbonated Apatites." *J. Colloid Interface Sci.*, **167**: 414-423.
21. J. Hsu, J. L. Fox, G. L. Powell, M. Otsuka, W. I. Higuchi, D. Yu, J. Wong and R. Z. LeGeros, 1994, "Quantitative Relationship between Carbonated Apatite Metastable Equilibrium Solubility and Dissolution Kinetics." *J. Colloid Interface Sci.*, **168**: 356-372.
22. W. E. Brown, 1973, Solubilities of Phosphate and Other Sparingly Soluble Compounds, *Enviromental Phosphorus Handbook*, New York, Wiley and Sons.
23. W. E. Brown, P. R. Patel and L. C. Chow, 1975 "Formation of $CaHPO_4 \cdot 2H_2O$ from Enamel Mineral and its Relationship to Caries Mechanism", *J. Dent. Res.*, **54**: 475.
24. P. R. Patel and W. E. Brown, 1975, "Thermodynamic Solubility Product of Human Tooth Enamel: Powdered Sample." *J. Dent. Res.*, **54**(4): 728-736.
25. M. J. Larsen and S. J. Jensen, 1989, "The Hydroxyapatite Solubility Product of Human Dental Enamel as Function of pH in the range 4.6-7.6 at 20 °C." *Archs. Oral Biol.*, **34**: 957-961.
26. H. Mika, L. C. Bell and B. J. Kruger, 1976, "The Role of Surface Reactions in the Dissolution of Stoichiometric Hydroxyapatite." *Arch. Oral Biol.*, **21**: 697-701.
27. L. C. Bell, H. Mika and B. J. Kruger, 1978, "Synthetic Hydroxyapatite Solubility Product and Stoichiometry of Dissolution." *Arch. Oral Biol.*, **23**: 329-336.

28. R. M. H. Verbeek, H. Steyaer, H. P. Thun and F. Verbeek, 1980, "Solubility of Synthetic Calcium Hydroxyapatite." *J.C.S. Faraday I.*, **76**: 209-219.

29. G. J. Levinskas and W. F. Neuman, 1955, "The Solubility of Bone Mineral. I. Solubility Studies of Synthetic Hydroxylapatite." *J. Phys. Chem.*, **59**: 164-168.

30. V. K. LaMer, 1962, "The Solubility Behavior of Hydroxyapatite." *J. Phys. Chem.*, **66**: 973-978.

31. H. M. Rootare, V. R. Deitz and F. G. Carpenter, 1962, "Solubility Product Phenomena in Hydroxyapatite Water Systems." *J. Colloid Sci.*, **17**: 179-206.

32. M. D. Francis, 1965, "Solubility Behavior of Dental Enamel and Other Calcium Phosphates." *Ann. New York Acad. Sci.*, **131**: 694-712.

33. R. Z. LeGeros, 1967, "Crystallographic Studies on the Carbonate Substitution in the Apatite Structure." Ph.D., (New York University).

34. R. Z. LeGeros, J. P. LeGeros, O. R. Trautz and E. Klein, 1970, Spectral Properties of Carbonate in Carbonate-Containing Apatites, *Developments in Applied Spectroscopy*, New York, Plenum Press.

35. R. Z. LeGeros, J. P. LeGeros, O. R. Trautz and W. P. Shirra, 1971, Conversion of Monetite, CaHPO4, to Apatites: Effect of Carbonate on the Crystallinity and the Morphology of the Apatite Crystallites, *Advances in X-ray Analysis.*, New York, Plenum Press.

36. D. McConnell, 1973, *Applied Mineralogy*, New York, Springer-Verlag.

37. D. G. A. Nelson, 1981, "The influence of Carbonate on the Atomic Structure and Reactivity of Hydroxyapatite." *J. Dent. Res,*, **60**(C): 1621-1629.

38. M. Okazaki, Y. Noriwaki, T. Aoba, Y. Doi and J. Takahashi, 1981, "Solubility Behavior of CO3-Apatites in Solutions at Physiological pH." *Caries Res.*, **15**: 477-483.

39. D. G. A. Nelson, J. D. B. Featherstone, J. F. Duncan and T. W. Cutress, 1982, "Paracrystalline Disorder of Biological and Synthetic Carbonate-Substituted Apatites." *J. Dent. Res.*, **61**: 1274-1281.

40. M. Okazaki, Y. Noriwaki, T. Aoba, Y. Doi, J. Takahashi and H. Kimura, 1982, "Crystallinity Changes in CO_3-Apatites in Solutions at Physiological pH." *Caries Res.*, **16**: 308-314.

41. J. D. B. Featherstone, C. P. Shields, B. Khaderazad and M. D. Oldershaw, 1983, "Acid Reactivity of Carbonated Apatites with Strontium and Fluoride Substitutions." *J. Dent. Res.*, **62**(10): 1049-1063.

42. R. Z. LeGeros and M. S. Tung, 1983, "Chemical Stability of Carbonate- and Fluoride-Containing Apatites." *Caries Res.*, **17**: 419-429.

43. D. G. A. Nelson, J. D. B. Featherstone, J. F. Duncan and T. W. Cutress, 1983, "Effect of Carbonate and Fluoride on the Dissolution Behaviour of Synthetic Apatites." *Caries Res.*, **17**: 200-211.

44. J. A. Budz, M. LoRe and G. H. Nancollas, 1987, "Hydroxyapatite and Carbonated Apatite as Models for the Dissolution Behavior of Human Dental Enamel." *Adv. Dent. Res.*, **1**(2): 314-321.

45. D. Yu. ,1991, "Laser Irradiation of Tooth Enamel." Ph.D. Thesis, (University of Utah).

46. J. Hsu, 1993, "Dissolution Mechanisms of Carbonate - Containing Apatites." Ph.D. Thesis, (University of Utah).

47. J. D. B. Featherstone, S. Pearson and R. Z. LeGeros, 1984, "An Infrared Method for Quantification of Carbonate in Carbonated Apatites." *Caries Res.*, **18**: 63-66.

48. E. J. Conway, 1957, Microdiffusion Analysis and Volumetric Error, London, Crosby Lockwood & Son Ltd.

49. J. O. Bockris and A. K. N. Reddy, 1977, *Modern Electrochemistry*, New York, Plenum Press.

50. 1994, *EQUIL.*, Version 2.12, MicroMath Scientific Software, Salt Lake City, UT.

51. J. N. Butler, 1982, *Carbon Dioxide Equililbria and their Application*, Reading, MA, Addison-Wesley.

52. M. S. Wu, W. I. Higuchi, J. L. Fox and M. Friedman, 1976, "Kinetics and Mechanism of Hydroxyapatite Crystal Dissolution in Weak Acid Buffers Using the Rotating Disk Method." *J. Dent. Res.*, **55**: 496-505.

53. Z. Wang, 1994, "Dissolution Kinetics of Carbonate Apatite and Effects of Solution Ions". Ph.D. thesis, (University of Utah).

54. B. Hodes, 1972, "Dissolution Rate Studies of Some Calcium Phosphates Using the Rotating Disk Method." Ph.D. thesis, (University of Michigan).

55. D. Yu, J. L. Fox, J. Hsu, G. L. Powell and W. I. Higuchi, 1993, "Computer Simulation of Surface Temperature Profiles During CO_2 Laser Irradiation of Human Enamel." *Optical Engineering*, **32**(2): 298-305.

56. S. J. Colby, A. Baig, Z. Wang, J. Hsu, J. L. Fox and W. I. Higuchi, "Metastable Equilibrium Solubility of Carbonate Apatite in the Presence of Fluoride." *A.A.D.R. Annual Meeting*, **1995**, Abstract #1462.

57. A. Barry, H. Zhuang, Z. Wang, A. Baig, J. Hsu, J. L. Fox, W. I. Higuchi, R. Z. LeGeros and M. Otsuka, "MES Distribution of CAP as a Function of Solution Composition." *A.A.D.R. Annual Meeting*, **1995**, Abstract #1680.

58. A. Baig, J. Hsu, J. L. Fox, M. Otsuka, W. I. HIguchi, G. L. Powell and R. Z. LeGeros, "Dependence of MES Distribution on Carbonate Content and Crystallinity of CAP's." *I.A.D.R. Annual Meeting*,**1994**, Abstract #1601.
59. A. Baig, J. Hsu, Z. Wang, J. L. Fox, M. Otsuka, W. I. Higuchi and R. Z. LeGeros, "Relationship Between Crystallinity and MES of CAP's: Trace Fluroide Effect." *A.A.D.R. Annual Meeting* ,**1995**, Abstract #1461.

20

THE CALCITE - HYDROXYAPATITE SYSTEM CRYSTAL GROWTH STUDIES IN AQUEOUS SOLUTIONS

P. G. Klepetsanis, P. Drakia, and P. G. Koutsoukos

Institute of Chemical Engineering and High Temperature Chemical
    Processes
P.O. Box 1414
Department of Chemical Engineering
University of Patras
GR-26500 Patras, Greece

INTRODUCTION

Scale formation usually takes place on heat transfer surfaces because they provide favourable conditions for the precipitation processes. A number of sparingly soluble salts may be formed depending on the local chemical composition of the aqueous films in contact with the heat transfer surfaces. The rate of formation of insoluble scale is very important as it determines the layers of scale deposited.[1] Phosphate has long been used for water treatment and as an agent against calcium carbonate scale formation. The extremely low solubility however of the calcium phosphate salts may promote their formation especially in high temperature alkaline water, encountered in boilers. Mixed carbonate-phosphate scale may be formed in cases in which supersaturation with respect to these two salts is exceeded. Calcium hydroxyapatite ($Ca_5(PO_4)_3OH$, HAP) is the most stable calcium phosphate phase and may readily be formed at low supersaturations[2] or through transformation from other precursor phases at high supersaturations.[3,4,5,6] Calcium carbonate polymorphs forming in aqueous supersaturated solutions include in the order of increasing solubility calcite, aragonite and vaterite. It should be noted that the solubility product for the calcium carbonate is approximately six orders of magnitude higher than that of HAP. It is therefore anticipated that the presence of low phosphate and calcium concentrations (<0.5mM at alkaline pH) may cause the spontaneous precipitation of HAP, while the precipitation of calcium carbonate is inhibited by the presence of the phosphate ions[7] and the threshold for spontaneous precipitation at these conditions is at considerably higher concentrations.[8] The presence however of HAP crystalline deposits in an aqueous medium supersaturated with respect to calcium carbonate may selectively induce the precipitation of one of the calcium carbonate polymorphs by providing the appropriate template for crystal growth.

Mineral Scale Formation and Inhibition, Edited by Zahid Amjad
Plenum Press, New York, 1995

The present work was undertaken in order to investigate the possibility of the selective formation of calcium carbonate onto HAP substrates, and the measurement of the kinetics of the preferred overgrowth. A prerequisite to this study was the experimental identification of the stability domain of the calcium carbonate system in order to preclude spontaneous precipitation. All experiments were done at constant supersaturation.[9,10]

EXPERIMENTAL

The experiments were done in a 250 ml double walled thermostatted Pyrex® vessel at $25.0 \pm 0.1°C$. Thermostatting of the double walled vessel was done by water circulating through a thermostat. The stock solutions were prepared from solid, reagent grade chemicals (Merck, Pro analisi) dissolved in triply distilled water and were filtered through membrane filters (0.2 μm, Schleicher and Schuell). Calcium chloride and sodium chloride stock solutions were standardized by ion chromatography (Metrohm, IC 690 with conductivity detector). The sodium carbonate and sodium bicarbonate stock solutions were prepared freshly for each experiment by exact weighing the amounts of the respective solids and dissolution in triply distilled water.

The preparation of HAP seed crystals was done by mixing calcium chloride and sodium dihydrogen phosphate at pH 10, 70°C.[11] The crystalline solid was aged, filtered, washed and dried. Chemical analysis gave a molar calcium:phosphate ratio of 1.67 ± 0.01 and was characterized further by physicochemical methods including powder x-ray diffraction (Philips 1840/ 30), scanning electron microscopy (SEM, JEOL JSM 5200), infrared spectroscopy (Nicolet FTIR 740) and specific surface area measurements (multiple point nitrogen adsorption BET). The powder x-ray diffraction spectrum of the crystalline HAP preparation coincided with that of the reference material.[12] The specific surface area of the seed crystals, was found equal to 35 m^2g^{-1}.

The supersaturated solutions were prepared by mixing equal volumes, 100 ml each of calcium chloride and sodium bicarbonate solutions. The ionic strength of the supersaturated solutions was adjusted by addition of appropriate amount of sodium chloride. The pH was next adjusted to 8.50 by the addition of a standard (0.1 N) sodium hydroxide solution (Merck, Titrisol). The pH measurements were done by a combination glass / saturated calomel electrode (Metrohm) standardized before and after each experiment with NBS Buffer Solutions (pH 7.41 and 9.18 at 25.0°C).[13]

The stability of the supersaturated solutions was checked by monitoring the solution pH. The onset of the precipitation was accompanied with a pH decrease in the solution. The time lapsed between the pH adjustment of the pH of the supersaturated solution and the observed pH decrease was taken as the induction time.

The investigation of the overgrowth of calcium carbonate on HAP crystals was done in solutions which were stable for at least one week at 25°C. Following the verification of the stability of the supersaturated solutions employed, about 50 mg weighed dry of HAP crystals were suspended under constant magnetic stirring. The pH of the supersaturated solutions remained constant for a time period, after which precipitation started, accompanied by a pH-decrease due to the liberation of hydrogen ions due to the precipitation of calcium carbonate:

$$Ca^{2+} + xH_2CO_3 + yHCO_3^- + zCO_3^{2-} = CaCO_{3\,(s)} + uH^+ \qquad (1)$$

A shift of the solution pH as small as 0.05 pH units triggered the addition of titrant solutions from two mechanically coupled burets of an automatic titrator (TITRATOR Type 11 in combination with AUTO-BURET Type ABU1C, Radiometer). The added volume of titrants was recorded with a recorder (TITRIGRAPH type SBR2C, Radiometer) the pen of which was coupled with the pistons shaft.

The titrant solutions in the two burets were calcium chloride (Titrant A) and mixture of sodium carbonate and sodium bicarbonate (Titrant B) having a molar ratio of calcium to carbonate equal to 1:1, matching the stoichiometry of the precipitating solid. The supersaturated solution contained sufficient concentration of inert electrolyte (Sodium Chloride) so as to maintain the solution ionic strength, which would otherwise change, due to the release of sodium and chloride ions from the precipitating calcium and carbonate ions respectively. The titrant solutions were made as follows:

$$\text{Buret 1: } 10C_1 + 2C_1M \tag{i}$$

$$\text{Buret 2: } C_2M + 2C_3M \tag{ii}$$

where C_1 and C_3 are the total calcium and total sodium bicarbonate concentrations in the working solution and C_2 is the Na_2CO_3 concentration equal to $10C_1M$.

During the precipitation process, samples were withdrawn and filtered through membrane filters. The filtrates were analyzed for calcium by ion chromatography. The solid precipitates on the filters were characterized by physicochemical methods in order to identify the phase forming. The rates of calcium carbonate formation spontaneously and on the introduced seed crystals, R_g, were in all cases computed from the volume of titrants added:

$$R_{sp} = \frac{dV}{dt} C_t \ (\text{mol min}^{-1}) \tag{2}$$

$$R_g = \frac{dV}{dt} C_t \frac{I}{A_s} \ (\text{mol min}^{-1}) \tag{3}$$

where dV/dt is the rate of addition of titrants of concentration C_t mol l^{-1} and A_s the total surface area of the seed crystals used to inoculate the supersaturated solution.

RESULTS AND DISCUSSION

The driving force for the formation of a crystalline phase A_mB_n (n=m+n) is the average change in Gibbs free energy, per ion, ΔG, for the transition from the supersaturated solution to equilibrium and is given by:

$$\Delta G = -\frac{R_gT}{v} \times \ln \frac{(a_{A^{n+}})^M \times (a_{B^{m-}})^n}{K_s^o} = \frac{-R_gT}{v} \ln \Omega \tag{4}$$

where the parentheses denote ionic activities, T is the absolute temperature, R_g is the gas constant, K_s^o the thermodynamic solubility product of the precipitating solid phase, and Ω is the supersaturation ratio:

$$\Omega = \frac{(a_{A^{n+}})^m (a_{B^{m-}})^n}{K_s^o} \tag{5}$$

The following values were used for the thermodynamic solubility products for the calcium carbonate polymorphs : calcite, $K_s^o = 3.8 \times 10^{-9}$ M^2, aragonite, $K_s^o = 6.03 \times 10^{-9}$ M^2, and vaterite $K_s^\circ = 1.23 \times 10^{-8} M^2$.[14] The relative supersaturation, σ, with respect to each of the three polymorphs which may form in a supersaturated solution is defined as:

$$\sigma = \Omega^{1/2} - 1$$

$$= \frac{[(Ca^{2+})(CO_3^{2-})]^{1/2}}{K_s^\circ} - 1 \qquad (6)$$

where K_s° is the thermodynamic solubility product of the corresponding phase. As may be seen for the calculation of the relative supersaturation, calculation of the activities of the ionic species is needed. The ionic speciation in our system took into account all equilibria including phosphate in equilibrium with the HAP seed crystals. The equilibria considered and the corresponding stability constants are summarized in Table 1.

For the calculations the HYDRAQL program was used[16] written in FORTRAN 77 and transferred to a Vax 11/750 computer. The program begins with an initial guess (provided by the user) of the free concentration in solution of the components that define the system investigated. The Gibbs free energy is minimized using equilibrium constants for relevant reactions regardless of mass balance constraints. During subsequent iterations the solution is improved until mass balance for each component is satisfied. When solids are present, the species in solution must be in equilibrium with the solid phase(s) present. In this case, mass balance for all components must be satisfied and no solubility products may be exceeded. Mathematically, simultaneous consideration of all mass balance constraints results in a system of non-linear algebraic equations which is solved with the Newton-Raphson iterative method. At each iteration the Jacobian of the system is formed and the resulting system of linear equations is solved using the Gaussian elimination-substitution method.

Table 1. Equilibria and corresponding thermodynamic formation
constants for the $Ca(OH)_2$ - H_3PO_4 - H_2O system; 25°C[15]

Equilibrium	LogK°
$H^+ + HCO_3^- = H_2CO_3$	6.35
$H^+ + CO_3^{2-} = HCO_3^-$	10.33
$Ca^{2+} + CO_3^{2-} = CaCO_3$	3.15
$Ca^{2+} + HCO_3^- = CaHCO_3^+$	1.00
$H^+ + PO_4^{3-} = HPO_4^{2-}$	12.18
$H^+ + HPO_4^{2-} = H_2PO_4^-$	7.25
$H^+ + H_2PO_4^- = H_3PO_4^\circ$	2.21
$Ca^{2+} + PO_4^{3-} = CaPO_4^-$	6.54
$Ca^{2+} + HPO_4^{2-} = CaHPO_4^\circ$	1.50
$Ca^{2+} + H_2PO_4^- = CaH_2PO_4^+$	2.83
$Ca^{2+} + OH^- = CaOH^+$	1.32
$H^+ + OH^- = H_2O$	13.997

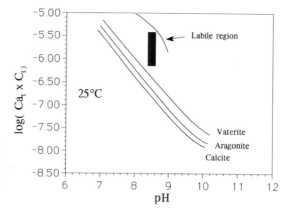

Figure 1. Solubility isotherms for calcium carbonate polymorphs showing the metastable zones (stable and labile) for pH ranges between pH 8.0 - 9.0, and the ranges of experimental conditions (shades areas).

The calcium carbonate supersaturated solutions were selected so that they were stable. All experiments were done at pH 8.50 and the supersaturation, defined by the total carbonate, C_t, correspond to the shaded area, shown in Figure 1, in which the experimental conditions with respect to the equilibrium (solubility) isotherms for the calcium carbonate polymorphs are presented. The line separating the labile region from the stable supersaturated solutions area is also shown over the pH range between 8.0-9.0 and it was defined as the locus of points corresponding to calcium carbonate solutions in which precipitation took place past induction times as long as a few days.

Our experiments have shown that the inoculation of the calcium carbonate supersaturated solutions with HAP seed crystals resulted in the formation of calcite past the lapse of induction periods as long as 65 hours. The experimental conditions and the results obtained are presented in Table 2.

As may be seen, the induction times measured were inversely proportional to the solution supersaturation. A semiempirical relationship between the induction times and the initial concentration of the solute in the supersaturated solutions is given by Christiansen and Nielsen[17]:

$$t = k_p C_o^{1-p} \tag{7}$$

In eq. (7) k_p is a constant C_o the initial solute concentration and p an integer corresponding to the number of growth units in the critical nucleus, i.e. to the critical cluster size. Plots of the logarithm of the measured induction times as a function of the logarithm of the initial

Table 2. Experimental conditions, induction times and subsequent crystal growth rates for the crystallization of calcite on HAP seed crystals at constant supersaturation. pH 8.50, 25°C

Exp #	Ca_t $\times 10^{-3}$ M	C_t $\times 10^{-4}$ M	σ	Induction time hr	R_g $\times 10^{-9}$ mol min^{-1} m^{-2}
1	3.92	1.50	0.420	3.83	12.0
2	3.44	1.31	0.341	15.30	10.0
3	3.05	1.17	0.258	13.30	5.1
4	2.95	1.13	0.238	21.00	4.0
5	1.98	0.76	-0.0983	55.00	1.4
6	1.78	0.68	-0.188	65.00	1.0

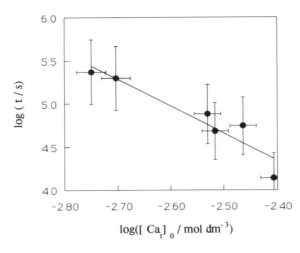

Figure 2. Kinetics of heterogeneous nucleation of calcite on HAP seed crystals at constant solution supersaturation; pH 8.5, 25°C. Plot of the logarithm of induction times as a function of the initial calcium concentration in the supersaturated solutions. Total calcium = Ca_t, total carbonate = C_t.

calcium concentration in the supersaturated solutions is shown in Figure 2. From the slope of this line a value of p=4 was computed for the size of the critical nucleus of the calcite crystals growing onto HAP. The same value has been reported for the heterogeneous crystallization of calcium carbonate monohydrate on sulfonated polystyrene and polystyrene divinylbenzene polymeric substrates.[18]

It is important for heterogeneous nucleation to proceed, to exibit a satisfactory lattice compatibility between the substrate and the overgrowth.[19] The lattice mismatch, d, between two crystal faces is given by:

$$d = \frac{a_s - a_o}{a_o}$$

(8)

where a_s and a_o are the stress-free lattice parameters of the substrate and of the overgrowth respectively.[20] Equation 8 is also applicable for the integer multiples of the stress-free lattice parameters. The (1010) face of calcite e.g. with net dimensions at 29.92A(90°) gave a misfit with the (1010) face of HAP (linear dimensions 28.26A x 34.4A (90°)) of 5.9% and 0.8% respectively.[21] Moreover comparison with the dimensions of 110 face of HAP 27.52A x 32.62A (90°) yielded a mismatch of 4.4% and 8.1% respectively. The favourable lattice matching is perhaps one of the factors suggesting that calcite may be a favourable substrate for HAP[22] and vice versa. The selective deposition of calcite on the HAP substrate was demonstrated by experiments in which the amount of the seed crystals was varied. Thus using 100 mg of HAP instead of 50 mg to inoculate the supersaturated solutions, the induction times remained the same while the rates were doubled, suggesting a constant number of active sites. A difficulty presented in the experimental work which added to the uncertainty of the measured kinetics parameters was the long induction times and the very slow rates of calcite overgrowth. Extension at the induction times and reduction of the rates was anticipated from earlier seeded crystal growth studies of calcite which have shown that the presence of very low concentrations of orthophosphate had a strong inhibitory effect.[7]

The calcite overgrowth of the HAP seed crystals was identified by the powder x-ray A comparison diffraction spectra shown in Figure 3 and the scanning electron micrographs presented in Figure 4 (a and b). As may be seen the rhombohedral calcite crystals outgrow the microcrystalline (<1 μm) HAP crystals.

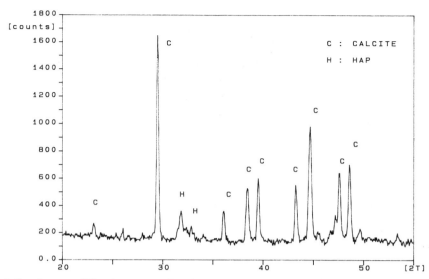

Figure 3. Powder x-ray diffraction spectra of calcite grown on HAP at constant supersaturation. pH 8.50, 25°C.

As may be seen from the data of Table 1, the rates of calcite formation showed a marked dependence on the solution supersaturation. The rates measured showed a parabolic dependence:

$$R_g = K_g \sigma^2 \tag{9}$$

where K_g is the rate constant. The kinetics plot is shown in Figure 5. Similar results have been obtained by other researchers from the seeded growth of calcite[23] and have been interpreted as a spiral growth mechanism for the heterogeneous growth of calcite.

It is interesting to note that the rates of calcite crystal growth were considerably lower (2-3 orders of magnitude) in comparison with the rates obtained for the seeded growth of calcite in the presence of orthophosphate (0-0.1 μM). This may be attributed to the higher (about 1 μM) levels of inorganic orthophosphate in equilibrium with the HAP seed crystals and also to the lattice matching factors. In any case, in combination with the independence of the measured kinetics parameters on the amount of the inoculating HAP crystals, it may be concluded that the calcite growth takes place selectively on the HAP surface. Should the precipitation be due to crystallization on suspended foreign particles (e.g. dust) the rates would be considerably higher and independent on the total surface area of the inoculating seed crystals.

CONCLUSIONS

The experiments in which stable supersaturated solutions were inoculated with HAP seed crystals showed that calcite may form selectively onto this substrate which provided a favourable lattice matching. The overgrowth was preceded by long induction times inversely proportional to the solution supersaturation while the subsequent crystallization rates showed a parabolic dependence on the relative solution supersaturation suggesting a surface controlled spiral growth mechanism.

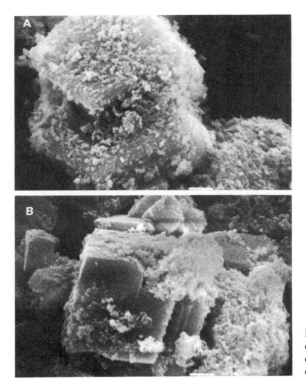

Figure 4. Scanning electron micrographs of calcite rhombohedral crystals growing on microcrystalline HAP seed crystals at constant supersaturation.

ACKNOWLEDGMENTS

Partial support of this work by the General Secretariat for Research and Technology, Hellas and of the Commission of the European Union, Contract No JOU2-CT92-0108 is gratefully acknowledged.

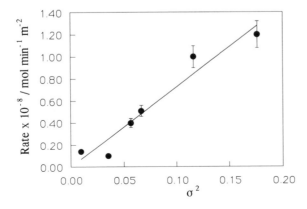

Figure 5. Kinetics of crystallization of HAP on calcium carbonate seed crystals at constant supersaturation. pH 8.50, 25°C. Plot of the rate of crystal growth as a function of the relative supersaturation.

REFERENCES

1. Betz Handbook of Industrial Water Conditioning, 8th Ed. Trevose PA. 1980, Betz Laboratories p. 86.
2. P.G. Koutsoukos, Z. Amjad, M.B. Tomson, G.H. Nancollas (1980). Crystallization of calcium phosphate. A constant composition study, *J. Amer. Chem. Soc.*, 102: 1553-1557.
3. G.H. Nancollas, B. Tomazic (1974), Growth of calcium phosphate on hydroxyapatite crystals. Effect of supersaturation and Ionic Strength medium, *J. Phys. Chem.*, 78: 2218-2225.
4. G.H. Nancollas, B. Tomazic, M.B. Tomson (1976), The precipitation of calcium phosphates in the presence of magnesium, *Croat Chim. Acta.*, 48; 431-438.
5. W. Kibalczyr, I.V. Melikhov, V.F. Komarov (1982) in Industrial Crystallization 81, p. 291, eds S.J. Jancic and E.J. de Jong, North Holland, Amsterdam.
6. Feenstra TP, de Bruyn P.L., (1979), Formation of calcium phosphates in moderately supersaturated solutions. *J.Phys. Chem.*, 83:2235-2240.
7. E.K. Giannimaras, P.G. Koutsoukos (1987), The crystallization of calcite in the presence of orthophosphate, *J.Colloid & Interface Sci.*,116: 423-432.
8. A.G. Xyla, E.K. Giannimaras, P.G. Koutsoukos (1991), The precipitation of calcium carbonate in aqueous solutions, *Colloids & Surfaces*, 53: 241-255.
9. M.B. Tomson, G.H. Nancollas (1978). Mineralization kinetics: A Constant Composition Approach, *Science*, 200: 1059-1061.
10. T.F. Kazmierczak, M.B. Tomson, G.H. Nancollas (1982), Crystal growth of carbonate. A controlled composition kinetic study, *J. Phys. Chem.*, 86: 103-108.
11. G.H. Nancollas, M.S. Mohan (1970), The growth of hydroxyapatite crystals, *Arch. Oral Biol.*, 15: 731-745.
12. JCPDS ASTM Card No 9-77.
13. R.G. Bates, (1973) Determination of pH, Theory and practice, 2nd ed. J. Wiley, New York, 479pp.
14. Plummer N.L., Wigley T.M.L., Parkhurst D.C. (1978), The Kinetics of Calcite Dissolution in CO_2-Water Systems at 5° and 60°C and 0.0 to 1.0 atm CO_2, *Amer. Journal of Science*, 278:179-216.
15. Martel A.C., Smith R.M. (1981), Critical Stability constants, Plenum, vol. 4, New York.
16. Papelis C., Hayes K.F. Leckie, O.J. (1988), HYDRAQL: A program for the computation of chemical equilibrium composition of aqueous batch systems including surface complexation modeling and ion adsorption at the oxide / solution interface, Tech, Report No 306, Stanford University, Stanford CA, 130 pp.
17. Mullin J.W. (1972), Crystallization 2nd ed. CRC Press p. 482.
18. Dalas E., Kallitsis J., Koutsoukos P.G. (1988), The crystallization of calcium carbonate on polymeric substrates, *J. Crystal Growth*, 89: 287-294.
19. Myerson A.S., Ginde R (1993), Crystals, crystal growth and nucleation, in Handbook of Industrial Crystallization H.S. Myerson Editor, Butterworth - Heinemann, Boston pp 33-63.
20. Newkirk J.B., Turubull D. (1955), Nucleation of ammonium iodide crystals from aqueous solutions, *J. Appl. Physics*, 26: 579-583.
21. Lonsdale K., (1968), The solid state epitaxy as a growth factor urinary calculi and gallstones, *Nature*, 217: 56-58.
22. Koutsoukos P.G., Nancollas G.H., (1981), Crystal growth of calcium phosphates-epitaxial considerations, *J. Crystal Growth*, 53: 10-19.
23. Söhnel O., Garside J. 1992, Precipitation Basic Principles and Industrial Applications, Butterworth-Heinemann, Oxford, pp 318-323.

21

ASSOCIATION BETWEEN CALCIUM PHOSPHATE AND CALCIUM OXALATE CRYSTALS IN THE DEVELOPMENT OF URINARY STONES

S. R. Khan

Department of Pathology
College of Medicine
University of Florida
Gainesville, Florida

INTRODUCTION

Urinary stones consist of crystals and matrix with crystals accounting for up to 98% of the stone weight.[1] Calcium oxalate (CaOx) and calcium phosphate (CaP) are the two most common calcific crystals in human urinary stones and usually occur together.[2,3,4] Since human urine is not sufficiently supersaturated for homogeneous nucleation of CaOx, suggestions have been made that CaP acts as the substrate for heterogeneous nucleation of CaOx. Such a possibility is investigated in this chapter.

COMPOSITION OF URINARY STONES

CaOx exists in three different forms; calcium oxalate monohydrate (COM), calcium oxalate dihydrate (COD) and calcium oxalate trihydrate (COT). All three forms have been seen in human urinary stones.[5] COM is the most common while COT the rarest. In most urinary stones, CaP is present as apatite which occurs in two forms, hydroxylapatite (HAP) and carbonate apatite. Hydroxylapatite is generally encountered in stones developed in sterile urine while carbonate apatite is found in stones formed in association with infection. CaP is sometimes present as brushite (calcium hydrogen phosphate dihydrate) and on rare occasions as whitlokite (b-tri-calcium phosphate) or octacalcium phosphate.

Results of various studies of mineral composition of urinary stones from different countries and regions indicate a presence of some 35 different crystalline components.[2,3,4,6] COM is the most common crystal and a major component of majority of urinary stones. Apatite is the second most common crystal while COD is third. Approximately 2/3 of the stones contain more than one type of crystal and CaOx/HAP is the most common combina-

Mineral Scale Formation and Inhibition, Edited by Zahid Amjad
Plenum Press, New York, 1995

tion. Otnes et. al. analyzed 245 stones from Norwegian patients[2] and found 30% with CaOx alone, 56% with CaOx and CaP and 3% with only apatite. They also found that pure CaOx stones were more common in men (40%) than in women (7%), while pure CaP stones were more frequent in women (11%) than men (2%). CaOx stones were also more common in upper urinary tract while CaP stones frequented the bladder.

Both HAP and carbonate apatite occur as small spherulitic aggregates of needle-shaped or plate-like crystals.[7] COM has a plate-like tabular habit. COM crystals are radially organized inside the stones and seen as dumbbell-shaped aggregates in urine. COD has a distinctive bipyramidal tetragonal form. Smaller fine-grained powdery crystals of apatite commonly appear in the spaces between larger crystals of COM and COD (Figure 1). Since stones with COD crystals have many interstices their CaP contents are generally higher than that of COM stones. Similar crystals of apatite have also been seen occupying the core of mixed CaOx/CaP stones. Less common is the habit where apatite forms thin solid laminations between layers of COM. Such apatite layers appear harder and less brittle.

NEPHROCALCINOSIS

Nephrocalcinosis refers to CaP deposition in renal parenchyma, is common in humans[8,9] and can be found in almost all the kidneys. The deposits are located in the interstitium and basement membrane or renal epithelial cells. CaP deposits in tubular lumen

Figure 1. Fractured surface of a human CaOx stone showing powdery CaP in association with more solid COM. Plate-like habit of COM crystals (arrows) is also evident. Magnification = 450X.

are often called intranephronic calculi. Such deposits have been suggested to precede the development of kidney stones.[10] Malek and Boyce examined fifty two kidneys from CaOx stone patients and found intranephronic calculi in every one of them.[11] Randall found calcification plaques in renal papillae of 19.6% of 1154 pairs of kidneys he examined.[12] 28.5% of 500 spontaneously passed stones examined by Cifuentes-Delatte contained nidi of calcified renal tubules.[13]

CRYSTALLURIA

Crystalluria refers to urinary excretion of crystals. Human urine is often super-saturated beyond the metastable limit with respect to a variety of stone forming salts including calcium phosphate and calcium oxalate.[1,14] As a result crystals of both calcium phosphate and calcium oxalate are common in urinary voidings from both normal and stone forming people. CaOx is generally encountered in urine with a pH between 5.0 and 6.2 whereas at a pH above 6.2 crystalluria particles are predominantly calcium phosphate.[15] Urinary crystals of CaOx are either COM or COD while CaP crystals are generally hydroxylapatite (HAP) or with amorphous morphology.[16] Urinary HAP crystals have poor crystallinity.[17] CaP exists in many phases whose precipitation is critically dependent upon pH of the milieu.[18] Forms such as brushite and octacalcium phosphates are preferentially formed in more acidic circumstances such as found in human urine while HAP is more likely under basic conditions. Brushite is rare as crystalluria particle but has been seen in the urine from stone patients with primary hyperparathyroidism or idiopathic calcium urolithiasis. It is also interesting to note that only the urine from calcific stone formers is supersaturated for brushite.[19,20] Brushite was the first precipitate to appear at a pH below 6.9, when urinary supersaturation was experimentally increased by addition of calcium chloride.[21] Since brushite is rarely found in urinary stones it was proposed that brushite nucleates CaOx and then transforms into the more stable HAP. Octacalcium phosphate, a very rare constituent of urinary stones, has similarly been proposed to nucleate CaOx and then transform into apatite. Because brushite does not form at a pH of 5.6 or less, CaOx stone disease might be prevented by lowering the urinary pH; but acidification of CaOx stone formers urine did not produce desired results.[22,23]

Many investigators have reported the presence of CaOx and CaP crystals in urine from both normal and stone formers but clear differences in crystalluria between the two groups have not been demonstrated. Studies of Werness et. al.[16,20] however, revealed that stone formers as a group void more crystals in their urine and more often than normal controls and that hydroxyapatite is the most common crystal in urine from both normal controls and stone formers. In one study they examined urinary crystals in 4,835 total voidings obtained from 16 normal controls, 45 patients with primary hyperparathyroidism, 12 with primary hyperoxaluria and 89 with idiopathic calcium urolithiasis.[16] Seventy four percent of voidings from normal controls, 21% from stone patients with primary hyperparathyroidism, 8% from primary hyperoxaluria patients and 34% from patients with idiopathic urolithiasis were negative for crystals. Fifty percent urinary voidings with crystalluria of normal controls, 70% of stone patients with primary hyperparathyroidism and 30% of patients with idiopathic calcium urolithiasis contained HAP as the only crystal. All patients with hyperoxaluria had COM as the major crystal usually mixed with small amount of COD crystals. Patients with primary hyperoxaluria produced only CaOx crystals. In another study Werness et al. compared crystalluria between 96 idiopathic stone formers and 20 normal controls.[20] Sixty three percent of the normal voidings had no crystals as compared to 37% of voidings from stone patients. HAP was present in urine from 80% of normal controls and 96% of stone patients. It was a major crystal in urine from 75% of normal controls and 83% of stone patients and was the only crystal in urine from 60% of normal controls and 32% of stone patients. 45% of the stone patients had both HAP and CaOx together in at least one voiding. Only 3% of the normal controls showed such a mixed crystalluria.

EPITAXIAL RELATIONSHIPS BETWEEN CALCIUM PHOSPHATE AND CALCIUM OXALATE

Since most urinary stones contain more than one mineral phase and oriented growth appears common, it has been suggested that epitaxy may be a major mechanism involved in development and growth of urinary stones of mixed composition.[24] Studies of actual faces in contact to determine epitaxy is a difficult problem. Most of the research has been limited to calculations of lattice misfits for the common stone components. It has been calculated that a lattice misfit of no greater than 10-20% is favorable for epitaxial growth.[25] Studies of Mandel and Mandel have demonstrated that large number of matches are possible at 15% misfit but are dramatically reduced at 10%, nearing 0 for most stone components at 5%.[26] To understand the relationship between CaOx and CaP, the most common combination in urinary stones, a number of experimental in vitro studies have been carried out. Using solution depletion seeded crystallization techniques, Meyer et. al. showed that addition of HAP seeds to a metastable solution of CaOx induced crystallization of COM on the seed.[27] Brushite was also shown to induce nucleation of COM but induction time was longer than with HAP.[28] On the other hand COM was able to induce crystallization of brushite but not of HAP from a metastable solution of CaP in conditions suitable for their respective nucleation. Recent studies using constant composition crystallization system, confirmed the ability of HAP to induce nucleation of COM and inability of COM to induce nucleation of HAP.[29] Calcium oxalate trihydrate was however shown capable of nucleating HAP. A very interesting study was performed by Ebrahimpour et. al. who pretreated the HAP seeds with human serum albumin and then incubated them in a metastable CaOx solution at a constant supersaturation.[30] Pretreatment of seeds resulted in reduction in crystallization induction time and increase in rate of subsequent COM crystallization. HAP surfaces covered with albumin appear to be more effective in inducing formation of COM.

Above mentioned studies clearly demonstrate that many of the seed crystals of calcium containing urinary stone salts can provoke crystallization of a second crystalline phase. But the studies do not prove that epitaxy was involved. Only that heterogeneous nucleation took place.

Recently Achilles et. al. evaluated the ability of CaP crystals to induce nucleation of CaOx crystals utilizing a dynamic flow model of crystallization in gels.[31] First a metastable solution of calcium phosphate was coursed over the surface of 1 wt % agar gel at 37° for 4-8h. This resulted in the formation of about 200 μm diameter hemispherulitic particles of CaP in the gel matrix. These particles contained poorly crystalline carbonate apatite in the center and well crystallized radially arranged plate-like crystals of octacalcium phosphate on the periphery. CaOx crystals were then generated by flowing a metastable solution of calcium oxalate over the gel surface containing the CaP crystals. Scanning electron microscopic examination showed them to be plate-like crystals of COM which appeared to grow in continuation with the plate-like crystals of octacalcium phosphate. This study is one of the few to actually demonstrate epitaxially mediated growth of COM on a CaP crystal.

CRYSTAL/ORGANIC MATRIX ASSOCIATION IN URINARY STONES

All urinary stones contain a ubiquitous organic matrix which is intimately associated with the crystalline phases. Demineralization of stones and crystalluria particles by EDTA treatment in the presence of aldehyde fixatives causes dissolution of crystals resulting in the appearance of crystal ghosts.[32,33,34] Despite the total loss of mineral contents, stones and crystals for the most part maintain their shapes because matrix material is pervasive throughout the crystals. COM ghosts appear tabular, COD ghosts have pyramidal shape and apatite keep their spherulitic morphology. CaOx crystal ghosts are delineated by an organic outer coat while ghosts of spherulitic CaP show organic material radially organized around a central space. Thus CaOx/CaP crystal interface is actually occupied by organic matrix material associated with the two types of crystals.

CALCIUM PHOSPHATE/CALCIUM OXALATE ASSOCIATION IN ANIMAL MODELS OF NEPHROLITHIASIS

Cut surface of a mammalian kidney shows two basic zones, an outer cortex and inner medulla. Inner zone of the medulla is called papilla. Architectural and functional unit of the kidney is called nephron and includes glomerulus and its tubule. Each tubule starts from the glomerulus in the cortex as proximal tubule, and follows a convoluted path before becoming straight. Then the tubule narrows to a thin descending limb and crosses into the medulla where it forms a U-shaped loop of Henle. After forming a loop it crosses back into the cortex, becomes wider and later becomes distal tubule. Distal tubules of adjacent nephrons give rise to collecting ducts. A series of collecting ducts join to make ducts of Bellini which have their openings at the papillary tips.

Nephrolithiasis or kidney stone formation refers to deposition of crystals in the kidneys and starts in lumina of renal tubules. Studies of CaOx nephrolithiasis in rats[35] have demonstrated that hyperoxaluria alone can induce CaOx crystalluria and nephrolithiasis without the assistance of any CaP phase. Acute hyperoxaluria promotes the diffuse deposition of CaOx crystals throughout the kidney while a sustained low level chronic hyperoxaluria causes preferential deposition of CaOx crystals in the renal papilla.[36] Cellular degradation products in the form of membranous vesicles are consistently seen in association with crystal deposits.

CaP nephrolithiasis can be experimentally induced in rats by various treatments which cause increased urinary excretion of calcium[37] or phosphate[38] or decreased excretion of magnesium.[39] CaP preferentially deposits in the renal tubules at the cortico-medullary junction and crystals are closely associated with cellular degradation products.[40]

To further our understanding of the relationship between CaP and CaOx crystal deposition in the kidneys we studied the phenomenon in male and female rats.[41] Renal deposition of CaP was induced by nutritional modifications giving rats a semi-purified diet, AIN-76A. This diet meets nutritional requirements but results in CaP deposition in kidneys of female rats. CaOx deposition was produced by administration of ethylene glycol, a hyperoxaluria-inducing lithogen, and ammonium chloride, a urine acidifying agent. It was hypothesized that induced CaP crystallization followed by hyperoxaluria would result in CaOx deposition on preexisting crystal deposits of CaP. One group of male and female rats was put on CaOx inducing regime and another on CaP inducing regime. A third group received CaP inducing diet for first 2 weeks followed by CaP inducing diet in association with hyperoxaluria inducing regime of ethylene glycol and ammonium chloride. Last group received CaP inducing diet followed by hyperoxaluria inducing protocol. Urine was analyzed for crystalluria and CaOx and CaP relative supersaturation. Kidneys were examined for crystal deposition.

No significant differences were noticed in CaOx relative supersaturation between male and female rats. CaP relative supersaturation was generally lower in males than females. Both male and female rats receiving CaOx inducing treatments produced CaOx crystals in their urine. Only female rats on CaP inducing protocol developed renal deposits of CaP and only male rats on CaOx inducing treatment developed renal deposits of CaOx. CaP deposits were located at the cortico-medullary junction of the kidneys (Figure 2) while CaOx deposits were located in collecting ducts of the renal medulla (Figure 3).

Results indicate an association between an individual's gender and nephrolithiasis. Under similar urinary CaOx and CaP relative supersaturation male rats were prone to form CaOx, while female rats were susceptible to produce CaP. Moreover CaP and CaOx did not deposit in the same location within the kidneys. It is obvious that CaP is not necessary for renal deposition of CaOx and even pre-existing CaP deposits do not provoke an overgrowth of CaOx. Gender dependent crystallization of CaOx or CaP may be the reason why more men develop CaOx stones while women produce CaP stones.[2,3] In a separate experiment a group of female rats received AIN diet and regular water for 4 weeks and then switched to regular diet and hyperoxaluria inducing treatment of 0.75% ethylene glycol and 2% ammonium chloride in drinking water for next 18 days. Both CaP and CaOx deposited in rat kidneys. Crystals of CaP were located at the cortico-medullary junction and were more

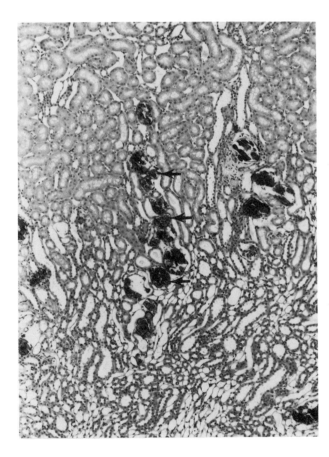

Figure 2. Light micrograph showing CaP deposits (arrows) in renal tubules at the cortico-medullary junction of a female rat kidney. Upper segment of the kidney is cortex and lower segment is medulla.

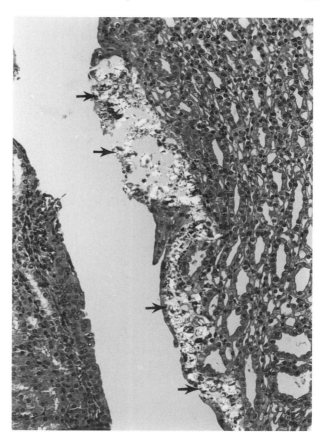

Figure 3. Light micrograph showing birefringent COM crystal deposits (arrows) in medullary collecting ducts of a male rat kidney.

abundant than CaOx crystals. The later were randomly dispersed as small deposits in the kidneys, but no direct association was evident between CaOx and CaP crystals. Thus even when CaP crystals were present in the kidneys to act as substrate for nucleation of CaOx and conditions were suitable for CaOx precipitation from a urine with higher urinary oxalate and obviously metastable for CaOx, CaOx crystals did not form in association with CaP.

CONCLUSION

Based on the many observations, that CaP is the most common crystal in human urine, is ubiquitous in human urinary stones, is seen at the center of mixed CaOx/CaP urinary stones and can induce CaOx precipitation from a metastable solution, it has been proposed that in the absence of appropriate CaOx supersaturation, CaP is the heterogenous nucleator of CaOx and nidus for renal stone formation. Experimental studies have demonstrated that CaP and CaOx crystals deposit at different locations in the kidneys, CaP at the cortico-medullary junction and CaOx in the renal papilla. Perhaps urinary environment is more suited for CaP precipitation in earlier segments of the renal tubules and for CaOx crystallization in the later segments. Evidence has been provided that urine in thin descending limb of the loop of Henle is metastable for CaP and can support crystallization of HAP.[42] Recent studies of Luptak et. al. have also indicated that urinary conditions in proximal and distal segments of the renal tubules may be more suitable for precipitation of CaP while those in collecting ducts more amenable for CaOx crystallization.[43] It has been proposed that HAP formed in earlier segments of the renal tubules migrates to the collecting ducts of the renal papilla where urine is more concentrated and conditions are suitable for heterogeneous nucleation of CaOx. Results of *in vivo* experiments however indicate that crystals formed in the proximal segments usually stay there. Moreover it should be realized that urine spends only a few minutes in the renal tubules which is not enough time for nucleation and retention of CaOx.[44] Formation of CaP substrate crystals and then heterogeneous nucleation and growth of CaOx will take much longer making it even more difficult for crystals to reach a size large enough for retention and development of a stone.

Crystals formed in the urine and seen either as crystalluria particles or as constituents of urinary stones are always coated with a layer of organic material. Thus only the organic matrix material covering the crystals appears accessible for heterogeneous nucleation. Therefore, even though a crystal may be able to act as substrate for nucleation as shown by *in vitro* studies, it may not be available to do so *in vivo*. Studies of Campbell et. al.[45] and Ebrahimpour et.al.[30] demonstrate that CaP crystals coated with certain macromolecules are better substrate for CaOx nucleation than the native crystals themselves, clearly indicating the importance of the organic material coating the crystals in heterogenous nucleation of CaOx.

Crystalluria is an indication of supersaturation and not necessarily of nephrolithiasis. It is common in both stone formers as well as normal people. Although all patients with hyperparathyroidism demonstrate hydroxyapatite crystalluria, only some produce kidney stones. Presence of CaP and CaOx crystals in the urine and urinary stones does not necessarily mean nucleation of one by the other, only that urine can be appropriately supersaturated with respect to the two types of crystals. CaP crystals present in the CaOx stones may simply represent crystalluria particles trapped between the larger CaOx crystals. The situation is similar to that existing in struvite stones where magnesium ammonium phosphate hexahydrate crystals are seen in association with carbonate apatite not because one nucleated the other but because urinary conditions were suited for crystallization of both during stone formation.

What about the presence of CaP in the stone nidus? In the stones with CaP as nidus, stone center may be formed by aggregation and retention of CaP crystals which are then enclosed by a mantle of organic matrix material. The retained mass of CaP crystals coated with organic material provide platform for further crystal deposition similar to foreign body encrustation.[46] Randall's plaques which are calcifications at the papillary tips may similarly act as encrustation platforms and become nidi for formation of renal stones. Organic matrix of urinary stones contain membranous as well as fibrillar and amorphous substances. Organic matrix,[47] its lipids as well as cellular membrane from the renal proximal tubular epithelial cells,[48] are all capable of inducing CaOx crystals from metastable solution of CaOx.

Development of multicrystalline kidney stones does not appear to be a result of heterogeneous nucleation of one crystal by the other but because conditions are suitable for precipitation of various constituent crystals. Heterogeneous nucleation of calcific crystals is most probably induced by membranous cellular degradation products which are abundant in the mammalian urine.

ACKNOWLEDGMENTS

This work was supported in part by National Institutes of Health grants #5PO1-DK20586 and 5RO1-DK41434. Prof. Raymond L. Hackett kindly reviewed the manuscript.

REFERENCES

1. Finlayson, B., 1974, Renal lithiasis in review, *Urol. Clin. North Am.* 1:181-212.
2. Otnes, B., 1983, Urinary stone analysis, methods, materials and value, *Scand. J. Urol. Nephrol.* Supplement 71: 1-109.
3. Leusmann, D.B., Blaschke, R., and Schmandt, 1990, Results of 5035 stone analyses: a contribution to epidemiology of urinary stone disease, *Scand. J. Urol. Nephrol.* 24:205-210.
4. Murphy, B.T., and Pyrah, L.N., 1962, The composition, structure and mechanism of the formation of urinary calculi, *Br. J. Urol.* 34:129-159.
5. Heijnen, W., Jellinghaus, W., and Klee, W.E., 1985, Calcium oxalate trihydrate in urinary calculi, *Urol. Res.* 13:281-283.
6. Khan, S.R., 1992, Structure and development of calcific urinary stones, in *Calcification in Biological Systems*, Bonucci, E., Ed., CRC press, Boca Raton, 345-363.
7. Khan, S.R., and Hackett, R.L., 1986, Identification of urinary stone and sediment crystals by scanning electron microscopy and x-ray microanalysis, *J. Urol.* 135:818-825.
8. Burry, AF, Axelson, R.A., Trolove, P., and Saal, J.R., 1971, Calcification in the renal medulla, a classification based on a prospective study of 2,261 necropsies, *Human Pathol.* 7:435-449.
9. Laberke, H.G., 1988, Pathological anatomy, etiology, and pathogenesis of nephrocalcinosis, in *Nephrocalcinosis, calcium antagonists and kidney*, Eds., Bichler, K.-H., and Strohmaier, W.L., Springer Verlag, Berlin, 67-104.
10. Drach,G.W., and Boyce, W.H., 1972, Nephrocalculosis as a source for renal stone nuclei. Observations on humans and squirrel monkeys and on hyperparathyroidism in the squirrel monkey, *J. Urol.*, 107: 897-904.
11. Malek, R.S., and Boyce W.H., 1973, Intranephronic calculosis: its significance and relationship to matrix in nephrolithiasis, *J. Urol.* 109: 551-555.
12. Randall, A., 1940, The etiology of primary renal calculus, *Intl. Abst. Surg.* 71: 209-240.
13. Cifuentes-Delatte, L., Minon-Cifuentes, J.L.R., and Medina J.A., 1985, Papillary stones, calcified renal tubules in Randall's plaques, *J. Urol.* 133: 490-494.
14. Robertson, W.G., Peacock, M., and Nordin, B.E.C., 1968, Activity products in stone-forming and non-stone-forming urine, *Cln. Sci.* 34:579-594.
15. Robertson, W.G., Peacock, M., and Nordin, B.E.C., 1969, Calcium crystalluria in recurrent renal-stone formers, *Lancet* 2:21-24.
16. Werness, P.G., Bergert, J.H., and Smith, L.H., 1981, Crystalluria, *J. Crystal Growth* 53:166-181.
17. Tozuka, K., Yonese, Y., Konjika, T., and Sudo, T., 1987, Study of calcium phosphate crystalluria, *J. Urol.* 138:326-329.
18. Nancollas, G.H., Lore, M., Perez, L., Richardson, C., and Zawacki, S.J., 1989, Mineral phases of calcium phosphate, *Anat. Rec.* 224:234-241.
19. Berland, Y., Boistelle, R., and Olmer, M., 1990, Urinary supersaturation with respect to brushite in patients suffering calcium oxalate lithiasis, *Nephrol. Dial. Transplant.* 5: 179-184.
20. Werness, P.G., Wilson, J.W.L., and Smith, L.H., 1984, Hydroxyapatite and its role in calcium urolithiasis, in *Urinary stone*, Eds., Ryall, R., Brockis, J.G., Marshall, V., and Finlayson, B., Churchill Livingstone, London, New York, 273-277.
21. Pak, C.Y.C., Eanes, E.D., and Ruskin, B., 1971, Spontaneous precipitation of brushite in urine: evidence that brushite is the nidus of renal stones originating as calcium phosphate, *Proc. Nat. Acad. Sci.* 68:1456-1460.
22. Prien, E.L., 1975, Calcium oxalate renal stones, *Ann. Rev. Med.* 26:173-179.
23. Finlayson, B., 1977, Calcium stones: some physical and clinical aspects, in *Calcium metabolism in renal failure and nephrolithiasis*, Ed., David, D.S., Wiley, New York, pp. 337-382.
24. Lonsdale, K., 1968, Epitaxy as a growth factor in urinary calculi and gallstones, *Nature* 217:56-58.
25. Frank, F.C., and Van der Merwe, J.H., 1949, One dimensional dislocations (I) static theory, (II) misfitting monolayers and oriented overgrowth, *Proc. R. Soc. Lond. [Biol.]* 198A:205-211.
26. Mandel, N., and Mandel, G., 1990, Epitaxis in renal stones, in *Renal Tract Stone, Metabolic Basis and Clinical Practice*, Eds., Wickham, J.E.A., and Buck, A.C., Churchill Livingstone, New York, 87-101.
27. Meyer, J.L., Bergert, J.H., and Smith, L.H., 1975, Epitaxial relationships in urolithiasis: the calcium oxalate monohydrate - hydroxyapatite system, *Clin. Sci. Mol. Med.* 49: 369-374.
28. Meyer, J.L., Bergert, J.H., and Smith, L.H., 1977, Epitaxial relationships in urolithiasis: the brushite - whewellite system, *Clin. Sci. Mol. Med.* 52:143-148.
29. Koutsoukos, P.G., Sheehan, M.E., and Nancollas, G.H., 1981, Epitaxial considerations in urinary stone formation, *Invest. Urol.* 18:358-363.

30. Ebrahimpour, A., Perez, L., and Nancollas, G.H., 1991, Induced crystal growth of calcium oxalate monohydrate at hydroxyapatite surfaces. The influence of human serum albumin, citrate, and magnesium, *Langmuir* 7:577-583.

31. Achilles, W., Jockel, U., Schaper, A., Burk, M., and Riedmiller, H., 1994, In vitro formation of urinary stones: generation of spherulites of calcium phosphate in gel and overgrowth with calcium oxalate using a new flow model of crystallization, *Scann. Microsc.*, In press.

32. Khan, S.R., Hackett, R.L., 1984, Microstructure of decalcified human calcium oxalate urinary stones, *Scann. Elect. Microsc.* 2:935-941.

33. Khan, S.R., Hackett, R.L., 1993, Role of organic matrix in urinary stone formation: an uktrastructural study of crystal matrix interface of calcium oxalate monohydrate stones, *J. Urol.* 150:239-245.

34. Khan, S.R., Hackett, R.L., 1993, Backscattered electron imaging of urinary stone matrix, *Scanning* 15, Suuplement III:84-85.

35. Khan, S.R., Hackett, R.L., 1985, Calcium oxalate urolithiasis in the rat: is it a model for human stone disease? A review of recent literature, *Scann. Elect. Microsc.* 2:759-774.

36. Khan, S.R., 1991, Pathogenesis of oxalate urolithiasis: lessons from experimental studies with rats, *Am. J. Kid. Dis.* 17:398-401.

37. Duffy, J.L., Suzuki, Y., and Churg, J., 1971, Acute calcium nephropathy, early proximal tubular changes in the rat kidney, *Arch. Path.* 91:340-351.

38. Ritskes-Hoittinga, J., Lemmens, A.G., Danse, L.H.J.C., and Beynen, A.C., 1989, Phosphorus-induced nephrocalcinosis and kidney function in female rats, *J. Nutr.* 119:1423-1431.

39. Oliver, M.D., MacDowell, M., Whang, R., and Welt L.G., 1966, The renal lesions of electrolyte imbalance, IV. The intranephronic calculosis of experimental magnesium depletion. *J. Exp. Med.* 124: 263-277.

40. Nguyen, H.T., and Woodard, J.C., 1980, Intranephronic calculosis in rats, *Am. J. Pathol.*100: 39-56.

41. Khan, S.R., and Glenton, P.A., 1995, Deposition of calcium phosphate and calcium oxalate crystals in the kidneys. *J. Urol.*, 153:811-817.

42. DeGanello, S., Asplin, J., and Coe, F.L., 1990, Evidence that tubule fluid in the thin segment of the loop of Henle normally is supersaturated and forms a poorly crystallized hydroxyapatite that can initiate renal stones. *Kidney Intl.* 37: 472.

43. Luptak, I., Bek-Jensen, H., Fornander, A.M., Hojgaard, I., Nilsson, M.A. and Tiselius, H.G., 1994, Crystallization of calcium oxalate and calcium phosphate at supersaturation levels corresponding to those in different parts of the nephron. *Scann. Microsc.* 8: 47-51, 1994.

44. Finlayson, B., and Reid, F. 1978, The expectation of free and fixed particles in urinary stone disease. *Invest. Urol.* 15: 442-447.

45. Campbell, A.A., Ebrahimpour, A., Perez, L., Smesko, S.A., Nancollas, G.H., 1989, The dual role of polyelectrolytes and proteins as mineralization promoters and inhibitors of calcium oxalate monohydrate, *Calcif. Tissue Itl.* 45:122-128.

46. Khan, S.R., and Hackett, R.L., 1985, Developmental morphology of calcium oxalate foreign body stones in rats, *Calcif. Tissue Intl.* 37:165-173, 1985.

47. Khan, S.R., Shevock, P.N., and Hackett, R.L., 1988, In vitro precipitation of calcium oxalate in the presence of whole matrix or lipid components of the urinary stones, *J. Urol.* 139:418-422.

48. Khan, S.R., Whalen, P.O., and Glenton, P.A., 1993, Heterogeneous nucleation of calcium oxalate crystals in the presence of membrane vesicles, *J. Crystal Growth* 134:211-218.

22

COMPOSITION AND SOLUBILITY PRODUCT OF A SYNTHETIC CALCIUM HYDROXYAPATITE

Chemical and Thermal Determination of Ca/P Ratio and Statistical Analysis of Chemical and Solubility Data

M. Markovic,[1] B.O. Fowler,[2] M.S. Tung,[1] and E.S. Lagergren[3]

[1] American Dental Association Health Foundation
   Paffenbarger Research Center
[2] National Institute of Dental Research, NIH
   Bone Research Branch Research Program
[3] Statistical Engineering Division
   Computing and Applied Mathematics Laboratory
[1,2,3] National Institute of Standards and Technology
   Gaithersburg, Maryland 20899

INTRODUCTION

Synthetic calcium hydroxyapatites (HA's) have, in general, excellent biocompatibility with tooth and bone tissues, and they are frequently used as biomedical materials.[1,2,3,4] The properties of synthetic HA's depend strongly on preparative conditions. They are well known to be variable[5] and the composition deviates from the stoichiometric HA formula, $Ca_{10}(PO_4)_6(OH)_2$, that comprises the unit-cell content. In order to properly characterize these HA's, accurate determinations of complete chemical composition, unit-cell parameters, solubility, surface area, crystallinity, etc., are needed. The focus of this paper is on the procedures and methods for accurate and precise determinations of two important parameters, the Ca/P molar ratio and solubility product, of a HA that was prepared and characterized in our group[6] for use as a HA reference material (HA-RM).

The Ca/P molar ratio reflects HA stoichiometry and purity. The Ca/P values determined from chemical analyses of calcium and phosphorus contents in HA-RM were compared with Ca/P values independently determined from thermal-product analyses of the heated HA-RM. The solubility of the HA-RM was determined at 37°C in suspensions with a large solid/solution ratio of 20 g l^{-1} and an initial concentration of phosphoric acid of ~5 mmol l^{-1}. The ionic strengths of equilibrated solutions were $I \approx 0.01$ mol l^{-1}. The calculated thermodynamic solubility product, K_{sp}, for HA-RM was compared with a previously published solubility product of HA determined under similar conditions.

Detailed statistical analyses of the experimentally determined data (calcium and phosphorus contents in HA-RM, pH, calcium and phosphate concentrations in equilibrated solutions) and of the calculated data (Ca/P molar ratio and K_{sp}) were performed. The statistical analysis was also utilized to assess the homogeneity of the whole batch of this HA-RM powder.

MATERIALS AND METHODS

Preparation

Calcium hydroxyapatite was synthesized by solution reaction of calcium oxide and phosphoric acid in accordance with the procedure of McDowell et al.[7] The precipitated HA-RM powder (~900 g), ~0.1 to 0.5 μm in crystal size[6], was dried at 105°C for 1 day. The associated water content of the dried HA-RM was 1.59 ± 0.05 mass % as determined by thermogravimetric analyses.[6]

Sampling Plan and Measurement Strategy

The HA-RM was placed in 180 bottles, each containing 5 g samples, and stored under ambient conditions. Ten bottles were randomly sampled from a population of 180 by first stratifying the bottles (in sequence) into 10 blocks of 18 bottles each and randomly selecting one bottle from each block. This was done to provide a representative set of bottles from the entire sequence. Two samples from each bottle were taken so that a test of material heterogeneity could be performed. Two samples from each of ten bottles were analyzed for calcium and phosphorus content and two samples from each of six bottles were used for solubility determination. In order to reduce confounding between possible material differences among bottles and from differences due to sample preparation and measurement drift, the order of sample preparation and measurement were each separately randomized.

Chemical Analysis of Ca and P Content

Calcium was determined by atomic absorption spectroscopy, AAS, with a Perkin-Elmer Model 603 spectrophotometer by using an air-acetylene flame and the 422.7 nm wavelength line. Analyses were performed directly on 200 ± 0.1 cm³ of ~0.02 mol l⁻¹ hydrochloric acid solution that contained 1000 ppm $LaCl_3$ and weighed amounts of ~2.9 mg (relative standard uncertainty, r_i = 0.11%) of HA sample. Standard calcium solutions (500 ± 0.2 cm³) contained HCl (~0.02 mol l⁻¹), 1000 ppm $LaCl_3$ and weighed amounts of ~7.0 mg (r_i = 0.043%) of calcium carbonate (NIST SRM 915, dried at 250°C for 2 h). These concentrations of sample and standard calcium solutions gave absorbances of ~0.500 (instrumental uncertainty of ± 0.001 absorbance unit). The absorbance was determined from four to twelve consecutive measurements (integration times, 10 or 16 s) for each sample and standard solution. The concentration of calcium in each sample was obtained by interpolation between two known calcium standards bracketing the unknown sample concentration within ± 3%.[8] The absorbances of these two standards were determined before and after absorbance determination of each sample.

Phosphorus was determined colorimetrically[9] as the phosphovanadomolybdate complex with a Cary Model 219 spectrometer and a wavelength of 420 nm; differential absorbance measurement was not used.[9] The phosphate solutions (100 ± 0.08 cm³) contained vanadomolybdate reagent and weighed amounts of ~8.0 mg (r_i = 0.038%) of HA-RM sample. Standard phosphate solutions (100 ± 0.08 cm^3) contained vanadomolybdate reagent and weighed amounts of ~6.5 mg (r_i = 0.046%) of potassium dihydrogenphosphate (Baker Ultrex Reagent, dried at 105°C for 2 h). Absorbances were obtained at essentially the same time of color development (~1 h) in the same quartz cell versus a water blank. These concentrations of sample and standard phosphorus solutions gave absorbances of ~0.8000 (instrumental uncertainty of ± 0.0008 absorbance units). The concentration of phosphorus in each sample was obtained by interpolation between two known phosphorus standards bracketing the unknown sample concentration within ± 3%.[8] The Ca/P molar ratio was calculated from the experimentally determined values of mass % of Ca and P with the values of their relative atomic masses 40.08 and 30.974, respectively.

Thermal-product Analysis

Samples of HA-RM were heated in three different ways: (a) at 850°C in air at ~50% relative humidity (r.h.) for times ranging from 16 to 309 h, (b) at 1000°C in air (~50% r.h.) for 24 h and then continued at 850°C in air (~50% r.h.) for 130 h, and (c) at 1000°C in a steam atmosphere (100 kPa) for 10 h. These heated samples were examined by X-ray diffraction to identify and quantify possible additional phases of α-, β-$Ca_3(PO_4)_2$ (α- and β-TCP) and CaO that may be present with the HA phase.[10]

Solubility

The saturated solutions with respect to HA were obtained by dissolution of HA-RM crystals in aqueous solutions of phosphoric acid. The phosphoric acid solution (5.026 mmol dm^{-3}) was prepared by diluting a stock solution (~0.5 mol l^{-1} H_3PO_4) with freshly boiled double-distilled water. About 400 mg of HA-RM was added to 20 cm^3 of 5.026 mmol l^{-1} phosphoric acid solution. This suspension was equilibrated in a capped 25 cm^3 polyethylene bottle completely wrapped in parafilm and kept at 37.0 ± 0.1°C for 60 days in a continuously shaken water bath. Twelve samples were prepared and equilibrated in this way.

The pH of each equilibrated suspension was obtained from emf measurements (in mV) performed with a Metrohm combined glass electrode connected to a Metrohm/Brink-mann pH meter Model 104. The emf measurements were carried out in the suspensions in original bottles shaken in a water bath at 37°C; the electrode was held in the bottleneck by means of a tight-fitting rubber stopper that excluded air. The measurements in standard buffer solutions were done in the same manner, and they were carried out before and after sample measurements. The emf reached a constant value ~5 min after the beginning of measurement. Two to four emf measurements were obtained for each suspension and buffer solution; the uncertainty in emf values was ± 0.1 mV. Two commercial buffer solutions (Fisher; pH = 4.63 ± 0.02 and 5.00 ± 0.01 at 25°C) and a buffer 0.05-molal solution of potassium hydrogen-phthalate, NIST SRM, pH standard (pH = 4.025 ± 0.005 at 37°C) were used for calibration. A straight line was fitted by least squares from the mean emf values versus pH values of the corresponding buffer solutions; this line had a slope of -59.30 and correlation coefficient of -0.99998, and the residual standard deviation was 0.252 based on 3 points.

The calcium and phosphate concentrations were determined in diluted supernatants by AAS and colorimetry, respectively, generally in the manner explained in the section on Chemical Analysis of Ca and P content. The supernatant was separated from the suspension as follows: the suspension was centrifuged at 10,000 rpm and the liquid phase was filtered through a 0.1 μm Millipore filter.

The solubility product, K_{sp}, of HA-RM defined as

$$K_{sp}(HA) = a^5 (Ca^{2+})\, a^3(PO_4{}^{3-})\, a(OH^-) \tag{1}$$

was calculated with a computer program[11] from measured equilibrium calcium and phosphate concentrations and pH values as input data. The dissociation constants of phosphoric acid (equilibria 1 to 3), the dissociation constant of water (equilibrium 4), and stability constants of the complex species (equilibria 5 to 8) used in computation[7,11] are listed in Table 1. The ionic activity coefficients, γ_i, were calculated from the extended Debye-Hückel equation

$$\log \gamma_i = - A\, z_i^2\, \sqrt{I} / (1 + B\, b_i\, \sqrt{I}) \tag{2}$$

where $A = 0.5232$ and $B = 0.3316$ at 37°C[12], z_i and b_i are valence charge and interatomic distance parameter[13] of ion i, and I is ionic strength of the solution, defined as $I = 0.5\Sigma c_i z_i^2$, where c_i is the concentration of corresponding ion.

Statistical Analysis

Heterogeneity Test. A test of material heterogeneity was performed for each of the estimated parameters (Ca and P content and solubility product) by means of analysis of variance (ANOVA).[14] This analysis assesses whether the parameter value differs across bottles by comparing the variation in measurements between bottles to variation in measurements within bottles. If variation in measurements between bottles is suitably greater (as determined by a statistical test of significance) than variations in measurements within bottles, then it is concluded that the material is heterogeneous, otherwise there is no evidence to suspect heterogeneity.

Assessment of Uncertainties. Uncertainties were assessed by use of the CIPM (International Committee for Weights and Measures) approach.[15] The specific quantity subject to measurement is called the measurand. Generally the result of a measurement will be an estimate of the measurand and therefore has uncertainty associated with it. The uncertainty of a measurement result typically consists of several components. These components may be grouped into one of two categories, named A and B. Type A components of uncertainty are those that can be evaluated by statistical means, i.e., from measurement data. Type B components of uncertainty are evaluated by other means, examples of which will be given in Results and Discussion. An estimated standard deviation called a standard uncertainty, u_i, is used to represent each component of uncertainty. An example of a type A standard

Table 1. Thermodynamic equilibrium constants[a] used in computation

No.	Equilibrium	K ($I = 0$, 37°C)
1.	$H_3PO_4 \rightleftharpoons H^+ + H_2PO_4^-$	6.22×10^{-3}
2.	$H_2PO_4^- \rightleftharpoons H^+ + HPO_4^{2-}$	6.58×10^{-8} (K_2)
3.	$HPO_4^{2-} \rightleftharpoons H^+ + PO_4^{3-}$	6.84×10^{-13} (K_3)
4.	$H_2O \rightleftharpoons H^+ + OH^-$	2.40×10^{-14} (K_w)
5.	$Ca^{2+} + H_2PO_4^- \rightleftharpoons CaH_2PO_4^+$	6.99
6.	$Ca^{2+} + HPO_4^{2-} \rightleftharpoons CaHPO_4^0$	355
7.	$Ca^{2+} + PO_4^{3-} \rightleftharpoons CaPO_4^-$	2.90×10^6
8.	$Ca^{2+} + OH^- \rightleftharpoons CaOH^+$	20

[a]References 7 and 11.

uncertainty (for component i) is the standard deviation (or standard error) of the sample mean, $s(\overline{X}_i)$, where

$$s(\overline{X}_i) = s_i/\sqrt{n_i} \tag{3}$$

and

$$s_i = \sqrt{[\Sigma(X_{ik} - \overline{X}_i)^2/(n_i - 1)]} \tag{4}$$

where s_i is the standard deviation of the n_i sample measurements, having $n_i = n - 1$ degrees of freedom (the denominator of equation 4). Here $u_i = s(???_i)$. Examples of type B uncertainty calculations are given in reference 15.

A combined standard uncertainty, u_c, of a measurement result is intended to represent, at the level of one standard deviation, the combined effects of all components of uncertainty. It is computed by the method of propagation of uncertainties (propagation of errors)[15,16], which is briefly described.

Often the measurand depends on quantities $X_1, \dots, X_N$, estimated by $x_1, \dots, x_N$, through the functional relationship

$$y = f(x_1, \dots, x_N) \tag{5}$$

where y is the estimate of the measurand. In the case of independent X_i's, the method of propagation of uncertainties states that

$$u_c^2 = \Sigma c_i^2 u_i^2 = \Sigma[u_i(y)]^2 \tag{6}$$

where $c_i = \partial f/\partial x_i$ evaluated at the x_i's and $u_i(y) = |c_i| u_i$. Note that $u_i(y)$ can also be viewed as a standard uncertainty, but as a standard uncertainty in y due to x_i. In this paper this method is applied to functions of the form

$$y = x_1^{p_1} \cdot x_2^{p_2} \cdot \dots \cdot x_n^{p_n} \tag{7}$$

So, for example, if $y = x_1^3/x_2$, then the powers of x_1 and x_2 are $p_1 = 3$ and $p_2 = -1$, respectively. For equation 7 the $u_i(y)$ can be expressed as

$$u_i(y) = |p_i| r_i y \tag{8}$$

where $r_i = u_i/\overline{X}_i$ is the relative standard uncertainty.

Additionally, the CIPM approach recommends giving an expanded uncertainty, U, as

$$U = 2u_c. \tag{9}$$

RESULTS AND DISCUSSION

Ca/P Molar Ratio

Chemical Analyses. The test of material heterogeneity performed for Ca and P contents determined in twenty samples (two from each of ten bottles of HA-RM) gave no

Table 2. Components of uncertainty for calcium content

No. (i)	Source	Standard uncertainty (mass %) $u_i(y)$	$\lvert p_i \rvert$	Relative standard uncertainty r_i (%)	Degrees of freedom (v_i)	Type A or B
1.	Replicate measurements	0.04181	1	0.107	19	A
2.	Uncertainty in mass for calcium standard	0.01684	1	0.043	29	A
3.	Uncertainty in volume for calcium standard solution	0.01566	1	0.040	∞	B
	Combined standard uncertainty, u_c	0.04772				
	Expanded uncertainty, $U = 2u_c$	0.09544				

evidence that the HA-RM batch was heterogeneous. As a result the samples were treated as being taken from a large batch of HA-RM, irrespective of bottle, so that determinations of Ca and P contents could be treated as twenty independent replicate measurements.

The estimate of the calcium content in HA-RM, 39.15 mass %, was the mean of 20 measurements. The sources of uncertainty in this estimate are listed in Table 2. The standard uncertainty for the sample mean of 20 replicate measurements, obtained with equation 3, was the largest source of uncertainty at 0.042 mass % ($r_i = 0.107\%$). The relative standard uncertainties, r_i, for the mass and volume measurements used for estimating the concentration of the calcium standards were 0.043% and 0.040%, respectively. The former was estimated from 30 measurements (type A), while the latter was given as the flask specification (type B). Since the mass concentration is the ratio of mass and volume, from equation 7, $p_i = 1$ and -1, respectively; then, the standard uncertainty, $u_i(y)$, for the mass and volume measurement was calculated with equation 8 where $\lvert p_i \rvert = 1$ and $y = 39.15$ mass %. The combined standard uncertainty, u_c, and expanded uncertainty, U, were obtained as 0.048 and 0.095 mass %, respectively (Table 2), with equations 6 and 9. Therefore the value for Ca content, expressed as mean ± U, was 39.15 ± 0.10 mass %.

A similar statistical analysis of the data was performed for mass % of phosphorus in HA-RM. The estimate of phosphorus content was the mean of 20 sample measurements, 18.18 mass %. The r_i's for the mass and volume measurements for phosphorus standards were 0.046% (based on 30 measurements, type A) and 0.08% (100 ± 0.08 cm^3 from the flask specification, type B), respectively (Table 3). The largest source of uncertainty was the uncertainty in volume of phosphorus standard solutions, with $u_i(y) = 0.015$ mass %. The value for P content in HA-RM, expressed as mean ± U, was 18.18 ± 0.04 mass %.

Table 3. Components of uncertainty for phosphorus content

No. (i)	Source	Standard uncertainty (mass %) $u_i(y)$	$\lvert p_i \rvert$	Relative standard uncertainty r_i (%)	Degrees of freedom (v_i)	Type A or B
1.	Replicate measurements	0.00745	1	0.041	19	A
2.	Uncertainty in mass for phosphorus standard	0.00836	1	0.046	29	A
3.	Uncertainty in volume for phosphorus standard solution	0.01454	1	0.080	∞	B
	Combined standard uncertainty, u_c	0.01836				
	Expanded uncertainty, $U = 2u_c$	0.03672				

Table 4. Components of uncertainty for Ca/P molar ratio

No. (i)	Source	Standard uncertainty u_i (y)	$\lvert p_i \rvert$	Relative standard uncertainty r_i (%)	Degrees of freedom (v_i)	Type A or B
1.	Uncertainty in calcium determination	0.002028	1	0.122	∞	B
2.	Uncertainty in phosphorus determination	0.001680	1	0.101	∞	B
	Combined standard uncertainty, u_c	0.002634				
	Expanded uncertainty, $U = 2u_c$	0.005268				

The Ca/P molar ratio of HA-RM obtained from the mean values for Ca and P content was 1.664. The uncertainty in this value is due to the uncertainties in the calcium and phosphorus content determinations. The relative standard uncertainties, r_i, for Ca and P contents were 0.12% and 0.10%, respectively (Table 4). These r_i's are relative combined uncertainties calculated from the combined standard uncertainties (Tables 2 and 3) and mean values of calcium and phosphorus content. The expanded uncertainty, U, for the Ca/P molar ratio was 0.005 (Table 4).

The accuracy of Ca and P determinations was inferred from the mass sum of all constituents detected in HA-RM.[6] The mass percents of all constituents with expanded uncertainties (except expanded uncertainties for SiO_3 and cationic impurities) were as follows and they sum to 100.0 ± 0.2 mass %: Ca (39.15 ± 0.10), PO_4 (55.16 ± 0.15), HPO_4 (0.592 ± 0.030), H_2O (1.59 ± 0.05), CO_3 (0.032 ± 0.002), SiO_3 (0.041), cationic impurities (0.018) and OH (3.37 ± 0.12, calculated from charge balance). This total sum of 100.0 ± 0.2 mass % indicated a high degree of accuracy for the calcium and phosphate values that together account for 94.9% of the total mass.

From the chemical analyses data, the number of Ca and H atoms (molar content of H is equal to molar content of HPO_4) normalized to 6 PO_4 groups per unit cell, was 9.985 ± 0.026 and 0.063 ± 0.003, respectively, and the OH content, calculated from the charge balance difference, was 2.033 ± 0.070. The content of impurities (CO_3, SiO_3, and cationic impurities) was neglected in this calculation. By use of these data the composition of HA-RM can be expressed with the general formula

$$Ca_{10-x}H_{2x-y}(PO_4)_6(OH)_{2-y} \tag{9a}$$

where $x = 0.015 \pm 0.026$ and $y = -0.033 \pm 0.070$. The content of the associated water molecules is not included in the general formula.

Thermal-Product Analyses. The thermal reaction of the calcium-deficient HA with the general formula $Ca_{10-x}H_{2x-y}(PO_4)_6(OH)_{2-y}$, which decomposed into HA, TCP and H_2O, can be described by the equation:

$$Ca_{10-x}H_{2x-y}(PO_4)_6(OH)_{2-y} \rightarrow (1-x)Ca_{10}(PO_4)_6(OH)_2 + 3xCa_3(PO_4)_2 + (2x-y)H_2O. \tag{10}$$

By using this equation, the Ca/P molar ratio can be calculated from the value of the experimentally determined mass fraction of TCP, w_{TCP}, in the thermal-product mixture:

$$\text{Ca/P molar ratio} = (10 - x)/6 \tag{11}$$

Table 5. Ca/P Molar ratios of calcium hydroxyapatites (HA) determined
by chemical and thermal-product analyses

Sample	Thermal analysis			Chemical analysis Ca/P
	Products	Mass %	Ca/P	
HA-RM[a]	HA	99.0 ± 0.3	1.6649 ± 0.0005	1.664 ± 0.005
	β-TCP	1.0 ± 0.3		
HA-N[b]	HA	82 ± 3	1.635 ± 0.005	1.631 ± 0.008 1.638 ± 0.01
	α-, β-TCP	18 ± 3		1.659 ± 0.003

[a]This paper. [b]Reference 20

where

$$x = 1.0796 \, w_{TCP} / (1 + 0.0796 \, w_{TCP}). \tag{12}$$

The contents of β-TCP determined in thermal-product mixtures after heating HA-RM samples at different conditions were as follows[10]: (a) The HA-RM heated at 850°C for 16 to 19 h in air at ~50% r.h. contained a very minor second phase of β-TCP (0.3 ± 0.1 mass %); heating for longer times (20 to 309 h) at 850°C did not cause any significant increase in β-TCP content. HA is stable on heating at 850°C in air at 50% r.h.[17] At 1000°C in air at 50% r.h., HA partially dehydroxylates to form oxyhydroxyapatite[10,17]; however, HA can be stabilized and dehydroxylation prevented at temperatures > 1000°C by increasing water vapor pressure.[18,19] In an effort to maximize the β-TCP content, a higher temperature, 1000°C, was used to increase diffusion in the sample. (b) A sample was heated at 1000°C in air at ~50% r.h. for 24 h, a treatment which caused slight dehydroxylation. This sample was reheated at 850°C in air at ~50% r.h. for 130 h; this treatment caused rehydroxylation of the HA. The x-ray analyses showed 1.1 ± 0.2 mass % β-TCP in this sample. (c) Another sample was heated at 1000°C in a water vapor atmosphere (100 kPa) for 10 h; this sample showed no dehydroxylation and contained 0.9 ± 0.2 mass % β-TCP. The average value of β-TCP content obtained from the higher temperature (1000°C) treatments, (b) and (c), was 1.0 ± 0.3 mass %.

The Ca/P molar ratio of heated HA-RM (Table 5), calculated from the average value of 1.0 ± 0.3 mass % β-TCP (and 99.0 ± 0.3 mass % HA) and equations 11 and 12, was Ca/P = 1.6649 ± 0.0005 (the original HA-RM, dried at 105°C, is expected to have this same molar ratio). The Ca/P ratio from the thermal-product data of HA-RM agrees well with the chemically determined value of 1.664 ± 0.005 for the original HA-RM (Table 5). This close agreement in values obtained by the two independent methods attests to the accuracy of both methods.

Comparison with Literature Data. To increase confidence in the accuracy of Ca/P ratios, it is desirable to obtain close agreement between values determined by different chemical methods and values independently determined by thermal-product analyses. In Table 5 are listed Ca/P molar ratios determined by thermal and chemical analyses for our HA-RM and for another hydroxyapatite, HA-N, that was prepared by J. and M.R. Christof-fersen and characterized in detail in different institutes.[20] For HA-N there is good agreement between the Ca/P values determined by independent chemical analyses in two institutes,

Table 6. Composition of solutions equilibrated with HA-RM (60 d, 37°C) and calculated thermodynamic solubility products, K_{sp}

No.	pH	$c(Ca)_{tot}$ (mmol l^{-1})	$c(PO_4)_{tot}$ (mmol l^{-1})	K_{sp} x 10^{59}
1.	4.688	3.422	7.059	1.754
2.	4.684	3.605	7.198	2.168
3.	4.691	3.521	7.037	2.065
4.	4.682	3.614	7.281	2.183
5.	4.686	3.495	6.991	1.803
6.	4.684	3.564	7.256	2.104
7.	4.687	3.572	7.099	2.109
8.	4.688	3.533	7.041	2.004
9.	4.683	3.559	7.187	2.009
10.	4.687	3.540	7.133	2.051
11.	4.683	3.574	7.290	2.118
12.	4.677	3.612	7.250	1.986

Ca/P = 1.631 ± 0.008 and 1.638 ± 0.01, and by thermal-product analysis, Ca/P = 1.635 ± 0.005 (Table 5). The third chemically determined value, Ca/P = 1.659 ± 0.003, which is ~1.5% higher than the others, corresponds to a calculated thermal-product composition of 4.3 mass % of TCP and 95.7 mass % of HA, as compared to the experimentally determined values of 18 ± 3 mass % TCP and 82 ± 3 mass % HA that give the Ca/P ratio of 1.635 ± 0.005.[20] The combined chemical and thermal data indicate that the Ca/P ratio of 1.659 is higher than the true value.

From the Ca/P molar ratio of 1.635 for HA-N (the average value of two independent chemical analyses, 1.631, 1.638, and thermal analyses, 1.635) and from the value of 2.34 mole % of HPO$_4$ (0.14 HPO$_4$ ion per 5.86 PO$_4$ ions, or 0.14 H atom per total 6 PO$_4$ groups) in HA-N determined by one of us (BOF), all above data published in reference 20, the values of x = 0.19 and y = 0.24 were obtained for the general formula, Ca$_{10-x}$H$_{2x-y}$(PO$_4$)$_6$(OH)$_{2-y}$, as follows. The x value was calculated from equation 11 and the Ca/P molar ratio, and the y value from this x value and number of H atoms, N(H) = 0.14, per total 6 PO$_4$ groups; y = 2x - N(H). From y = 0.24, the calculated (2 - y) number of OH ions per formula unit was 1.76; this value is in agreement with that of 1.66 ± 0.10 OH ions per total 6 PO$_4$ groups experimentally determined for HA-N by infrared analysis.[20]

Solubility Product

The pH values and the total concentrations of soluble calcium and phosphate in the HA-RM suspensions equilibrated for 60 d at 37°C are listed in Table 6. The ionic strengths of equilibrated solutions were $I = 0.0105 ± 0.0001$ mol dm^{-3}. From these data, the thermodynamic solubility products, K_{sp}(HA), were calculated (Table 6); the mean value of these 12 determinations was K_{sp}(HA) = a^5(Ca^{2+}) a^3(PO$_4^{3-}$) a(OH$^-$) = 2.03 x 10^{-59}.

The sources of uncertainty in this mean are listed in Table 7. From equation 3 the standard uncertainty in the mean value of K_{sp} of the twelve replicate measurements was 0.04 x 10^{-59} (Table 7, source 1). In order to evaluate contributions to the uncertainty in K$_{sp}$ from other sources, a relationship for K_{sp} as a function of these sources was developed.

The contributions to the combined uncertainty that derived from the equilibrium constants (Table 1) used to calculate K_{sp} were determined as described below. The calculated molar fractions of different species in total calcium and phosphate concentrations were evaluated. From the calculated concentrations (activities) of different calcium species

Table 7. Components of Uncertainty for Solubility Product

No. (i)	Source	Standard uncertainty $(10^{-59})u_i(y)$	$\|p_i\|$	Relative standard uncertainty r_i (%)	Degrees of freedom (v_i)	Type A or B
1.	Replicate measurements	0.03836	1	1.890	11	A
2.	Uncertainty in mass for calcium standard	0.00436	5	0.043	29	A
3.	Uncertainty in volume forcalcium standard solution	0.00406	5	0.040	∞	B
4.	Uncertainty in mass for phosphorus standard	0.00280	3	0.046	29	A
5.	Uncertainty in volume for phosphorus standard solution	0.00487	3	0.080	∞	B
6.	Uncertainty in $a(H^+)$ from emf measurements on pH standards	0.08382	7	0.59	1	A
7.	Uncertainty in $a(H^+)$ from pH standard buffer solution	0.19605	7	1.38	∞	B
8.	Uncertainty in K_w	0.02804	1	1.38	∞	B
9.	Uncertainty in K_2	0.02804	3	0.46	∞	B
10.	Uncertainty in K_3	0.28039	3	4.61	∞	B
	Combined Standard Uncertainty, u_c	0.35664				
	Expanded Uncertainty, $U = 2u_c$	0.71328				

present in the equilibrated solutions under our experimental conditions, it was evident that the molar fraction of complex ions $CaH_2PO_4^+$, $CaHPO_4^0$ and $CaPO_4^-$ in the total soluble calcium concentration was less than 5%; therefore, the uncertainties in corresponding equilibrium constants of calcium complex species did not contribute significantly to the combined uncertainty of K_{sp}. The dominant calcium species was ionic calcium, Ca^{2+}, whose activity was close to the activity of total calcium in solution:

$$a(Ca^{2+}) \approx a(Ca)_{tot}. \tag{13}$$

Among the phosphate species, $H_2PO_4^-$ was dominant (~97.5% of total PO_4):

$$a(H_2PO_4^-) \approx a(PO_4)_{tot}. \tag{14}$$

The $a(H_2PO_4^-)$ can be expressed as $a(H_2PO_4^-) = a(PO_4^{3-}) \, a^2(H^+) / K_2 \, K_3$ (Table 1, equilibria 2 and 3) and, therefore

$$a(PO_4)_{tot} \approx a(PO_4^{3-}) \, a^2(H^+) / K_2 \, K_3. \tag{15}$$

From equations 1, 13 and 15, the following equation, used for propagation of uncertainty in K_{sp}, was derived:

$$K_{sp} \approx a^5(Ca)_{tot} \, a^3(PO_4)_{tot} \, a^{-7}(H^+) \, K_w \, K_2^3 \, K_3^3. \tag{16}$$

Given the relative standard uncertainties, r_i, the standard uncertainties $u_i(y)$ for sources 2 to 10 in Table 7 were obtained by equation 8 with powers, p_i's, from equation 16. The r_i's for sources 2 to 5 (Table 7) were taken from Tables 2 and 3. The relative standard uncertainties in $a(H^+)$, r_i's for sources 6 and 7 (Table 7), were computed from the emf versus pH calibration data and arose from emf measurements on the pH standards used to fit the calibration line (source 6) and from uncertainty in the certified values of the pH standards (source 7). For source 6, the standard uncertainty in the pH calibration prediction at the mean pH of 4.685 ($n = 12$) was 0.00255 (for calculation of the uncertainty in a calibration prediction, see reference 14). Since pH = $-\log_{10}a(H^+)$, by the method of propagation of uncertainties[15,16] it can be shown that $r_6 \approx 0.00255$ x ln(10) = 0.0059 or 0.59%. For source 7, r_7 was computed from the estimated u_c's in the pH standards and the fitted calibration line. The manufacturer's stated uncertainty (see Solubility section) for the NIST pH standard of 4.025 at 37°C was $U = 0.005$ and uncertainties for the Fisher pH standards of 4.63 and 5.00 at 25°C were given as ranges of ± 0.02 and ± 0.01, respectively. Regarding all of these stated uncertainties as U's, where $U = 2u_c$, the estimated u_c is ≤ 0.01 for each pH standard; consequently, $u_c = 0.01$ was used for computing r_7. The standard uncertainty in the pH calibration prediction at the mean pH of 4.685 ($n = 12$) arising from $u_c = 0.01$ in the pH standards was 0.00601 and so $r_7 \approx 0.00601$ x ln(10) = 0.0138 or 1.38%. The r_i's for K_w, K_2 and K_3 (sources 8, 9 and 10, respectively) were computed from uncertainties in pK_w, pK_2 and pK_3 of 0.006, 0.002 and 0.02, respectively, estimated from literature data.[21,22] The calculated r_i's for K_w, K_2 and K_3 were ≈ 0.0138 or 1.38%, ≈ 0.0046 or 0.46%, and ≈ 0.0461 or 4.61% respectively, and corresponding u_i's were 0.028 x 10^{-59}, 0.028 x 10^{-59}, and 0.280 x 10^{-59} respectively.

From the ten u_i's (Table 7, sources 1 to 10) the calculated combined uncertainty for K_{sp} was $u_c = 0.356$ x 10^{-59}. The largest contributions were from the uncertainties in $a(H^+)$ and K_3 values (sources 7 and 10). The expanded uncertainty, U, was 0.71 x 10^{-59}; thus, the thermodynamic $K_{sp}(HA)$ at 37°C, expressed as mean $\pm U$, was (2.03 $\pm$ 0.71) x 10^{-59} and $pK_{sp}(HA)$ was 58.69 $\pm$ 0.15.

This $K_{sp}(HA)$ value of (2.03 $\pm$ 0.71) x 10^{-59} is in very good agreement with the literature value of (2.36 $\pm$ 0.28) x 10^{-59} determined under similar conditions by McDowell et al.[7] The direct comparison of these two K_{sp} mean values is valid because the same values of all constants and parameters were used in both calculations. The comparison of uncertainties for these two K_{sp} values is not valid because the uncertainties were calculated by different statistical methods. Other published values of $K_{sp}(HA)$ that were calculated with parameters and constants different from those used in this paper (See Solubility section and Table 1) cannot be directly compared without recalculation, which requires the original solubility data.

Certain commercial materials and equipment are identified in this paper to specify the experimental procedure. In no instance does such identification imply recommendation or endorsement by the National Institute of Standards and Technology, the National Institutes of Health, or the ADA Health Foundation or that the material or equipment identified is necessarily the best available for the purpose.

ACKNOWLEDGMENT

This work was supported in part by ADAHF, FDA and NIST. M. Markovic is on leave of absence from the Faculty of Science, University of Zagreb, Zagreb, Croatia.

REFERENCES

1. LeGeros, R.Z., 1991, Calcium phosphates in oral biology, Karger, Basel.

 2. Bioceramics, 1993, Volume 6, Proc. of the 6th International Symposium on Ceramics in Medicine, Philadelphia, November 1993, Ducheyne, P., and Christiansen, D., Eds., Butterworth-Heinemann Ltd, Oxford.
 3. Hydroxyapatite and related materials, 1994, Brown, P.W., and Constantz, B., Eds., CRC Press, Boca Raton.
 4. Elliott, J.C., 1994, Structure and chemistry of the apatites and other calcium orthophosphates, Elsevier, Amsterdam.
 5. Young, R.A., and Holcomb, D.W., 1982, Variability of hydroxyapatite preparations, *Calcif. Tissue Int.* 34:517-532.
 6. Markovic, M., Fowler, B.O., and Tung, M.S., 1995, Calcium hydroxyapatite. 1. Preparation and characterization of hexagonal form, *J. Res. Natl. Inst. Stand. Technol.*, in preparation.
 7. McDowell, H., Gregory, T.M., and Brown, W.E., 1977, Solubility of $Ca_5(PO_4)_3OH$ in the system $Ca(OH)_2-H_3PO_4-H_2O$ at 5, 15, 25 and 37°C, *J. Res. Natl. Bur. Stand.* 81A:273-281.
 8. Markovic, M., Fowler, B.O., and Brown, W.E., 1993, Octacalcium phosphate carboxylates. 2. Characterization and structural considerations, *Chem. Mater.* 5:1406-1416.
 9. Gee, A., and Deitz, V.R., 1953, Determination of phosphate by differential spectrophotometry, *Anal. Chem.* 25:1320-1324.
10. Markovic, M., Fowler, B.O., and Tung, M.S., 1995, Calcium hydroxyapatite. 2. Characteristics of monoclinic form, *J. Res. Natl. Inst. Stand. Technol.*, in preparation.
11. Gregory, T.M., Moreno, E.C., and Brown, W.E., 1970, Solubility of $CaHPO_4 \bullet 2H_2O$ in the system $Ca(OH)_2-H_3PO_4-H_2O$ at 5, 15, 25, and 37.5°C, *J. Res. Natl. Bur. Stand.* 74A:461-475.
12. Bates, R.G., 1973, Determination of pH: Theory and practise, John Wiley & Sons, New York, p. 48.
13. Kielland J., 1937, Individual activity coefficients of ions in aqueous solutions, *J. Amer. Chem. Soc.* 59:1675-1678.
14. Neter, J., Wasserman, W., and Kutner, M. H., 1990, Applied linear statistical models, 3rd ed., Irwin, Boston.
15. Guide to the expression of uncertainty in measurement, 1993, ISBN 92-67-10188-9, 1st ed., ISO, Switzerland.
16. Ku, H.H., 1966, Notes on the use of propagation of error formulas, *J. Res. Natl. Bur. Stand.* 70C:263-273.
17. Fowler, B.O., unpublished data.
18. Fowler, B.O., 1974, Infrared studies of apatites. II. Preparation of normal and isotopically substituted calcium,strontium, and barium hydroxyapatites and spectra-structure-composition correlations, *Inorg. Chem.* 13:207-214.
19. Riboud, P.V., 1975, Composition et stabilite des phases a structure d'apatite dans le systeme $CaO-P_2O_5$-oxyde de fer-H_2O a haute temperature, In: Physico-chimie et crystallographie des apatites d'interet biologique, *Colloques internationaux C.N.R.S.* (Paris) 230:473-480.
20. Arends, J., Christoffersen, J., Christoffersen, M.R., Eckert, H., Fowler, B.O., Heughebaert, J.C., Nancollas, G.H., Yesinowski, J.P., and Zawacki, S.J., 1987, A calcium hydroxyapatite precipitated from an aqueous solution. An international multimethod analysis, *J. Crystal Growth* 84:515-532.
21. Sillen, L.G., and Martell, A.E., 1964, Stability constants of metal-ion complexes, Spec. Publ. No. 17, The Chemical Society, London.
22. Smith, R.M., and Martell, A.E., 1977, Selected stability constants, Vol. 4: Inorganic Complexes, Plenum Press, New York, London.

23

FOURIER TRANSFORM INFRARED SPECTROSCOPY OF SYNTHETIC AND BIOLOGICAL APATITES

A Review

Sergio J. Gadaleta,[1] Richard Mendelsohn,[1] Eleftherios L. Paschalis,[2]
Nancy P. Camacho,[2] Foster Betts,[2] and Adele L. Boskey[2]

[1] Rutgers University
Newark New Jersey
[2] The Hospital for Special Surgery
New York, New York

INTRODUCTION

Infrared (IR) spectroscopy has been widely used to characterize the mineral phase in physiologically calcified tissues. These and other ultrastructural studies[1,2,3] have shown this material to be a poorly crystalline, carbonate containing analogue of the naturally occurring mineral hydroxyapatite, HA, $Ca_{10}(PO_4)_6(OH)_2$. The IR spectrum of calcified tissues are characterized by absorbances attributable to water, the organic matrix, and from phosphate and carbonate constituents of the mineral. By monitoring the changes in the ν_1, ν_3 phosphate contour, a region accessible with detectors used on FT-IR microscopes[4], important information can be elucidated about the crystal size and perfection (crystallinity) as well as changes in the molecular environment of the mineral as it matures *in vitro* and *in vivo*.[4,5] Recently, Fourier transform infrared microscopy (FT-IRMS) has been applied to the characterization of mineralized tissues.[6,7,8,9] FT-IRMS provides a method to probe heterogenous biological mineralized tissues with minimal sample preparation at spatial resolution to 10μm. This method provides a more detailed analysis of the spatial variations in the molecular environment of calcified tissues, as well as providing information about the protein conformation[10] and orientation of the organic matrix[6] at a resolution comparable to the size of significant biological structures. Although earlier IR analyses have identified the mineral phase in bones[11,12], teeth[13], tendon[6], calcified cartilage[14], and certain dystrophic calcifications[15] as apatitic, detailed analyses of the underlying components in the phosphate and carbonate bands reveal that there are significant site dependent variations in parameters such as crystallinity and acidic phosphate content, which are not deducible from the raw absorbance spectra.

The IR spectrum of a material arises from the absorption of infrared radiation corresponding to the frequency of one or more of its interatomic vibrations.[16] The isolated PO_4 ion has T_d

symmetry.[16,17] It is expected to have 3N-6 (N = number of atoms) vibrational modes of which, IR active are: a triply degenerate asymmetric stretch, v_3, at 1082 cm^{-1}, a doubly degenerate asymmetric O-P-O bend, v_2, at 363 cm^{-1} and a triply degenerate asymmetric O-P-O bend, v_4, at 515 cm^{-1}.[17] Nevertheless, distortions due to vacancies, substitutions (such as CO_3^{2-} and HPO_4^{2-}), and/or correlated motions of the phosphate vibrations (factor group splitting), may split the degenerate modes[19,20,21,22] and introduce additional modes. For large, highly crystalline/perfect HA crystals, these distortions are few and the corresponding spectra have been interpreted with minimal spectral manipulation.[19,20,21,22] However, assignments for the poorly crystalline biological apatites and their synthetic analogues have been more difficult to make.

To accomplish qualitative and quantitative determination of the underlying components in the broad contours of poorly crystalline HA, three data reduction techniques have routinely been used, namely, Fourier self-deconvolution[24,18], second derivative spectroscopy[6,5,23,24] and curve fitting.[25] Fourier self-deconvolution requires the subjective choice of line narrowing parameters such as the full width at half height (FWHH) of the underlying bands being deconvolved and the ratio of band widths before and after deconvolution. The values of these parameters vary according to the material being analyzed and are often different even among closely related compounds. Nevertheless, if properly employed, Fourier self-deconvolution provides information about the positions and intensities of the underlying bands. Second derivative spectroscopy has the advantage of being a mathematically objective method which allows consistent processing of spectral data. Peaks in the derivative spectrum provide an estimate of the number and position of the underlying bands in a contour. The observed frequencies provide information about the existing molecular species and are sensitive to environmental changes. However, it has been empirically shown that the signal to noise ratio decreases by a factor of 3-5 with each level of derivation.[26,27,28] Curve fitting, which provides the most information about the underlying bands in a contour, requires choice of several input parameters such as the number, position, and shape of the bands as well as selection of criteria for the "goodness of fit".[25] Knowledge of these parameters is crucial to effectively implement this procedure. For broad contours such as the v_1, v_3 phosphate region of the biologically relevant calcium phosphates, determination of these values is an arduous task and to some degree subjective.

Recently, Gadaleta et al[5] have employed a combination of second derivative spectroscopy and curve fitting in the analysis of poorly crystalline synthetic and biological apatites. This method of analysis provided a more reliable procedure for the quantitation of the underlying bands in the phosphate contour of poorly crystalline HA. In this paper we review the use of this technique as applied to the study of synthetic apatites and biologic mineral, and show the applicability to the analysis of highly crystalline apatite.

METHODS

Two methods were used to synthesize the poorly crystalline HA for these experiments.[5] In each method, solutions of calcium chloride and sodium phosphate were prepared in *tris*-(hydroxymethyl)-aminomethane (Tris) buffer and mixed with constant stirring at room temperature.[29] The apatites synthesized by the first method were prepared at pH 7.4. The pH was monitored and maintained constant throughout the reaction by addition of dilute NH_4OH. In the second method of preparation, the concentration of sodium phosphate was increased and the pH of the solution was allowed to vary during the reaction. For both methods the total reaction time was 3 weeks. Aliquots were withdrawn at time 0, then every 30 minutes for the first 4 hours, at 24 hours, and at 3 weeks. Precipitates were filtered through a medium porosity fritted glass filter, washed with ammoniated (pH 10) water, dried with spectroscopic grade acetone and lyophilized.

Highly crystalline HA was prepared according to the method of Ebrahimpour et al.[30] In brief, a solution of 0.125M $CaHPO_4$ and 0.13M H_3PO_4 was slowly added over a period of 10 hours to three liters of double distilled water saturated with $Ca(OH)_2$. The precipitate was removed after 36 hours in solution, filtered and lyophilized. Octacalcium phosphate (OCP) was prepared by seeded growth using the constant composition method in Dr. George Nancollas's laboratory.

For the experiments discussed in this study, synthetic HA crystals made by methods 1 and 2[5], and the highly crystalline HA and OCP were analyzed as KBr pellets. Data were collected on a Bio-Rad FTS-40 infrared spectrometer (Cambridge, MA). The data (256 interferograms) were co-added, and the resultant interferogram Fourier transformed. The crystallinity of the HAs was evaluated by x-ray diffraction (XRD), using a Siemens powder diffractometer (Siemens, Iselin, NJ) scanned in 0.05°(2θ) intervals between 20°-37° (2θ). Calcium to phosphate ionic ratios were determined, and carbonate to phosphate ratios were calculated from the spectra as described elsewhere.[5]

Segments of turkey leg tendons, ranging from unmineralized to heavily calcified, were cut into pieces 3/8" in length, and sliced into 5 µm sections. The sections were placed on a BaF_2 window and analyzed by FT-IRMS.[6] FT-IRMS was performed on a UMA 500 Fourier transform infrared microscope coupled to an FTS-40 infrared spectrometer, both from Bio-Rad (Cambridge, Mass). The aperture of the microscope was initially centered near an area devoid of mineral. A 500 µm by 500 µm area was mapped at 50 µm increments. Each spectrum was obtained by the co-addition of 256 interferograms at 4 cm⁻¹ resolution.

The derivative and curve fitting calculations were performed with software developed at the National Research Council of Canada, Ottawa or with GRAMS/386 (Galactic Software, Salem, NH) as described elsewhere.[5] The X-ray diffraction patterns in the region analyzed consisted of a single peak and a broad intrinsically overlapping contour. The half width of the single peak associated with the c-axis (002) reflection, where 002 represents the hkl Miller indices , was determined directly. The half widths of the overlapping 211, 112, 300, 210, 102 and 202 reflections were calculated after curve-fitting these peaks keeping their positions constant at the standard American Society for Testing and Materials (ASTM) values. The FWHH of an x-ray reflection is inversely proportional to the particle size along that axis.[31] For FT-IR data, second derivative spectra were calculated to obtain the number and position of the underlying bands used to define the initial parameters in the curve-fitting algorithm. Correlations between calculated particle size for each of the x-ray reflections and the percent areas of FT-IR subbands were calculated by linear regression.

RESULTS

A summary of previously reported[5] spectral changes accompanying the maturation of HA synthesized by the two different methods follows. Figure 1a illustrates a typical spectrum showing the v_1, v_3 (900 cm⁻¹ - 1200 cm⁻¹) phosphate region of the calcium phosphate mineral formed immediately after mixing the reagents according to method 1 described above. This spectrum is characteristic of amorphous calcium phosphate, the precursor to HA in this experiment. After 30 minutes, the ACP had converted into poorly crystalline hydroxyapatite (Figure 1b). As the mineral matured in solution, few changes were seen in the spectral characteristics of the v_1, v_3 phosphate contour. The spectrum of the mineral at 3 weeks is shown in Figure 1c. The apatite spectra exhibit three broad regions at 900 cm⁻¹ - 980 cm⁻¹, 980 cm⁻¹ -1080 cm⁻¹, and 1080 cm⁻¹ - 1200 cm⁻¹. Second derivative spectroscopy was used to obtain more information concerning chemical and crystallographic changes occurring during the maturation (Figure 2a,b,c). The second derivative spectra for all time points after the conversion of ACP in method 1 revealed peaks at 960 cm⁻¹, 985 cm⁻¹, 1028 cm⁻¹, 1055 cm⁻¹, 1075 cm⁻¹, 1097 cm⁻¹, 1118 cm⁻¹ and 1145 cm⁻¹, although the broad single peak at 985 cm⁻¹ developed into two distinct peaks centered at 982 cm⁻¹ and 999 cm⁻¹ as the conversion progressed.

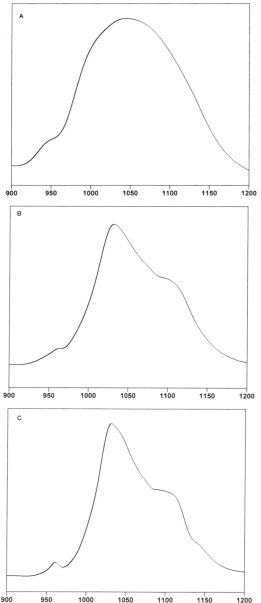

Figure 1. Typical FT-IR spectral changes in the v_1, v_3 (900 cm^{-1} - 1200 cm^{-1}) phosphate region during the formation and maturation of hydroxyapatite formed at constant pH (method 1). a) Spectrum of the calcium phosphate mineral formed immediately after mixing calcium and phosphate at pH 7.4. b) Spectrum of the poorly crystalline apatite formed at 30 minutes. c) Spectrum of the poorly crystalline apatite matured for 3 weeks. Adapted from reference 5.

Although the raw absorbance spectra for the apatite synthesized by method 2 appeared qualitatively similar to those prepared by method 1, second derivative spectra of both methods reveal dramatic differences in the positions of the peaks at early time points of the conversion (Figures 3 and 4). For example, at 30 minutes, peaks occurred at 959 cm^{-1}, 985 cm^{-1}, 1020 cm^{-1}, 1038 cm^{-1}, 1055 cm^{-1}, 1075 cm^{-1}, 1112 cm^{-1}, and 1127 cm^{-1}. However, after the first 3 hours of the reaction the positions of the bands had shifted

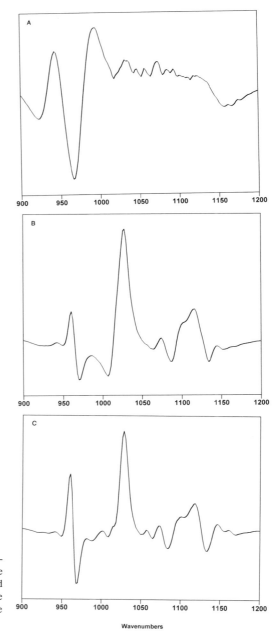

Figure 2. Second derivative spectra of the contours shown in Figure 1. a) Calcium phosphate formed immediately after mixing calcium and phosphate at pH 7.4. b) Poorly crystalline apatite formed at 30 minutes. c) Poorly crystalline apatite matured for 3 weeks. Adapted from reference 5.

to within ± 2 cm^{-1} of the apatites in method 1. The bands at 1020 cm^{-1}, 1127 cm^{-1}, 1112 cm^{-1}, and 1038 cm^{-1} were attributable directly to HPO$_4$ or to PO$_4$ in an acid phosphate environment. For each method, Ca:P ratios increased with time, although the samples in method 2 had a lower initial value (1.38), and exhibited greater variability than the samples prepared by method 1. The CO$_3^{2-}$:P ratio as determined by FT-IR increased as a function of time in both methods.

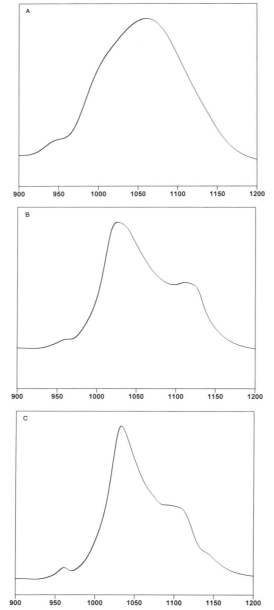

Figure 3. Typical FT-IR spectral changes in the n_1, n_3 (900 cm^{-1} - 1200 cm^{-1}) phosphate region during the formation and maturation of hydroxyapatite. (method 2). a) Spectrum of the calcium phosphate mineral formed immediately after mixing calcium and phosphate solutions. b) Spectrum of the poorly crystalline apatite formed at 30 minutes. c) Spectrum of the poorly crystalline apatite matured for 3 weeks. Adapted from reference 5.

The size/perfection of the HA crystals, determined by x-ray diffraction as described above was correlated with the infrared spectral parameters. In particular, the line width at half maximum of the underlying components attributable to the 002 and 300 reflections in the XRD spectra were inversely correlated ($r \geq 0.7$) with the percent areas of the 1117 cm^{-1} (002), 1074 cm^{-1} and 1053 cm^{-1} (300) bands in the FT-IR data.[16] Thus, it was suggested that combining data derived from the raw absorbance spectra (e.g. carbonate content, mineral to

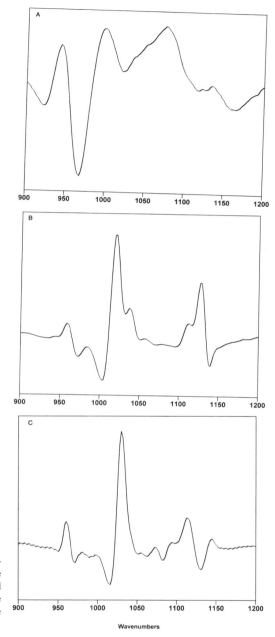

Figure 4. Second derivative spectra of the contours shown in Figure 3. a) Calcium phosphate formed immediately after mixing calcium and phosphate solutions. b) Poorly crystalline apatite formed at 30 minutes. c) Poorly crystalline apatite matured for 3 weeks. Adapted from reference 5.

matrix ratio) with information about the underlying peaks would provide the most detailed characterization of the material.

Based on the positions of the second derivative peaks and their assignments in the literature[19,32,33,22,25], it was concluded that the poorly crystalline apatites described above contained both HPO_4 and PO_4. Since this study was the first application using a combination of second derivative spectroscopy and curve fitting in the analysis of poorly crystalline apatites, we now have also analyzed two highly crystalline calcium phosphates using this technique for which band assignments have been made.[19,33,25] The spectra of highly crystalline hydroxyapatite (Ca:P = 1.66±.05) and octacalcium phosphate (Ca:P = 1.32±.02) examined by this method are reported for the first time in this review. Although the crystallographic parameters of these materials differ significantly from those of biologic materials, the highly crystalline hydroxyapatite served as a model of PO_4 in an environment free of acid phosphate, while the OCP provided a model for the study of PO_4 in an acid phosphate environment. Figures 5a-c show the raw, second-derivative, and curve-fit spectra of the well crystallized HA. Similar data for OCP are presented in Figure 6a-c. The peak positions obtained from these experiments are shown in Table I. To validate the results of the curve-fitting process, the second derivative of the curve fit spectra was calculated and compared to those obtained from the raw absorption spectra (Figures 5d and 6d). This comparison provides a unique means of validation of the proposed technique.

FT-IRMS spectra of the mineral phase in calcified turkey leg tendon have also been reported.[6] Second derivative spectra of newly mineralized tendon were found to be comparable to that of the least mature of the poorly crystalline synthetic HA prepared in method 2, exhibiting peaks at 960 cm^{-1}, 982 cm^{-1}, 1018 cm^{-1}, 1038 cm^{-1}, 1055 cm^{-1}, 1075 cm^{-1}, 1112

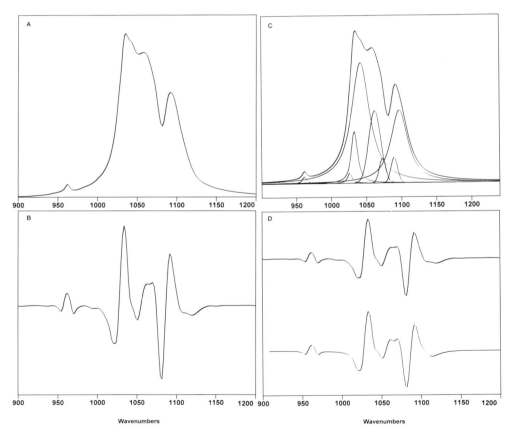

Figure 5. The v_1, v_3 (900 cm^{-1} - 1200 cm^{-1}) phosphate region spectra of highly crystalline HA. a) the raw spectrum. b) second-derivative spectrum. c) curve-fit spectrum. d) second derivative of the original raw absorbance spectrum compared with the second derivative of the curve fit analysis.

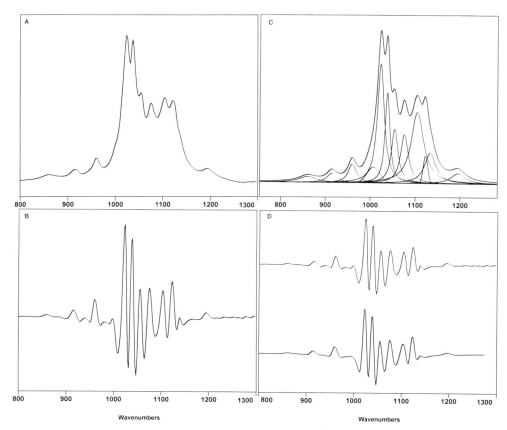

Figure 6. The v_1, v_3 (900 cm^{-1} - 1200 cm^{-1}) phosphate region of OCP. a) the raw spectrum. b) second-derivative spectrum. c) curve-fit spectrum. d) second derivative of the original raw absorbance spectrum compared with the second derivative of the curve fit analysis.

cm^{-1}, and 1127 cm^{-1} agreeing within ± 2 wavenumbers with peak positions of the poorly crystalline HA. Second derivative spectra of older mineral in the calcified tendon showed peaks at 960 cm^{-1}, 986 cm^{-1} with a shoulder at ~1000 cm^{-1}, 1030 cm^{-1}, 1055 cm^{-1}, 1075 cm^{-1}, 1099 cm^{-1}, and 1118 cm^{-1} which agree within ± 2 wavenumbers with peak positions of most mature poorly crystalline HA prepared by method 2. It is important to note that the biologic minerals all had peak positions and percent areas significantly different from the highly crystalline synthetic materials.

DISCUSSION

The results presented in this review demonstrate the capabilities of second derivative spectroscopy coupled with curve-fitting in the analysis of both poorly and highly crystalline apatites. The inherent objectivity of second derivative spectroscopy provides a method of analysis relatively free of subjective factors. Combining second derivative spectroscopy with a curve fitting algorithm provides a more reliable starting point in the quantitative analysis of broad contours. An advantage of this technique is that the validity of the curve fitting

Table 1. Underlying band positions obtained from curve fitting the FT-IR spectra of poorly crystalline HA, highly crystalline HA and octacalcium phosphate

Poorly crystalline HA method 1* (cm⁻¹)	Highly crystalline HA (cm⁻¹)	Octacalcium phosphate (cm⁻¹)
		861
		915
960	962	960
985		
999		1008
	1027	1023
1030	1033	
	1042	1039
1055	1062	1055
1075	1073	1076
	1090	
1099	1098	1105
1117		
		1123
		1134
1145		
		1197

*Reprinted from reference 16.

protocol can be demonstrated by the similarities between the second derivative spectra of the curve fit and the original raw absorbance data. As described previously[6,5], the relatively constant peak positions in the second derivative spectra of the poorly crystalline HA synthesized in method 1 can all be assigned to phosphate vibrations in the apatitic environment of the highly crystalline HA prepared by the method of Ebrahimpour.[30]

The constancy of positions of the second derivative peaks during maturation also indicates that in the HA prepared by method 1, the environment of the phosphate species did not change as the crystals developed. In contrast, certain second derivative peaks of the early (< 210 min) poorly crystalline HA synthesized by method 2 change during maturation.[5] The frequencies (1020 cm⁻¹, 1038 cm⁻¹, 1127 cm⁻¹, and 1112 cm⁻¹) in these early samples were attributed to phosphate vibrations in non-stoichiometric environments as these frequencies are seen in the other acid phosphate containing calcium phosphate phases such as octacalcium phosphate (OCP), monetite, and brushite.[19,32,33,29,22,25,18] The detection of these bands in the apatites synthesized by method 2 at early stages of maturation was reported to indicate the presence of a phosphate moiety in a non-apatitic or acid phosphate environment.

Studies by Rey et al[18], whose results were based on Fourier self-deconvolution of a series of apatitic and non-apatitic calcium phosphates, assigned bands at 970 cm⁻¹, 990 cm⁻¹, and 1145 cm⁻¹ to HPO_4^{2-}, 1110 cm⁻¹ and 1125 cm⁻¹ to non-apatitic PO_4, and 1020 cm⁻¹ and 1030 cm⁻¹ to non stoichiometric and stoichiometric PO_4, respectively. Here, non-stoichiometric refers to the presence of HPO_4^{2-}, CO_3^{2-}, or vacancies in the lattice. Leung et al[23], studying highly crystalline HA, assigned the subbands in the 900 cm⁻¹ - 1200 cm⁻¹ region to symmetry elements of the PO_4 species of the apatite. The second derivative peak positions for the poorly crystalline HA synthesized at variable pH in method 2 correspond to the assignments suggested by Rey et al[18], with subbands due to the non stoichiometric and acid phosphate contributions decreasing with time. The data for the poorly crystalline HA synthesized by method 1 and the highly crystalline HA , calculated using the combination of second derivative and curve fitting, agree both with Rey's assignments for stoichiometric apatites and with Leung's assignments. The only differences are that the material synthesized by method 1 also exhibits peaks at 985, 999, 1118, and 1145 cm^{-1}, while the highly crystalline HA has a peak at 1027 cm^{-1}. which was assigned to stoichiometric apatite by Rey[18], and a peak at 1098 cm^{-1}. The advantage of combining second derivative spectroscopy and curve fitting is that it allows both detection and quantitation of "non-apatitic" species.

The correlations between x-ray diffraction and FT-IR data reported in this review confirm and complement those published by Pleshko et al.[25], relating the percent area of the FT-IR 1060 cm^{-1} subband with the half-width of the 002 reflection in the x-ray diffraction pattern. These are useful in the analyses of the mineral in calcified tissues as described in detail elsewhere.[5]

Although numerous studies have been reported on the arrangement, distribution, size, and shape of the crystalline phase in sections of calcified tissues[34,35,3,36,38], FT-IRMS, as reviewed in the present study, is a unique technique able to address the spatial variations in the mineral phase in a quantitative and qualitative manner, on the scale of significant biological structures with minimal sample preparation. Furthermore, the raw spectra provide information about the type and extent of carbonate substitution[37], along with information about the organic matrix. Mineralized and non-mineralized regions of the spectra can be analyzed using second derivative spectroscopy coupled with curve-fitting, facilitating the enhancement and identification of subtle changes in broad IR contours. As demonstrated in this review, the combination of these methods provides information about the phosphate environment not previously accessible.

ACKNOWLEDGMENTS

This work was supported by NIH grant AR 41325.

REFERENCES

1. Blumenthal, N.C., Betts, F. and Posner, A.S., 1975, Effect of Carbonate and Biological Molecules on Formation and Properties of Hydroxyapatite, *Calcif. Tissue. Res.*, 18:81-90.
2. Posner, A.S. and Betts, F., 1975, Synthetic Amorphous Calcium Phosphate and its Relation to Bone Mineral Structure, *Accts. Chem. Res.*, 8:273.
3. Termine, J.D. and Eanes, E.D., 1972, Comparitive Chemistry of Amorphous and Apatitic Calcium Phosphate Preparations, *Calcif. Tissue Res.*, 10:171-197.
4. Boskey, A.L., Pleshko, N., Doty, S.B. and Mendelsohn, R., 1992, Applications of Fourier Transform Infrared (FT-IR) Microscopy to the Study of Mineralization in Bone and Cartilage, *Cells and Mat.*, 2:209-220.
5. Gadaleta, S.J., Paschalis, E.P., Betts, F., Camacho, N.P., Mendelsohn, R. and Boskey, A.L., 1994, Second Derivative FT-IR Spectroscopy of the Solution Mediated Conversion of Amorphous Calcium Phosphate to Hydroxyapatite: New Correlations between XRD and FT-IR Data, *Calcif. Tissue Int.*, (Submitted)
6. Gadaleta, S.J., Camacho, N.P., Mendelsohn, R. and Boskey, A.L., 1994, Application of Fourier Transform Infrared Microspectroscopy to Calcified Turkey Leg Tendon, *Calcif. Tissue Int.*, (Submitted)
7. Mendelsohn, R., Hassankhani, A., DiCarlo, E. and Boskey, A.L., 1989, FT-IR Microscopy of Endochondral Ossification at 20 Spatial Resolution, *Calcif. Tissue Int.*, 44:20-24.
8. Pleshko, N.L., Boskey, A.L. and Mendelsohn, R., 1992, An Infrared Study of the Interaction of Polymethyl Methacrylate with the Protein and Mineral Components of Bone, *J. Histo.*, 40:1413-1417.
9. Pleshko, N.L., Mendelsohn, R. and Boskey, A.L., 1992, Developmental Changes in Bone Mineral: an FT-IR Microscopy Study, *Calcif. Tissue Int.*, 51:72-77.
10. Payne, K.J. and Veis, A., 1988, Fourier Transform IR Spectroscopy of Collagen and Gelatin Solutions: Deconvolution of the Amide I Band for Conformational Studies, *Biopolym.*, 27:1749-1760.
11. Baud, C.A., Very, J.M. and Courvois, B., 1988, Biophysical Study of Bone-Mineral in Biopsies of Osteoporotic Patients Before and After Long-Term Treatment with Fluoride, *Bone*, 9:361-365.

12. Cohen, L. and Kitzes, R., 1981, Infrared Spectroscopy of Magnesium Content of Bone Mineral in Osteoporotic Women, *Isrl J. Med. Sci.*, 17:1123.

13. LeGeros, R.Z., 1977, Apatites from Aqueous and Non-Aqueous Systems: Relation to Biological Apatites, *IMPHOS* 347-360.

14. Rey, C., Griffin, R. and Glimcher, M.J., 1991, Structural Studies of the Mineral Phase of Calcifying Cartilage, *J. Bone and Min. Res.*, 6:515-525.

15. McCarty, D.J., Lehr, J.R. and Halverson, P.B., 1983, Crystal Populations in Human Synovial Fluid-Identification of Apatite, Octacalcium Phosphate, and Tricalcium Phosphate, *Arthr. Rheum.*, 26:1220-1224.

16. Bailey, R.T. and Holt, C. Fourier Transform Infrared Spectroscopy and Characterisation of Biological Calcium Phosphates. In: *Calcified Tissue*, edited by Hukins, D.W. London: Macmillan Press, 1989, p. 93-119.

17. Herzberg, G.H. *Molecular Spectra and Molecular Structure Volume II: Infrared and Raman Spectra of Polyatomic Molecules*, Malabar, Fl: Krieger Publishing Company, 1991. pp. 322.

18. Rey, C., Shimizu, M., Collins, B. and Glimcher, M.J., 1991, Resolution-Enhanced Fourier Transform Infrared Spectroscopy Study of the Environment of Phosphate Ion in the Early deposits of a Solid Phase Calcium Phosphate in Bone and Enamel and Their Evolution with Age: 2. Investigations in the v_3 PO_4 Domain, *Calcif. Tissue Int.*, 49:383-388.

19. Baddiel, C.B. and Berry, E.E., 1966, Spectra Structure Correlations in Hydroxy and Fluorapatite, *Spectrochim. Acta.*, 22:1407-1416.

20. Fowler, B.O., 1974, Infrared Studies of Apatites. II. Preparation of Normally and Isotopically Substituted Calcium, Strontium, and Barium Hydroxyapatites and Spectra-Structure Correlations, *Inorg. Chem.*, 13:207-214.

21. Fowler, B.O., 1974, Infrared Studies of Apatites. I. Vibrational Assignments for Calcium, Strontium, and Barium Hydroxyapatites Utilizing Isotopic Substitution, *Inorg. Chem.*, 13:194-206.

22. Fowler, B.O., Moreno, E.C. and Brown, W.E., 1966, Infra-red Spectra of Hydroxyapatite, Octacalcium Phosphate and Pyrolysed Calcium Phosphate, *Arch. Oral. Biol.*, 11:477-492.

23. Leung, Y., Walters, M.A. and LeGeros, R.Z., 1990, Second Derivative Spectra of Hydroxyapatite, *Spectrochim. Acta.*, 46A:1453-1459.

24. Rey, C., Shimizu, M., Collins, B. and Glimcher, M.J., 1990, Resolution-Enhanced Fourier Transform Infrared Spectroscopy Study of the Environment of Phosphate Ions in the Early Deposits of a Solid Phase of Calcium-Phosphate in Bone and Enamel, and Their Evolution with Age. I: Investigations in the v_4 PO_4 Domain, *Calcif. Tissue Int.*, 46:384-394.

25. Pleshko, N.L., Boskey, A.L. and Mendelsohn, R., 1991, Novel Infrared Spectroscopic Method for the Determination of Crystallinity of Hydroxyapatite Minerals, *Biophys. J.*, 60:786-793.

26. Hawkes, S., Maddams, W.F., Mead, W.L. and Southon, M.J., 1982, The Measurement of Derivative IR Spectra-II. Experimental Measurements, *Spectrochim. Acta.*, 38A:445-457.

27. Maddams, W.F. and Mead, W.L., 1982, The Measurement of Derivative IR Spectra-I. Background Studies, *Spectrochim. Acta.*, 38A:437-444.

28. Maddams, W.F. and Southon, M.J., 1982, The Effect of Band Width and Band Shape on Resolution Enhancement by Derivative Spectroscopy, *Spectrochim. Acta.*, 38A:459-466.

29. Boskey, A.L. and Posner, A.S., 1973, Conversion of Amorphous Calcium Phosphate to Microcrystalline Hydroxyapatite. A pH-Dependent, Solution Mediated, Solid-Solid Conversion, *J. Phys. Chem.*, 77:2313-2317.

30. Ebrahimpour, A., 1990, Thesis, *State University of New York at Buffalo*

31. Cullity, B.D. *Elements of X-ray Diffraction*, Reading, MA:Addison-Wesley Publ., 1967. pp. 99.

32. Berry, E.E. and Baddiel, C.B., 1967, The Infra-red Spectrum of Dicalcium Phosphate Dihydrate (Brushite), *Spectrochim. Acta.*, 23A:2089-2097.

33. Berry, E.E. and Baddiel, C.B., 1967, Some Assignments in the Infra-red Spectrum of Octacalcium Phosphate, *Spectrochim. Acta.*, 23A:1781-1792.

34. Krefting, E.R., Barkhaus, R.H. and Hohling, F., 1980, Analysis of the Crystal Arrangement in Collagen Fibrils of Mineralizing Turkey Tibia Tendon, *Cell Tissue Res.*, 205:485-492.

35. Landis, W.J., Moradian-Oldak, J. and Weiner, S., 1991, Topographic Imaging of Mineral and Collagen in the Calcifying Turkey Tendon, *Conn. Tiss. Res.*, 25:181-196.

36. Walters, M.A., Leung, Y.C., Blumenthal, N.C., LeGeros, R.Z. and Konsker, K.A., 1990, A Raman and Infrared Spectroscopic Investigation of Biological Hydroxyapatite, *J. Inorg. Biochem* 39:193-200.

37. Rey, C., Renugopalakrishnan, V., Collins, B., Glimcher, M.J., 1991, Fourier Transform Infrared Spectroscopic Study of the Carbonate Ions in Bone Mineral During Aging, *Calcif. Tissue.Int.*, 49:251-258.

38. Traub, W., Arad, T. and Weiner, S., 1989, Three-Dimensional Ordered Distribution of Crystals in Turkey Tendon Collagen Fibrils, *Proc. Natl. Acad. Sci.* 86:9822-9826.

24

DETERMINATION OF SOLUBILITY AND CALCIUM ION STABILITY CONSTANTS OF A PHOSPHONOALKYLPHOSPHINATE (PAP) AND BISPHOSPHONATES (BPs) SUCH AS EHDP, RISEDRONATE, ALENDRONATE, 3-pic AMBP, AND 3-pic AMPMP

A. Ebrahimpour,[1] F.H. Ebetino,[1] G. Sethuraman,[2] and G.H. Nancollas[2]

[1] The Procter & Gamble Co.
Miami Valley Laboratories
P.O. Box 538707, Cincinnati, Ohio 45253
[2] The State University of New York at Buffalo
Buffalo, New York 14214

INTRODUCTION

The first member of the bis- or diphosphonate (BP) analogous was first synthesized in 1897.[1] Over six decades later Blaser and Worms patented the first use of a BP[2], Ethylidene-1,1-hydroxy bisphosphonic acid (EHDP), for the soluble complexation of metal ions, primarily calcium and magnesium. Since that time many interesting structural variants of the BPs have been synthesized which have found extensive use in chemical, medical, and dental application. The first dental use of a BP was for the inhibition or blockage of calculus or tartar deposition. The principle involved was the adsorption of EHDP on the tooth surface and to any nuclei of calculus that might form on the tooth surface. The concentrations of topical solution and dentifrice were such that EHDP saturated the calcium phosphate crystal's growing surfaces thus, inhibiting deposition and accumulation of supragingival calculus. The first medical use of a BP was based on a principle similar to that of calculus inhibition involving the oral dosing of EHDP to a sixteen month old child with ossifying deposition (myositis ossificans progressiva). The principle again was to produce a saturating adsorption of BP on the sites of calcium phosphate deposition in the muscle tissues and so block the debilitating effects of the unwanted accretion. In a somewhat similar medical application for heterotopic ossification, etidronate has been used successfully for the inhibition of calcification of soft tissue resulting from hip replacement or spinal chord injury. Again the principle is the same, sites of rapidly forming calcium phosphate in muscle tissue can be saturated by

Mineral Scale Formation and Inhibition, Edited by Zahid Amjad
Plenum Press, New York, 1995

oral administration of etidronate and blocked from further deposition which destroys muscle tissue function.

For their *in vivo* application, BPs are best known for their ability to block bone resorption such as in Paget's disease, hypercalcemia of tumor origin or in other bone resorptive processes such as osteoporosis. Several hypotheses have been put forward concerning the mechanism of this antiresorptive effect. These include the inhibited solubility of BP adsorbed bone, an antiosteoclastic toxic effect due to phagocytosis of bone particles coated with adsorbed BP, and possibly site specific interaction of the BPs with the cell membrane of the osteoclast producing a functional inactivation.

Bone modeling and remodeling involves the process of resorption of previously mineralized bone matrix, and the deposition of organic matrix and its subsequent mineralization. Although many calcium phosphate phases such as dicalcium phosphate dihydrate (DCPD) and octacalcium phosphate (OCP) have been suggested as the possible precursors to bone mineral, the mature bone mineral is virtually all in the form of calcium phosphate's most thermodynamically stable phase, hydroxyapatite (HAP). Well-characterized physicochemical studies of the calcium phosphates and the BPs, bisphosphinates (BPIs), and phosphonoalkylphosphinate (PAPs) are bringing new learnings to this field.[3] These *in vitro* experiments include the kinetic studies of mineralization and demineralization which are conducted under physiologically relevant conditions while the concentration of every involved component is maintained constant throughout the experiment using the dual constant composition method.[4,5] Consequently, in these studies the rates of HAP growth and dissolution in the absence and in the presence of the drugs is measured. The affinity of these compounds for the mineral surfaces and for metal ions, as well as the solubility of metal ion salts of these compounds play an important role in their ability to modify the re- / demineralization processes. The present work focuses on quantifying the association constants of four BPs and a PAP hydrogen and calcium ions. In addition to the complexation constants, the solubility of calcium salts of these PAP and BPs were determined.

MATERIALS AND METHOD

Solutions were prepared in grade A glasswares using analytical grade chemical reagents and deionized, triply distilled water (TDW). They were filtered twice (0.22 mm membrane filters, Millipore) before use. Sodium hydroxide solution was prepared in boiled TDW and stored in carbon dioxide free environment.

The double walled reaction cells were maintained at 37°C by the circulation of thermostated water. Calcium and hydrogen ion activities were measured using Orion calcium ion selective electrode and Orion pH combination electrode, respectively. The calcium electrodes were calibrated with standard calcium chloride solution and pH electrodes were calibrated using two buffer solutions of pH 7.43 and 4.28.

Sodium hydroxide was standardized using standard potassium hydrogen phthalate. Calcium was determined by EDTA complexometric titration method and by atomic absorption spectroscopy (Perkin Elmer). Phosphonate concentration was determined by Hach phosphonate analysis method, in which phosphonate in the presence of persulfate was converted to orthophosphate by exposure to UV radiation for 10-15 minutes. Ammonium molybdate was added to the orthophosphate solution the color was developed and after two minutes the absorption was measured at 700 nm.

The protonation equilibria, calcium complex formation constants and solubility of the calcium salts of BPs have been studied. Chemical analysis, SEM, EDAX and X-ray measurements were made on the calcium BP salts. Experiments were carried out in a nitrogen

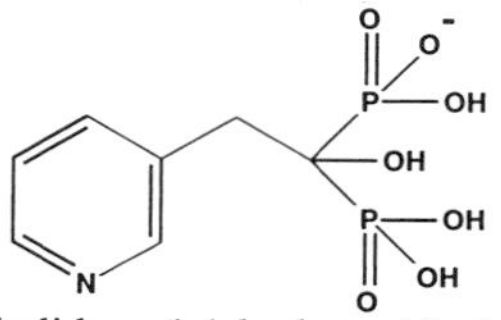

Risedronate
2-(3-pyridyl)ethylidene-1,1-hydroxy bisphosphonic acid
NE-58095

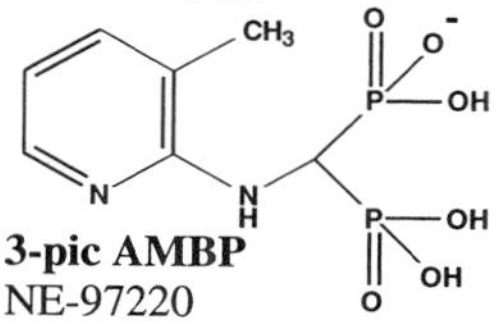

3-pic AMBP
NE-97220

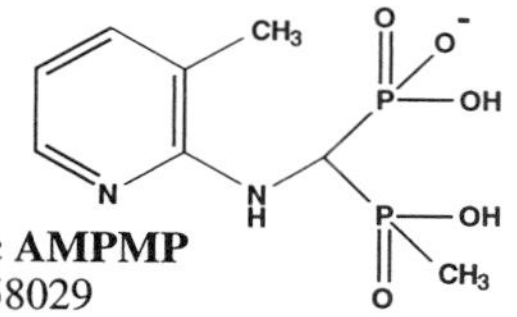

3-pic AMPMP
NE-58029

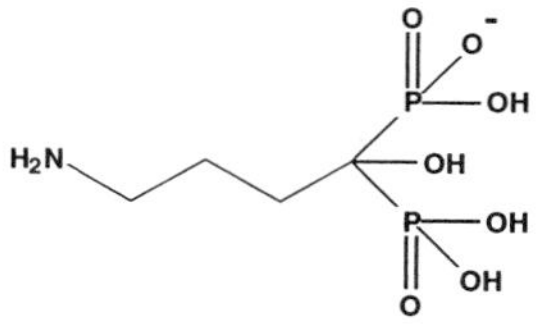

Alendronate
NE-10503
(4-amino-1-hydroxybutylidene)bisphosphonic acid

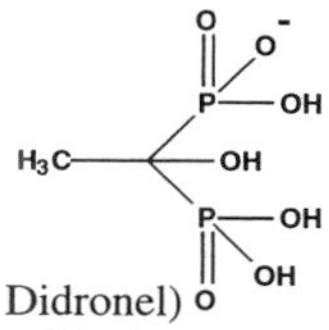

Figure 1. Bisphosphonates and phosphonoalkylphosphinate structures.

EHDP (Etidronate or Didronel)
Ethylidene-1,1-hydroxy bisphosphonic acid

atmosphere at 37°C and in the presence of 0.1 M sodium chloride background electrolyte. The BPs are shown in Figure 1.

In experiments for the determination of the formation constant of calcium bisphosphonate and PAP complexes known quantities of calcium chloride were added to the bisphosphonate and PAP solution (calcium to BP or PAP molar ratio was usually 1:1) in the presence of 0.1 M sodium chloride. Each resulting solution was titrated with sodium hydroxide solution. The stability constants of the bisphosphonates (CaH_2Phn^0 and $CaHPhn^-$ for Phn^{-4} ligand, and CaH_2Phn^+ and $CaHPhn$ for Phn^{-3} ligand) and PAP were calculated.

Preparation and Solubility of Calcium BP and PAP Salts

Calcium salts of BPs were prepared by adding calcium chloride solution to 50 mg of BP in 5 ml of solutions under nitrogen, and with constant stirring at 37°C. Samples of NE97220 and NE58029 precipitated at a pH of about 5.2 and precipitation was completed in less than an hour. In contrast, the calcium salts of NE58095 and NE 10503 did not precipitate under these conditions at pH 5. However, by adding additional base to the solution, a fine white precipitate was observed at a pH about 6.5. Precipitation appeared to be complete after about 12 hours. The filtered precipitates from each preparation was dried under vacuum and studied using SEM, EDAX and X-ray diffraction, as well as wet chemical analysis.

Solubility measurements of the calcium phosphonate salts were made by dispersing known amounts of solids in 50 ml of 0.15 M sodium chloride solutions at 37°C under nitrogen. The pH of the solution was maintained at 7.00 ± 0.02 by adding 0.100 M HCl and/or 0.110 M KOH. After 24 hours the solids remaining in the solution were filtered (0.22 mm Millipore filter membrane) and dried in a vacuum oven at 50°C. The concentrations of free calcium and phosphonate in the solids and in solution at equilibrium with these solids were analyzed as described above.

Method Used for the Determination of Proton Activity Constant

In the presence of excess strong acid the BPs have six pK values including the two nitrogen groups each of which has a pair of unshared electrons available for binding hydrogen ions from excess acid. The actual protonated species in dilute solutions are shown below. Figure 2 shows a typical stepwise deprotonation of the BP.

In the presence of excess strong acid reactions 1 through 6 must be taken into account:

$$H_6Phn^{2+} \rightarrow H_5Phn^+ + H^+ \tag{1}$$

$$H_5Phn^+ \rightarrow H_4Phn + H^+ \tag{2}$$

$$H_4Phn \rightarrow H_3Phn^- + H^+ \tag{3}$$

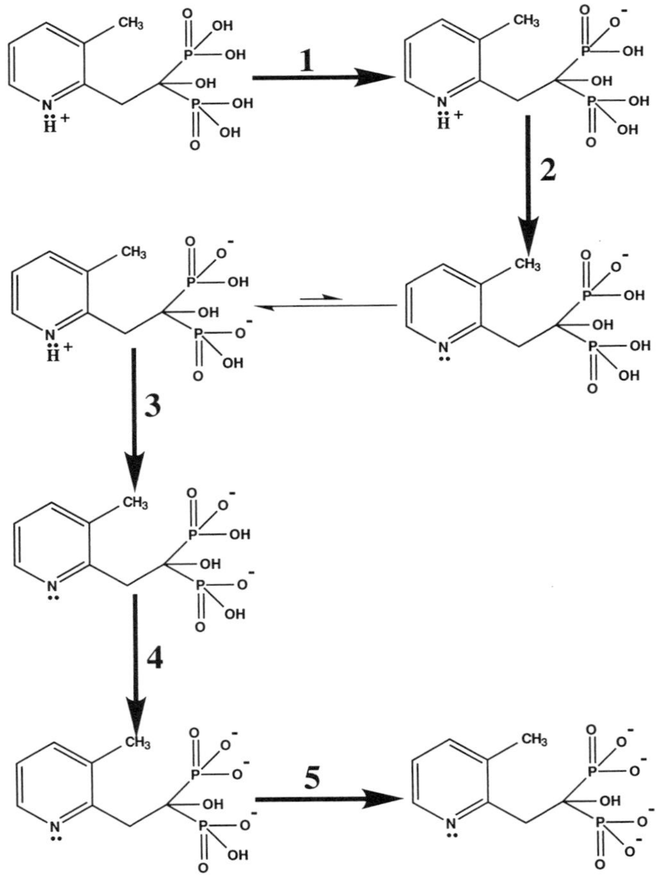

Figure 2. Stepwise deprotonation of BP.

$$H_3Phn^- \rightarrow H_2Phn^{-2} + H^+ \tag{4}$$

$$H_2Phn^{-2} \rightarrow HPhn^{-3} + H^+ \tag{5}$$

$$HPhn^{-3} \rightarrow Phn^{-4} + H^+ \tag{6}$$

In the absence of strong acid, the reactions are limited to equations 7 to 10:

$$H_4Phn \rightarrow H_3Phn^- + H^+ \qquad K1 = (H_3Phn^-)(H^+) / (H_4Phn) \tag{7}$$

(only in strong acid, not applicable in pH range of interest)

$$H_3Phn \rightarrow H_2Phn^{-2} + H^+ \qquad K2 = (H_2Phn^{-2})(H^+) / (H_3Phn^-) \tag{8}$$

$$H_2Phn^{-2} \rightarrow HPhn^{-3} + H^+ \qquad K3 = (HPhn^{-3})(H^+) / (H_2Phn^{-2}) \tag{9}$$

$$HPhn^{-3} \rightarrow Phn^{-4} + H^+ \qquad K4 = (Phn^{-4})(H^+) / (HPhn^{-3}) \tag{10}$$

Mass balance relationships involving total molar concentrations of the components are shown in equations 11 to 13:

$$T_{Phn} = H_4Phn + H_3Phn^- + H_2Phn^{-2} + HPhn^{-3} + Phn^{-4} \tag{11}$$

$$T_{Phn} = Phn^{-4}\{[(H^+)^3 / (K_2K_3K_4)] + [2(H^+)^2 / (K_3K_4)] + [3(H^+) / (K_4)]+1\} \tag{12}$$

$$Phn^{-4} = T_{Phn} / \{[(H^+)^3 / (K_2K_3K_4)] + [2(H^+)^2 / (K_3K_4)] + [3(H^+) / (K_4)]+1\} \tag{13}$$

Charge balance requirements are given in equations 14 to 18:

$$Na^+ + H^+ = H_3Phn^- + 2H_2Phn^{-2} + 3HPhn^{-3} + 4Phn^{-4} + OH^- \tag{14}$$

$$Na^+ + H^+ - [Kw / (H^+)] = H_3Phn^- + 2H_2Phn^{-2} + 3HPhn^{-3} + 4Phn\text{-}4 \tag{15}$$

$$R = Na^+ + H^+ - [Kw / (H^+)] \tag{16}$$

$$R = Phn^{-4} [(H^+)^3 / (K_2K_3K_4)] + [2(H^+)^2 / (K_3K_4)] + [3(H^+) / (K_4)]+ 4 \tag{17}$$

$$R = T_{Phn} \{[(H^+)^3 / (K_2K_3K_4)] + [2(H^+)^2 / (K_3K_4)] + [3(H^+) / (K_4)]+ 4\} / \{[(H^+)^3 / (K_2K_3K_4)] + [2(H^+)^2 / (K_3K_4)] + [3(H^+) / (K_4)]+1\} \tag{18}$$

multiplying by $(K_2K_3K_4) / (K_2K_3K_4)$

$$R = TPhn [(H^+)^3 + 2(H^+)^2 K_2 + 3(H^+)K_2K_3 + 4K_2K_3K_4] / [(H^+)^3 + (H^+)^2 K_2 + (H^+)K_2K_3 + K_2K_3K_4] \tag{19}$$

$$R [(H^+)^3 + (H^+)^2 K_2 + (H^+)K_2K_3 + K_2K_3K_4] = T_{Phn} [(H^+)^3 + 2(H^+)^2 K_2 + 3(H^+)K_2K_3 + 4K_2K_3K_4] \tag{20}$$

$$R (H^+)^3 + R (H^+)^2 K_2 + R (H^+)K_2K_3 + R K_2K_3K_4 = T_{Phn} (H^+)^3 + 2 T_{Phn} (H^+)^2 K_2 + 3 T_{Phn} (H^+)K_2K_3 + 4 T_{Phn} K_2K_3K_4 \tag{21}$$

separating the terms, will result in equation (22)

$$[R\ (H^+)^2 - 2T_{Phn}\ (H^+)^2]\ K_2 + [R\ (H^+) - 3T_{Phn}\ (H^+)]\ K_2\ K_3 + (R - 4T_{Phn})\ K_2\ K_3\ K_4 = -\ R$$
$$(H^+)^3 - T_{Phn}\ (H^+)^3 \tag{22}$$

$$a_{11}x_1 + a_{12}x_2 + a_{13}x_3 = b_1 \text{ (data Point from region 1 of Figure 3)} \tag{23}$$

$$a_{21}x_1 + a_{22}x_2 + a_{23}x_3 = b_2 \text{ (data Point from region 2 of Figure 3)} \tag{24}$$

$$a_{31}x_1 + a_{32}x_2 + a_{33}x_3 = b_3 \text{ (data Point from region 3 of Figure 3)} \tag{25}$$

$$x_1{}^* = (b_1 - 0 - 0)\ /\ a_{11} \tag{26}$$

$$x_2{}^* = (b_2 - a_{21}x_1 - 0)\ /\ a_{22} \tag{27}$$

$$x_3{}^* = (b_3 - a_{31}x_1 - a_{32}x_1)\ /\ a_{33} \tag{28}$$

$$x_1 = (b_1 - a_{12}x_2{}^* - a_{13}x_3{}^*)\ /\ a_{11} \tag{29}$$

$$x_2 = (b_2 - a_{21}x_1 - a_{23}x_3{}^*)\ /\ a_{22} \tag{30}$$

$$x_3 = (b_3 - a_{31}x_1 - a_{32}x_2)\ /\ a_{33} \tag{31}$$

$$K_2 = x_1$$

$$K_3 = x_2 - x_1$$

$$K_4 = x_3 - x_2$$

The Formation Constants Calculation Of Calcium Complexes: (For L^{-4} Ligand)

$$Ca^{+2} + H_2L^{-2} \rightarrow CaH_2L \qquad K_{CaH_2L} = (CaH_2L)\ /\ [(Ca^{+2})(H_2L^{-2})] \tag{32}$$

$$Ca^{+2} + HL^{-3} \rightarrow CaHL^- \qquad K_{CaHL^-} = (CaHL^-)\ /\ [(Ca^{+2})(HL^{-3})] \tag{33}$$

$$H_2L^{-2} \rightarrow HL^{-3} + H^+ \qquad K_3 = (HL^{-3})(H^+)(H_2L^{-2}) \tag{34}$$

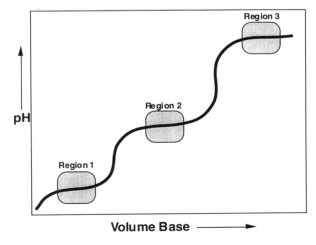

Figure 3. pH vs volume of base titration curve for BP.

$$(H_2L^{-2}) = (HL^{-3})(H^+) / K_3 \tag{35}$$

Mass balance relationship involving the total molar concentration of the components indicated are shown in equations 36 and 37

$$T_L = (CaH_2L) + (CaHL^-) + (H_2L^{-2}) + (HL^{-3}) \tag{36}$$

$$T_{Ca} = (CaH_2L) + (CaHL^-) + (Ca^{+2}) \tag{37}$$

$$T_{Ca} - (Ca^{+2}) = (CaH_2L) + (CaHL^-) \tag{38}$$

Equation 39 can be obtained by substituting equation 38 into equation 36:

$$T_L = [T_{Ca} - (Ca^{+2})] + (H_2L^{-2}) + (HL^{-3}) \tag{39}$$

Equation 40 can be obtained by substituting equation 35 into equation 39

$$T_L = [T_{Ca} - (Ca^{+2})] + [(HL^{-3})(H^+) / K3] + (HL^{-3}) \tag{40}$$

$$T_L = [T_{Ca} - (Ca^{+2})] + (HL^{-3})\{[(H^+) / K3] + 1\} \tag{41}$$

$$HL^{-3} = \{T_L - [T_{Ca} - (Ca^{+2})]\} / \{[(H^+)/K3] + 1\} \tag{42}$$

$$H_2L^{-2} = T_L - \{(HL^{-3}) + [T_{Ca} - (Ca^{+2})]\} \tag{43}$$

Charge balance requirement is shown in equation 44.

$$(Na^+) + (H^+) + 2(Ca^{+2}) = (CaHL^-) + 2(H_2L^{-2}) + (3HL^{-3}) + (OH^-) + (Cl^-) \tag{44}$$

$$(CaHL^-) = (H^+) + (Na^+) + 2(Ca^{+2}) - 2(H_2L^{-2}) + (3HL^{-3}) + (OH^-) + (Cl^-) \tag{45}$$

$$(CaH_2L) = T_L - [(CaHL^-) + (H_2L^{-2}) + (HL^{-3})] \tag{46}$$

K_{CaH_2L} and K_{CaHL^-} values are calculated using equation 32 and 33.

The Formation Constants Calculation Of Calcium Complexes (For L^{-3} Ligand)

$$Ca^{+2} + H_2L^- \rightarrow CaH_2L^+ \quad K_{CaH2L}{}^+ = (CaH_2L^+) / [(Ca^{+2})(H_2L^-)] \tag{47}$$

$$Ca^{+2} + HL^{-2} \rightarrow CaHL \quad K_{CaHL} = (CaHL) / [(Ca^{+2})(HL^{-2})] \tag{48}$$

$$H_2L^- \rightarrow HL^{-2} + H^+ \quad K_2 = (HL^{-2})(H^+) / (H_2L^-) \tag{49}$$

$$(H_2L^-) = (HL^{-2})(H^+) / K_2 \tag{50}$$

Mass balance relationship involving the total molar concentration of the components indicated are shown in equations (51) and (52)

$$T_L = (CaH_2L^+) + (CaHL) + (H_2L^-) + (HL^{-2}) \tag{51}$$

$$T_{Ca} = (CaH_2L^+) + (CaHL) + (Ca^{+2}) \tag{52}$$

$$T_{Ca} - (Ca^{+2}) = (CaH_2L^+) + (CaHL) \tag{53}$$

Equation 54 can be obtained by substituting equation 53 into equation 51

$$T_L = [T_{Ca} - (Ca^{+2})] + (H_2L^-) + (HL^{-2}) \tag{54}$$

Equation 55 can be obtained by substituting equation 50 into equation 54

$$T_L = T_{Ca} - (Ca^{+2}) + [(HL^{-2})(H^+) / K_2)] + HL^{-2} \tag{55}$$

$$T_L = T_{Ca} - (Ca^{+2}) + (HL^{-2})\{[(H^+) / K_2] + 1\} \tag{56}$$

$$HL^{-2} = \{T_L - [T_{Ca} - (Ca^{+2})]\} \{[(H^+) / K_2] + 1\} \tag{57}$$

$$H_2L^- = T_L - \{(HL^{-2}) + [T_{Ca} - (Ca^{+2})]\} \tag{58}$$

Charge balance requirement is shown in equation 59,

$$(Na^+) + (H^+) + (CaH_2L^+) + 2(Ca^{+2}) = (H_2L^-) + 2(HL^{-2}) + (OH^-) + (Cl^-) \tag{59}$$

$$(CaH_2L^+) = [(H_2L^-) + 2(HL^{-2}) + (OH^-) + (Cl^-)] - [(Na^+) + (H^+) + 2(Ca^{+2})] \tag{60}$$

$$(CaHL) = T_L - [(CaH_2L^+) + (H_2L^-) + (HL^{-2})] \tag{61}$$

K_{CaHL+} and K_{CaHL}- values are calculated using equations 47 and 48.

Equilibrium Constant Determination

Fifty milliliters of a solution containing a known concentration of phosphonate in 0.1 M sodium chloride solution, was equilibrated in the reaction cell under nitrogen at 25.0 ± 0.05°C. This solution was titrated with sodium hydroxide and the pH was recorded. After each addition the solution was allowed to equilibrate for 3 to 5 minutes until the pH of the solution reached a steady value (± 0.005). The protonation constants of the bisphosphonates were calculated using the steps shown in Figure 2. The data (pH and volume of base added) required for these calculations were chosen from three pH regions (as shown in Figure 3). For NE97220 the data were selected from the pH regions, 4-5.4 (region 1), 7-8.3 (region 2) and 9.5-10.5 (region 3); for NE58095, 2.6-3.2 (region 1), 5.5-6.9 (region 2), and 10.2-10.8 (region 3); for NE10503, 2.3-3.7 (region 1), 6.0-7.1 (region 2), and 10.0-11.0 (region 3); and for EHDP, 2.3-3.5 (region 1), 6.1-7.2 (region 2), and 10.0-11.0 (region 3).

RESULTS AND DISCUSSIONS

The protonation constants of all four bisphosphonates as well as a PAP are given in Table 1; each value is an average of 10 to 15 data points. EHDP was used as a model compound for comparison. Table 2 summarizes the calculated formation constants of calcium ion with the PAP and bisphosphonate anions.

In order to determine the chemical make up of the calcium salts of the PAP and bisphosphonate their corresponding crystalline phases where synthesized and analyzed. The stoichiometric ratio of calcium ion per BP or PAP anion for each salt is given in Table 3. Moreover, the apparent solubility products of the calcium PAP and BPs were determined using the corresponding synthesized calcium salt of each compound and the calculated

Table 1. Protonation constants of phosphonates

Sample	pK2	pK3	pK4
NE58095, (stdev)	2.77, (0.22)	6.79, (0.31)	10.45, (0.16)
NE97220, (stdev)	4.18, (0.52)	7.29, (0.16)	9.73, (0.39)
NE10503, (stdev)	2.35, (0.10)	6.55, (0.13)	10.09, (0.32)
ref. 6, (0.1 M KNO_3)	2.72	8.72	10.04
EHDP, (stdev)	2.41, (0.07)	6.95, (0.01)	10.38, (0.04)
ref. 7, (0.1 M KCl)	2.47	7.48	10.29
ref. 8, (1 M KCl)	2.34	6.44	10.08
ref. 9, (0.1 M KNO_3)	2.59	6.87	10.98
ref. 10	2.54	6.97	11.41
	pK1	pK2	pK3
NE58029, (stdev)	5.67, (0.09)	8.26, (0.10)	10.45, (0.04)

solubility products are given in Table 4. For comparative purposes the concentration of PAP^{-3} and BP^{-4} anions at their corresponding conditional solubility limits at physiological conditions of pH 7.4 and calcium concentration of 1.5 mM were also calculated and these are listed in Table 5.

CONCLUSIONS

Substitution of a hydroxyl group on the phosphorus atom of the BP yields the PAP derivative. The results in Table 5 compare the PAP^{-3} and BP^{-4} concentrations that are achievable in aqueous solutions at physiological conditions. Herein, it is shown that maximum concentration of a PAP ion (NE58029) is well over one hundred times less than its BP analog (NE97220). The decrease in polarity and the number of sites for hydrogen bonding of the PAP molecule compared with the its BP analog may have resulted in the PAP's decreases solubility.

Substitution of an amine group between the geminal carbon and the rest of the molecules, as is the case for NE97220, resulted in the marked increase in the solubility of this bisphosphonate compared with both risedronate and alendronate, as shown in Table 5. Compared with risedronate and alendronate the presence of an amine group, versus a methyl

Table 2. Stability constants for complexes of phosphonate and calcium ions

Sample	$logK_{CaH2L}$	$logK_{CaHL^-}$
Ca-NE58095, (stdev)	3.68, (0.05)	4.30, (0.13)
Ca-NE97220, (stdev)	4.32, (0.01)	5.46, (0.06)
Ca-NE10503, (stdev)	2.77, (0.35)	3.46, (0.31)
ref 6, in 0.1 M KC*l	5.37	6.01
Ca-HEDP, (stdev)	3.88, (0.05)	4.89, (0.11)
ref 9, in 0.1M KNO_3*	2.63	4.20
ref 11, in 0.02 M NaCl*	-	4.73
ref 12, in 0.1 M R4N.X*	-	3.12
	$logK_{CaHL}$	
Ca-NE58029, (stdev)	4.33, (0.07)	

*Literature Values, at 25°C

Table 3. Stoichiometry of calcium ions per BP or PAP molecule in the solids

	NE58095	NE97220	NE10503	NE58029
Ca / BP or PAP ratio	0.90	1.03	1.10	0.84
(stdv)	(± 0.05)	(± 0.05)	(± 0.09)	(± 0.06)
Morphology of the CaBP	flaky	flaky	flaky	flaky
or CaPAP solids				(plastic like)

group, on the geminal carbon not only resulted in the decreased rigidity of NE97220 at this position, but also the increased hydrophilicity of NE97220 and its greater solubility.

The calcium to BP and PAP ratio in the calcium salts of these agents ranged from 0.84 for the PAP NE58029, to 0.90, 1.03 and 1.09 for the bisphosphonates, risedronate,

Table 4. Concentration of calcium and phosphonate in solutions at equilibrium with BP and PAP solids. The solubility products of CaH_2BP and CaHPAP are given in the last column

Sample	pH	$(Ca^{2+})/10^{-5}$ M	Total BP/10^{-5} M	$(BP^{-4})/10^{-9}$ M	pKsp
NE58095	7.147	1.52	1.75	6.05	27.73
NE97220	7.150	1.52	3.50	38.6	26.93
NE10503	6.970	0.36	0.42	2.31	28.42
			Total PAP/10^{-5} M	$(PAP^{-3})/10^{-9}$ M	
NE58029	6.986	0.916	2.64	0.438	21.78

NE97220 and alendronate, respectively. As shown in Table 3, the calcium to PAP ratio is significantly lower for NE58029 than the calcium to BP ratio of its bisphosphonate analog, NE97220. NE97220 complexes the calcium ion in a tridentate manner through its hydroxyl and amine groups. As shown in Figure 1, a hydroxyl group on one of the phosphorus atoms of NE97220 was substituted with a methyl group, yielding NE58029. Thus, NE58029 can complex the calcium ion only in a bidentate manner through its hydroxyl and amine groups. The bidentate calcium complexation of NE58029 versus the tridentate calcium complexation of NE97220 may result in the decreased calcium ratio in the PAP versus the BP salt.

Table 5. The calculated concentration of PAP^{-3} and BP^{-4} anions at their corresponding conditional solubility limits in physiological solutions (pH 7.4 and calcium concentration of 1.5 mM)

Sample	pH	$(Ca^{2+})/mM$	$(BP^{-4})/10^{-11}$ M
NE58095	7.4	1.5	7.84
NE97220	7.4	1.5	49.4
NE10503	7.4	1.5	1.60
			$(PAP^{-3})/10^{-11}$ M
NE58029	7.4	1.5	0.28

REFERENCES

1. Von Beyer, H., and K.S. Hofmann, 1897, 1973, Acitodiphosphonige Saure, *Berichte*, 30.
2. Blaser, B., and K.H. Worms, with Henkel & Cie. G.m.b.H. Germany 1, 1960, Application of Organic Acylation Products of Phosphorous Acids or Their Derivatives as Complexing Agents for Metal Ions, *German Patent*, 080,235.
3. Sunberg, R.J., F.H. Ebetino, T.C. Mosher, and C.F. Roof, 1991, Designing drugs for stronger bones, *Chem. Tech*.
4. Zhang, J., A. Ebrahimpour, and G.H. Nancollas, 1992, Dual constant composition studies of phase transformation of dicalcium phosphate dihydrate into octacalcium phosphate, *J. of Colloid and Interface Science*, 152: 132-140.
5. Ebrahimpour, A., J. Zhang, and G.H. Nancollas, 1991, Dual constant composition method and its application to studies of phase transformation and crystallization of mixed phases, *J. of Crystal Growth*, 13: 83-91.
6. Kabachnik, M.I., et al., 1978, *Izv. Akad Nauk.* (USSR), 2, 433, (374).
7. Kabachnik, M.I., et al., 1967, *Proc. Acad. Sci.*, (USSR)., 177, 1060.
8. Mioduski, T., 1980, *Talanta*, 27,299.
9. Rizkalla, E.N., et al., 1980, *Talanta*, 27, 715.
10. Caroll, R.L., and R.R. Irani, 1967, *Inorg. Chem.*, 6, 1994.
11. Vasil'ev, V.P., G.A. Zaitsev, and I.N. Borisova, 1986, *Zhur. Neorg. Kim*, 31 (3)812, 462.
12. Claesssens, R.A.M. and J. van der Linden, 1984, *J. Inorg. Biochem.*, 21,73.

25

SOLUTION AND PRECIPITATION CHEMISTRY OF PHOSPHONATE SCALE INHIBITORS

M. Tomson, A. T. Kan, J. E. Oddo, and A. J. Gerbino

Rice University
Department of Environmental Science and Engineering
6100 S. Main - MS 317, Houston, Texas 77005

INTRODUCTION

Major progress has been made toward the understanding of phosphonate solution and precipitation chemistry and how it relates to the fate and transport of phosphonates. To prevent scale formation in gas and oil production chemicals are forced into a reservoir, allowed to react with the solid matrix and then slowly released into the production fluids. This is called a "squeeze". A polymer ionization model has been developed for modeling phosphonate speciation in water and a more stable calcium phosphonate solid phase has been identified, for the first time. In this chapter, phosphonate solution and precipitation chemistry will be discussed. Preliminary results have shown that the solution and precipitation chemistry can be used to explain the fate of phosphonates following an oil field squeeze. Modeling and prediction of an inhibitor squeeze life has began using this information. Most of this chapter will be devoted to a discussion on calcium phosphonate solution chemistry and a new precipitation mechanism.

Organophosphorus compounds are a group of widely used industrial chemicals. They are particularly important in a variety of water treatment applications, e.g., corrosion inhibition, threshold scale inhibition, chelating metal ions, deflocculation, crystal growth modifications and sludge modification. Therefore, it is important to understand the solution speciation and precipitation chemistry of this type of compound, but little information is available regarding the solution chemistry of phosphonates. Oddo and Tomson[1] reported on the precipitation chemistry of Ca-diethylenetriamine- N,N,N',N",N"-penta(methylenephosphonic acid) (DTPMP). However, the follow up work in that area has shown that their precipitate was not the thermodynamically most stable phase. The objective of this research has been to understand the fundamental solution chemistry of one representative organophosphorus compound (i.e., Ca-diethylene-triamine N,N,N',N",N"-penta(methylenephosphonic acid)) and the precipitation chemistry of a more stable, or less soluble Ca-DTPMP solid phase. DTPMP is a commonly used phosphonate. It is sold by Monsanto Chemical Company as a 50% (w/w) solution under the name Dequest® 2060.

Mineral Scale Formation and Inhibition, Edited by Zahid Amjad
Plenum Press, New York, 1995

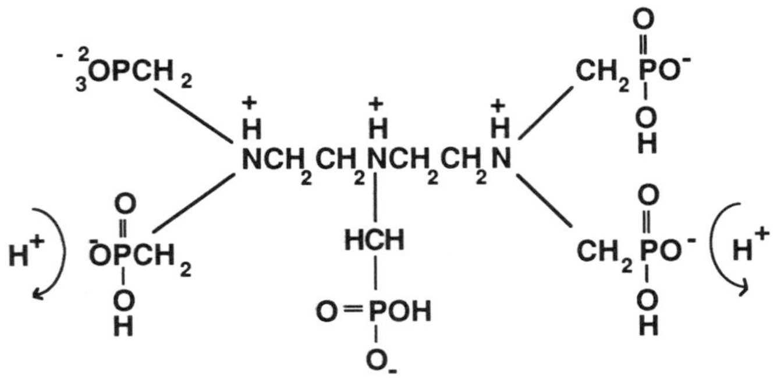

$$^{-2}_{3}OPCH_2 \qquad\qquad\qquad\qquad CH_2PO^-$$

(Figure 1 chemical structure of DTPMP)

Figure 1. Structure of diethylenetriamine penta(methlene phosphonic) acid, DTPMP, used in this study.

SOLUTION CHEMISTRY

In order to understand the retention and release of phosphonates from formations, a reliable model of the aqueous acid/base and complex formation behavior of metal phosphonates has been developed. This new model is based upon electrostatics and uses simplifications from modified theory of polymer ionization.[2] First, a short review of the status of phosphonate speciation will be presented.

The structure of DTPMP is shown in Figure 1. There are 10 ionizable protons on DTPMP and the distribution of species is strongly dependent upon pH.

The term, H_nPhn (n = 10 for DTPMP) will be used to represent an arbitrary phosphonate with a maximum of "n" ionizable protons. When DTPMP is fully ionized, the charge on the anion is minus ten, Phn^{10-}. Even neglecting possible chelate effects or covalent bond formation and considering only electrostatics, such highly charged anions should form stable complexes with di- and trivalent metal ions, such as Ca^{2+}, Mg^{2+}, Fe^{2+}, or Al^{3+}. The divalent metal ion Ca^{2+} will be used for illustration purposes. Formation of a stable complex between Ca^{2+} and Phn^{10-} could be represented by:

$$Phn^{10-} + Ca^{2+} \rightarrow Ca\,Phn^{8-} \tag{1}$$

The product in Equation 1, $Ca\,Phn^{8-}$, is still highly charged and would be expected to form additional complexes with four or five calcium ions.[3] In Figure 2 is written a possible scheme of DTPMP, or $H_{10}Phn$, ionization and concomitant calcium ion complexation.

To calculate the species distribution in Figure 2, it would be necessary to determine a total of thirty equilibrium constants, ten ionization constants and twenty metal complex constants. A similar scheme could be written for any polyprotic oligotrophic acid and metal ion. In principle it is possible to calculate all thirty constants from acid/base titration data in the absence and presence of various total concentrations of divalent metal ions. The fact that such a set of equilibrium constants has never been, and probably never will be, determined for any such polyprotic acid-metal combination is illustrated qualitatively by the following arguments. In Figure 3 is illustrated the overall shape of typical titration curves of DTPMP, determined by the authors, in the absence and presence of calcium.

The titrations were done at 70°C and one molar ionic strength. It is possible to determine ten ionization constants from the titration data in the absence of calcium, essentially by noting the pH at the midpoint of successive addition of 0.5, 1.5, 2.5 - - - - - 9.5 equivalents of base per mole of phospohonate. It should be noted that although this is essentially correct, the actual numerical and experimental procedures are considerably more

Diethylenetriaminepenta (methylene phosphonic) acid (DTPMP)

H_0Phn^{10-}
(0.00)

$\downarrow$ 12.58

H_1Phn^{9-}
(12.58)

$\downarrow$ 11.18

	5.04		3.78		2.52		1.26	
H_2Phn^{8-} (23.76)	$\rightarrow$	CaH_2Phn^{6-} (28.80)	$\rightarrow$	$Ca_2H_2Phn^{4-}$ (32.58)	$\rightarrow$	$Ca_3H_2Phn^{2-}$ (35.10)	$\rightarrow$	$Ca_4H_2Phn^0$ (36.36)

$\downarrow$ 8.30

	4.41		3.15		1.88		0.63	
H_3Phn^{7-} (32.06)	$\rightarrow$	CaH_3Phn^{5-} (36.47)	$\rightarrow$	$Ca_2H_3Phn^{3-}$ (39.62)	$\rightarrow$	$Ca_3H_3Phn^{1-}$ (41.51)	$\rightarrow$	$Ca_4H_3Phn^{+1}$ (42.14)

$\downarrow$ 7.27

	3.78		2.52		1.26	
H_4Phn^{6-} (39.33)	$\rightarrow$	CaH_4Phn^{4-} (43.11)	$\rightarrow$	$Ca_2H_4Phn^{2-}$ (45.63)	$\rightarrow$	$Ca_3H_4Phn^0$ (46.89)

$\downarrow$ 6.23

	3.15		1.88		0.63	
H_5Phn^{5-} (45.56)	$\rightarrow$	CaH_5Phn^{3-} (48.71)	$\rightarrow$	$Ca_2H_5Phn^{1-}$ (50.60)	$\rightarrow$	$Ca_3H_5Phn^{1+}$ (51.23)

$\downarrow$ 5.19

	2.52		1.26	
H_6Phn^{4-} (50.75)	$\rightarrow$	CaH_6Phn^{2-} (53.27)	$\rightarrow$	$Ca_2H_6Phn^0$ (54.53)

$\downarrow$ 4.15

	1.88		0.63	
H_7Phn^{3-} (54.90)	$\rightarrow$	CaH_7Phn^{1-} (56.79)	$\rightarrow$	$Ca_2H_7Phn^{1+}$ (57.42)

$\downarrow$ 3.11

	1.26	
H_8Phn^{2-} (58.01)	$\rightarrow$	CaH_8Phn^0 (59.27)

$\downarrow$ 2.08

	0.63	
H_9Phn^{1-} (60.09)	$\rightarrow$	CaH_9Phn^{1+} (60.72)

$\downarrow$ 1.04

$H_{10}Phn^0$
(61.13)

Figure 2. A representation of the DTPMP and calcium complexes considered in the present model. Stepwise association constants (logarithm base ten) are to the right or just above the respective arrows. Cumulative formation constants of each species, Ca_jH_iPhn, are in parentheses. For example, the concentration of $(Ca_2H_3Phn) = 10^{10.62}(Ca^{2+})^2\{H^+\}^3(Phn^{10-})$.

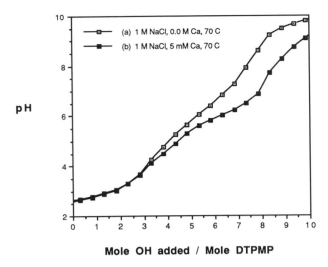

Figure 3. Potentiometric titration of DTPMP with NaOH where DTPMP solution is in 1M NaCl (a) and 1M NaCl, 5mM Ca (b).

complicated due to spontaneous ionization at pH values below about 4 pH and the presence of overlapping ionizations at higher pH values. The ten reported ionization constants of DTPMP at 25°C and one molar ionic strength are[4]:

$$
\begin{array}{ccccccccccc}
i & 1 & 2 & 3 & 4 & 5 & 6 & 7 & 8 & 9 & 10... \\
pK1 & (0.8) & (1.8) & 2.8 & 4.45 & 5.50 & 6.38 & 7.17 & 8.15 & 10.11 & 12.04
\end{array}
\tag{2}
$$

where the values in parentheses have been estimated by the present authors.

The presence of calcium tends to lower the measured pH at any specific amount of base added, Figure 3, lower curve. The difference in the shapes of the two curves in Figure 3 is typical of titration data and is not sufficient to uniquely determine the thirty constants indicated in Figure 2. For the DTPMP/Ca system one complex constant has been reported for 25°C and one tenth molar ionic strength ($K_{st} = 10^{7.11}$).[4] Such a single constant might represent the observed decrease in pH at higher pH values, but it can be shown that at lower pH values, typical of field brines, this does not reproduce the observed depression of pH. For example, at 5 pH and 1 mM total DTPMP the calculated free phosphonate, Phn^{10-}, would be too small to have an impact on the measured pH. Yet, the observed pH is lowered, as can be seen in Figure. 3.

To overcome the numerical and experimental difficulties discussed above, and to provide a framework for generalizing the results to more complicated brines and mixed interactions a new approach based upon notions from simple electrostatics and polymer theory has been developed. The derivation is based upon the assumption that the differences in all of the acid/base constants and metal-ligand complexation constants in e.g. Figure 2 can be modeled by a linear change in charge on the ligand or metal/ligand reactant species. Support for this approach is based upon four observations from the reported data in Smith and Martell.[4] First, when successive ionizations of 12 monophosphonates were examined, the average difference in the ionization constants was $\Delta pK = 5.20 \pm 0.16$, and this included orthophosphate, as suggested by Pauling.[5]

Secondly, the corresponding calcium complex stability constants could all be represented by log Kst = 1.55 ± 0.14, which is the expected electrostatic value.[3] Third, when pKi is plotted against charge in Figure 4 for DTPMP, a straight line is obtained with a correlation

Table 1. Comparison of the ionization constants and stability constants of various monophosphates

$R^\S$	pK_1	pK_2	pK_3	$\Delta pK^?$	log $Kst^¥$
H-	2.15	7.20		5.05	1.55
CH_3-	2.19	7.55		5.36	1.51
CH_3CH_2-	2.29	7.79		5.50	1.54
$ClCH_2$-	1.04	6.19		5.15	1.46
Cl_2CH-	(0.7)	5.21		(4.51)	(1.26)
Cl_3C-	(0.8)	4.48		(3.68)	(1.25)
$BrCH_2$-	1.15	6.24		5.09	1.34
ICH_2-	1.27	6.44		5.14	1.37
$HOCH_2$-	1.70	6.97		5.27	1.68
H_2NCH_2-	0.40	6.23	10.05	4.99	1.71
$H_2NCH_2CH_2$-	1.10	6.23	11.02	5.13	1.74
$H_2NCH_2CH_2CH_2$-	1.60	6/89	11/07	5.29	1.68
Mean				5.20	1.55
St. Dev				0.16	0.14

*Data are taken from Martell and Smith (1975) for 25 C and 0.1 M I.S.
§ R is the chemical structure attached to the function group $R-PO_3H_2$
$^?$ $\Delta K = pK1-pK2$
$^¥$ $K_{st} = (CaL)/(Ca)(L)$

coefficient of r = 0.99 and it will be shown later that the slope of this line is as expected from electrostatic theory.

Finally, when the reported metal complex constants are examined and log Kst vs. (q) is plotted for charges from one to six, (Figure 5) a similar straight line is obtained, with r = 0.99. The value of the DTPMP/Ca complex is not included because there is not enough reported data for charges from 6 to 10 and the value of log K = 7.11 may reflect a systematic error in the determination or a leveling effect due to charge distribution. Based upon these observations and based upon a need to develop a simpler approach for solution chemistry of phosphonates in brine, the following model is suggested and evaluated.

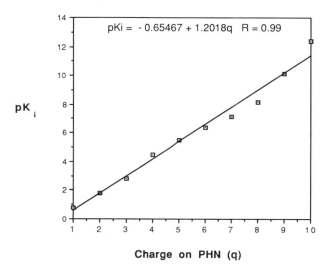

$$pKi = -0.65467 + 1.2018q \quad R = 0.99$$

Charge on PHN (q)

Figure 4. Plot of pKi (DTPMP) versus charge.

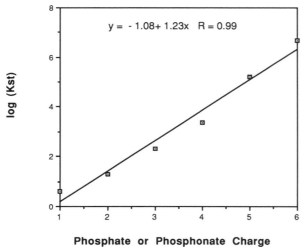

Figure 5. Plot of the logarithm of the Ca Phosphate ar Ca Phosphonate complex stability constants versus the charges of the phosphorus molecule.

y = - 1.08+ 1.23x   R = 0.99

log (Kst)

Phosphate or Phosphonate Charge

ELECTROSTATIC MODEL OF PHOSPHONATE/PROTON ASSOCIATION

To simply the notation and algebraic approach all equilibria will be written as stepwise association constants.[6] Tikhonova[7] showed that the first two protons associate with basic sp^3-hybridized nitrogens (see Figure 1). It will be shown below in the experimental section that the first two association constant, K_1^n and K_2^n, are $10^{12.58}$ and $10^{11.18}$, respectively.

The remaining proton association reactions and constants can be represented by the following equations:

$$H_{i-1}Phn^{(n-i+1)-} + H^+ \rightarrow HiPhn^{(n-i)-} \text{ for } i \geq 3 \tag{3a}$$

$$K_i = (H_iPhn^{(n-i)-}) / (\{H+\}(H_{i-1}Phn^{(n-i+1)-})) \tag{3b}$$

An arbitrary phosphonate species, $(H_iPhn^{(n-i)-})$, can be expressed in terms of equilibrium constants, free phosphonate concentration and pH:

$$(H_iPhn^{(n-i)-}) = K_1^N K_2^N K_3 \ldots K_i \{H^+\}^i (Phn^{n-}) \text{ for } i \geq 3 \tag{3c}$$

To simplify terms, the "beta" or cumulative formation constant notation will be used and charges will be dropped:

$$\beta_1 \equiv K_1^N; \tag{4a}$$

$$\beta_2 \equiv K_1^N K_2^N; \tag{4b}$$

and

$$\beta_i \equiv K_1^N K_2^N K_3 \ldots K_i, \text{ for } i \geq 3 \tag{4c}$$

A model can now be proposed to calculate the values of the K_i's, assuming that the primary difference between successive equilibrium constants is a consequence of decreasing charge. Numerous models based upon electrostatics have been proposed.[2] Most of these semiempirical models yield equations which are linear in increasing charge, as will be assumed herein (see Results and Discussion Section). The following model is proposed to calculate the values of K_i's for the proton association equilibria:

$$\log K_i = a_{H+} + b_{H+} |q_{i-1}| \tag{5a}$$

where a_{H+} and b_{H+} are constants to be determined from the titration data and $|q_{i-1}|$ represents the absolute value of the charge on the (i-1)th species, i.e., the species with one less proton, which is being protonated. From analysis of titration data presented herein, it has been shown that the value of α_{H+} in Equation 5a is statistically indistinguishable from zero and therefore Equation 5a can be written as follows:

$$\log K_i = b_{H+} |q_{i-1}| \tag{5b}$$

When values of the charges are written in terms of the integers, i and n, the b_i in Equation 4c can be shown to be:

$$\log \beta_i = \log K_1{}^N + \log K_2{}^N + \{i[n - \tfrac{1}{2}(i-1)] - 2n + 1\}b_{H+} \text{ for } i \geq 2 \tag{6}$$

The value of b_{H+} can be determined by combining the mass balance for total phosphonate with the electroneutrality condition and minimizes the sum squared error.

The procedure for calculating the stability constants of the calcium-DTPMP complexes using an electrostatic model is similar to that outlined above for the proton equilibria. Again, the speciation scheme depicted in Figure 2 will be used; specifically, it will be assumed that the two protons associated with the nitrogen atoms in DTPMP molecules are not displaced by calcium and therefore do not participate in the calcium-complex reaction scheme, i.e., the first metal complex to be formed is with the phosphonic acid species H_2Phn^{8-}. The general reaction and equilibrium constant for the formation of a calcium-DTPMP complex are as follows:

$$Ca_{j-1}H_iPhn + Ca^{2+} \rightarrow Ca_jH_iPhn \tag{7a}$$

$$K_{ij} = (Ca_jH_jPhn)/(Ca^{2+})(Ca_{j-1}H_iPhn) \text{ for } i \geq 2 \text{ and } j \geq 1 \tag{7b}$$

By repeated substitution, the concentration of (Ca_jH_iPhn) can be expressed as[16]:

$$(Ca_jH_iPhn) = \beta_iK_i1K_i2...K_{ij}(Ca^{2+})^j\{H^+\}^i(Phn) \tag{7c}$$

Again, the formation constants, K_{ij}, can be modeled with an equation similar to equation 5a, which assumes that the stability constant is proportional to the charge on the species with one less calcium ion:

$$\log K_{ij} = a_{Ca} + b_{Ca} |\theta_{2(j-1)+i}| \tag{8a}$$

The value of a_{Ca} would correspond to the association constant of calcium with an electronically neutral phosphonate molecule, i.e., H10Phn. The value of a_{Ca} determined from statistical analysis was not distinguishable from zero (0.00) and including it in the model did not improve the fit to the experimental data. Therefore, it was set to zero and the final model used was:

$$\log K_{ij} = b_{Ca} |q_{2(j-1)+i}| \tag{8b}$$

The cumulative formation constant for the metal complexes can be written in terms of the individual proton association constants and the individual calcium-phosphonate stability constants:

$$\beta_j \equiv K_1^N \ K^{2N} \ K_3...K_i K_{i1} K_{i2}...K_{ij} = \beta_i K_{i1} K_{i2}...K_{ij} \tag{9a}$$

By considering the correspondence of the charge to the ij-indexing, Equation 9a can be shown to be given by the following formula, which facilitates calculation of constants:

$$\log \beta_{ij} = \log K_1^N + \log K_2^N + \{i[n-\tfrac{1}{2}(i-1)]-2n+1\}b_{H+} + \{j(n-i-j+1)\}b_{Ca}, \text{ with } i\geq 2 \text{ and } j\geq 0 \tag{9b}$$

Equations 9a and b express all of the stability constants of DTPMP (Figure 2) in terms of only one additional constant, b_{Ca}. The value of b_{Ca} can be obtained using the method of least squares, analogous to the determination of b_{H+}.

EXPERIMENTAL

Potentiometric Titration

All titration experiments were done in a 250 ml Teflon covered water jacketed beaker (ACE Glass Inc.) connected to a circulation bath (Neslab, RTA-100). pH was monitored with a temperature compensated combination electrode (Ross Model 81-55). pH meter was calibrated with potassium biphthalate buffer (pH 4.01 at 25°C) and sodium borate (pH 9.18 at 25°C) solution at various temperature. Both the test solutions and calibration buffers were first flushed with N_2 gas and equilibrated at appropriate temperature before measuring pH. The stability of the electrode at 50, 70 and 90°C was confirmed prior to each experiment.

Before a titration experiment, a 100 ml solution and electrode was equilibrated at the appropriate temperature under N_2. The change in pH was observed with the addition of 0.050 ml increments of 1.000 M NaOH (Baker Diutit No. 4689). A total of 2 ml NaOH was added. The data was confirmed by back titration with 1.000 M HCl. Titration experiments were done to evaluate the effect of ionic strength, Ca ion complexation, and temperature on phosphonate ionization and calcium complexation. The effect of ionic strength was determined by titrating 1mM DTPMP solution in 0, 0.1, 0.3, 1.0 M NaCl solution. Each of these experiments was also done in the presence of 1×10^{-3} M Ca in the solution to observe the effect of calcium complexation at various ionic strength. Two additional titration experiments were done in the presence of 5×10^{-3} M Ca and 1 M NaCl. Finally, similar experiments were done in 1 M NaCl and 50, 70, 90°C temperature to study the temperature effect on ionization and stability constants.

Formation of the Low Solubility Ca-DTPMP Precipitate

A Ca-DTPMP precipitate was prepared by forming a less stable precipitate and washing the precipitate with excess of 0.1 M Ca solution to transform the initial material to a more stable phase. Figure 6 showed the schematic diagram of the experimental apparatus to prepare Ca-DTPMP precipitate. First, the less stable Ca-DTPMP salt was prepared in an Amicon stirred ultrafiltration cell, equipped with a 10,000 molecular weight cutoff ultrafiltration membrane (Amicon YM 10) as a reaction vessel. A 0.4 M DTPMP stock solution was prepared in DI water and neutralized to pH 6.8 with NaOH. The final solution contained 4.8

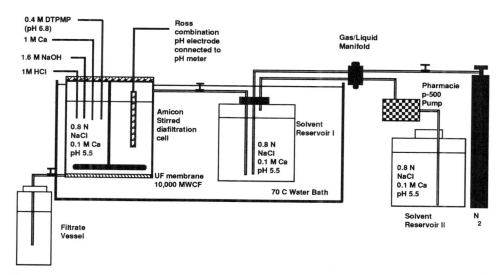

Figure 6. Schematic representation of the experimental apparatus for preparing the two calcium phosphate precipitates.

M NaCl after neutralization. Calcium solution (1.0 M) was prepared by dissolving 1 mole $CaCl_2 \times 2H_2O$ in a liter of deionized water. At the beginning of the reaction, 250 ml of synthetic brine (0.8 M NaCl, 0.1M $CaCl_2$, 1 x 10^{-3} M sodium acetate, pH 5.5) solution was added to the stirred cell with constant stirring by a magnetic stirrer and the solution was heated to 70°C in a water bath. The DTPMP and Ca solutions were titrated into the solution with a micrometer syringe (Gilmont GS-1200) in 0.050 ml increments. The pH was maintained at 5.5 pH with the addition of NaOH (1.6 N) or HCl (1 M). A total of 2.5 ml of DTPMP, 1.25 ml $CaCl_2$ and 0.31 ml NaOH was added to the solution. The final ionic strength of the resulting solution increased to 1.12 (0.84 M NaCl and 0.093 M $CaCl_2$). The reaction solution immediately formed a light precipitate upon injection of DTPMP and Ca. Equilibrium was assumed when the pH remained constant. Typically, a reaction was at equilibrium in 15 minutes. Following the reaction, the solution was continuously diafiltered with 8.5 l of synthetic brine at a 90 ml/hr diaflitration rate over 3 days. The synthetic brine was forced through the ultrafiltration membrane at constant flow rate by pressurizing the stirred cell with nitrogen at 5 psi. About 70% of Ca-DTPMP precipitate matured to a more stable solid phase. The resulting material was stored as suspension in a high density polyethylene bottle at 70°C for the following dissolution experiments.

Dissolution Experiments

Dissolution experiments were done in a wide span of solution compositions. The parameters tested included a pH span of 4.0-8.4, Ca concentration of 0.002 M to 0.3 M, ionic strength span of 0.2 - 3 M, and temperature span of 30-87°C. Batch dissolution experiments were done by preparing 90 ml of stock solution at the required ionic strength, pH, and Ca concentration. Sodium acetate (for pH 4 - 5.5) and sodium borate (pH 8.4) at 1 x 10^{-3} M was used as buffer. The solution was first warmed to a desired temperature in a water bath before adding 5 - 10 ml of Ca-DTPMP solid suspension. The Ca-DTPMP solid suspension was kept on constant stirring while sampling. The suspension contains about 1 mg solid as DTPMP per ml volume. The solution was stirred continuously at the desired temperature overnight and filtered through a 0.45 μm filter. The solution pH, Ca and phosphonate concentrations were determined by the methods described below.

Analysis of the Solid

Some Ca-DTPMP solid was allowed to mature for a week before being washed with deionized water and dried at 70°C for three day. The dried solid was analyzed with the scanning electron microscope (JEOL Model 5300) to study the crystal morphology. The crystallinity was studied by X-ray diffraction using a Phillips X-Ray powdered difractometer automated with DAPPLE automation software. The remaining solid following the dissolution experiment was washed with 30 ml water and dissolved in 0.5 ml of 1 N HCl. The solution was then diluted and neutralized before Ca and DTPMP concentration were determined.

Analytical Method

Calcium was measured by EDTA titration.[8] Phosphonate was first hydrolyzed to orthophosphate with persulfate in the presence of UV light. The orthophosphate concentration was then measured colorimetrically by the formation of molybdophosphorus complex and reduced to phosphomolybdenum blue complex with ascorbic acid.[9] The pH was determined using a Ross combination pH electrode (Model 81-5500).

RESULTS AND DISCUSSIONS

Phosphonate Ionization

Fig. 7 shows the potentiometric titration curve of DTPMP at four different ionic strengths, i.e. 0, 0.1, 0.316, and 1.0 M and 3 different temperatures. The pH of the corresponding solutions are plotted along the Y axis, and the ratio of the number of moles of base added to the solution to the number of moles of acid present in the solution is plotted along the X axis. Table 2 lists the calculated acidity constants from b_{H+} via Equation 6.

Using a nonlinear least squares procedure to match calculated versus experimental concentration of added sodium hydroxide in various titration curves the following curve fitted value for b_{H+} was obtained:

$$b_{H+} = 1.040 - 0.0047 \text{ (I.S., m)}^{1/2} - 5.97/T(K) \qquad (10)$$

Calcium-Phosphonate Complexation

Using additional data similar to that in Figure 7 the values of b_{Ca} were determined:

$$b_{Ca} = 1.109 + 0.021 \text{ (I.S., m)}^{1/2} - 176/T(K) \qquad (11)$$

Clearly, the values of b_{H+} and b_{Ca} are nearly independent of temperature and ionic strength for the ionic strengths above about 0.3 M as were used to determine the b-values.

Figure 8 plots the concentrations of each of the thirty species at pH 4 to 8. The X and Y axis are the number of hydrogen and calcium ions on a particular Ca-DTPMP species; concentrations are plotted as contour lines.

Each black square represent one of the 30 individual species. At each pH, there are 5 or 6 predominant species which compose 90% of all solution phase phosphonate. The

Table 2. Comparison of proton stability constants for DTPMP as determined in this study with those by Tikhonova reported in Smith and Martell.[4] The acidity constants determined by the polymer ionization theory compare favorably to the acidity constants reported in the literature

	$\log K_1^N$	$\log K_2^N$	$\log K_3$	$\log K_4$	$\log K_5$	$\log K_6$	$\log K_7$	$\log K_8$	$\log K_9$	$\log K_{10}$
Previous study:	12.04	10.10	8.15	7.17	6.38	5.50	4.45	2.8		
This study:	12.58	11.18	8.30	7.23	6.23	5.19	4.15	3.11	2.08	1.04

predominant Ca-DTPMP species are $Ca_{2,3}H_{4,5}DTPMP$ at pH 5.0 and $Ca_{2,3}H_{2,3}DTPMP$ at pH 7.0 and $Ca_{2,3,4}H_2DTPMP$ at pH 9.0. The results in Figure 8 are only qualitative.

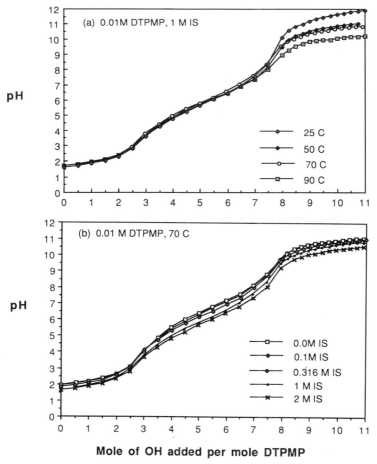

Figure 7. a) Plot of pH vs. reletive base added for 0.01M DTPMP solutions at 1M ionic strength and various temperatures and b) Plot of reletive base added for 0.01M DTPMP solutions and various ionic strengths.

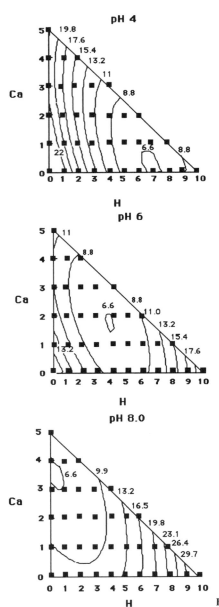

Figure 8. Distribution of the 36 Ca-DTPMP solution species.

Ca-DTPMP Precipitation Chemistry

Following the addition of 1 M Ca and 0.4 M DTPMP solution in the Amicon stirred cell at the beginning of the precipitation experiment, a light precipitate quickly forms. The resulting precipitate has a high solubility and the solution phase phosphonate concentration is about 3.45×10^{-4} M. This highly soluble material shares similar solubility properties of that of Oddo and Tomson.[1] The equilibrium solution phase phosphonate concentration of this less soluble material is in the range of 1×10^{-6} M.

Table 3. Comparison of calculated ion products for six oil and gas wells with the predicted K_{sp} values

Well	Ionic strength of brine (M)	Formation temperature (C)	Forma-tion pH	T_{Ca} (M)	T_{Phn} Initial (µM)	T_{Phn} Plateau (µM)	Ion[a] product initial	Ion[a] product plateau	Predicted pK_{sp} Crystalline	$\Delta pKsp$
Gladys McCall Grand Chenier, LA	1.9	148	4.13	0.1	351	0.18	51.4	54.7	54.3	-0.4
Pleasant Bayou No 2 Brazoria County, TX	2.7	152	4.32	0.24	602	0.13	50.7	54.4	54.5	0.1
Maxus ZEE-6 Offshore Indonesia	0.70	115	5.89	0.006	4	0.88	53.8	54.5	53.8	-0.7
O'Daniel No. 2 Hitchcock, TX	0.75	140	5.37	0.019	2840	2.0	50.5	53.6	54.1	0.5
N. R. Smith No. 4 Odem, TX	0.88	71	6.37	0.011	536	0.86	51.6	54.6	53.1	-1.5
Well A Offshore, North Sea	0.86	110	6.60	0.006	138	3.7	52.8	54.4	53.7	0.7

[a] Ion product is reported as $-\log_{10} (Ca^{2+})^3 \{H^+\}^4 (Phn^{10-})$.

Both the high solubility and low solubility Ca-DTPMP precipitates have been examined with Scanning electron microscope. The high solubility material appears to be amorphous and the low solubility material has a crystalline needle shape morphology. The X-ray diffraction pattern (Data not shown) of the low solubility Ca-DTPMP crystal shows a sharp crystalline structure. The X-ray diffraction pattern of the high solubility material confirms that the highly soluble material is amorphous in nature (Data not shown).

Table 4 lists the solution composition following the dissolution of Ca-DTPMP crystals at 70°C and 1.1 M ionic strength and the solid Ca/DTPMP stoichiometry of the remaining solids. In Table 4, T_{Ca} is the solution phase total Ca concentration and TDTPMP is the solution phase total DTPMP concentration. A total of 55 dissolution experiments have been done at 4.0-5.6 pH and Ca concentration of 0.002 - 0.3 M at 70°C and 1.1 M ionic strength. The Ca-DTPMP solid used in these experiments are from five replicate preparations and the data are very reproducible. The preliminary dissolution experiments have shown that the solution is not at equilibrium after 4 hours. No kinetic effect was observed when dissolution experiments are allowed to equilibrate from 17 to 96 hours. Most of the dissolution experiments were done by equilibrating at the appropriate temperature for 17 hours. In the dissolution experiments, the final solid to solution ratio ranged from 1 to 100 mg/l concentration and there was no dependence of solubility on solid to solution ratio. The solid Ca to DTPMP molar stoichiometric ratio is 3.07 (s = 0.08) for 21 experiments (Column 8, Table 4).

The solubility product of calcium DTPMP can be calculated from Ca^{2+}, H^+, and $DTPMP^{10-}$ concentrations by assuming a solid composition of Ca_3H_4DTPMP:

$$Ksp = [Ca^{2+}]^3[H^+]^4[DTPMP^{10-}] \qquad (12)$$

Table 4. The Ca and DTPMP concentrations, pH, and solubility product (K_{sp}), of the dissolution experiments and the solid stoichiometric coefficient (X)

Exp No.	pH	T_{Ca} (M)[¥]	Ca (M)[$]	T_{DTPMP}[¥] (μM)	DTPMP (M)[$]	pK_{sp}[*]	X[§]
1	4.10	0.051	0.051	10.10	1.27E-27	47.2	3.06
2	4.05	0.101	0.101	4.23	1.02E-28	47.2	3.07
3	4.02	0.151	0.151	1.68	1.55E-29	47.4	3.11
4	4.01	0.202	0.202	1.63	9.36E-30	47.2	3.06
5	4.18	0.029	0.029	23.80	1.63E-26	47.1	3.18
6	4.13	0.029	0.029	40.90	1.32E-26	47.0	3.20
7	4.09	0.020	0.020	107.00	2.89E-26	47.0	NA
8	4.09	0.026	0.026	66.40	1.41E-26	47.0	NA
9	4.09	0.097	0.097	1.42	6.95E-29	47.6	NA
10	4.06	0.198	0.198	0.30	3.07E-30	47.9	NA
11	4.02	0.293	0.293	0.44	1.27E-30	47.6	NA
12	4.44	0.052	0.052	1.57	1.86E-26	47.3	NA
13	4.41	0.107	0.107	0.95	2.44E-27	47.2	NA
14	4.37	0.157	0.157	0.81	6.45E-28	47.1	NA
15	4.36	0.216	0.216	0.72	2.71E-28	47.0	NA
16	4.40	0.029	0.029	5.57	7.95E-26	47.3	3.14
17	4.31	0.015	0.015	88.10	7.02E-25	46.9	NA
18	4.37	0.010	0.010	189.00	4.54E-24	46.8	3.21
19	4.33	0.010	0.010	114.00	1.54E-24	47.1	NA
20	4.83	0.050	0.050	0.65	9.66E-25	47.3	2.98
21	4.82	0.097	0.097	0.49	1.73E-25	47.1	2.93
22	4.78	0.152	0.152	0.42	3.97E-26	47.0	3.17
23	4.84	0.029	0.029	2.46	9.19E-24	47.0	3.13
24	4.67	0.01	0.010	36.90	5.68E-23	46.9	NA
25	4.73	0.005	0.005	114.00	6.37E-22	47.0	NA
26	4.64	0.015	0.015	19.90	1.43E-23	46.9	NA
27	4.77	0.018	0.018	1.98	6.16E-24	47.5	3.09
28	4.75	0.018	0.018	2.81	6.98E-24	47.4	3.03
29	4.73	0.019	0.019	5.70	1.02E-23	47.1	2.98
30	4.79	0.010	0.010	13.00	9.02E-23	47.1	2.98
31	4.79	0.010	0.010	12.30	8.74E-23	47.2	2.99
32	4.79	0.010	0.010	12.10	9.16E-23	47.2	3.07
33	4.76	0.010	0.010	13.20	6.37E-23	47.2	NA
34	4.77	0.010	0.010	16.80	9.22E-23	47.1	NA
35	4.77	0.010	0.010	11.50	5.98E-23	47.2	NA
36	4.77	0.011	0.011	15.30	7.80E-23	47.1	3.15
37	4.77	0.010	0.010	10.70	5.88E-23	47.3	NA
38	4.77	0.010	0.010	9.82	5.32E-23	47.3	NA
39	4.77	0.010	0.010	10.20	5.72E-23	47.3	NA
40	4.77	0.010	0.010	9.92	5.66E-23	47.3	NA
41	4.78	0.011	0.011	10.90	6.30E-23	47.2	3.01
42	4.79	0.048	0.048	0.80	7.48E-25	47.2	NA
43	4.82	0.097	0.097	0.46	1.65E-25	47.1	NA
44	4.75	0.050	0.050	1.25	7.13E-25	47.1	NA
45	4.85	0.010	0.010	10.60	1.70E-22	47.2	3.04
46	5.56	0.010	0.010	0.52	2.61E-20	47.9	NA
47	5.56	0.010	0.010	0.48	2.25E-20	47.9	NA
48	5.41	0.007	0.007	1.86	3.06E-20	47.7	NA
49	5.41	0.007	0.007	2.39	3.84E-20	47.5	NA
50	5.4	0.005	0.005	2.14	4.61E-20	47.9	NA
51	5.4	0.005	0.005	3.62	7.54E-20	47.6	NA
52	5.43	0.003	0.003	9.29	4.00E-19	47.6	NA
53	5.44	0.003	0.003	12.40	5.93E-19	47.5	NA
54	5.46	0.002	0.002	25.10	1.88E-18	47.5	NA

Table 4. (Continued)

55	5.46	0.002	0.002	25.40	1.96E-18	47.5	NA
56	8.4	0.012	0.012	0.73	1.19E-14	53.3	NA
MEAN						47.4	3.07
STD DEV						0.8	0.08

¥TCa = The solution phase Ca concentration and TDTPMP = The solution phase DTPMP concentration.

$Ca = Free Ca^{2+} concentration and DTPMP = Free $DTPMP^{10-}$ concentration.

* pKsp = -log $[Ca^{2+}]^3[H^+]^4[DTPMP^{10-}]$.

§ x = molar stoichiometric ratio of Ca versus DTPMP in the solid.

$$pK_{sp} = -\log (K_{sp})$$

The corresponding pK_{sp} values have been calculated for these 55 experiments (Column 7, Table 3). The temperature and ionic strength dependence of pK_{sp} is given by:

$$pKsp = 58.95 - 2084.5/T(K) + 0.048(I.S., m)^{1/2} \qquad (13)$$

Again, there is little dependence upon ionic strength, above about 0.3 M ionic strength. The solubility decreases with increasing temperature similar to that for $FeCO_3$[10], $CaCO_3$, and $CaHPO_4 \times 2H_2O$.[11]

RELATION TO INHIBITOR SQUEEZE LIFE

The low solubility Ca-DTPMP crystal is believed to be the stable form of phosphonate in a formation following a squeeze. Extensive research is in progress to simulate the inhibitor squeeze life using the precipitation model which will not be discussed in this chapter. To justify the hypothesis that precipitation controls phosphonate return following a squeeze, laboratory results will be compared with the field squeeze data in the following paragraph.

First, the solution chemistry of Ca-DTPMP dictate that the solution phosphonate concentration (TDTPMP) should remain in the range of 0.5-1.0 mg/l (or 0.5-2.0 μM) at pH 4.4 - 5.6 and Ca concentration between 0.05 - 0.2 M (see Table 4). The phosphonate flow back concentration in the field is typically in the range of 0.1-1 mg/l and the above pH values and Ca-concentrations are the typical concentrations of field brines. Secondly, one can compare the field data to the laboratory observation. Our research group has assisted in and implemented several inhibitor squeezes within the past few years. These squeezes were done by using aminotri (methylene phosphonic acid) (ATMP), another common phosphonate scale inhibitor. Two assumptions have been made in order to compare laboratory and field results: (1) the inhibitor in the brine is assumed to be at equilibrium with a Ca phosphonate solid phase in the formation; and (2) the orthophosphate group on ATMP is identical to that of DTPMP and the observed ATMP concentration can be converted to DTPMP in equal orthophosphate molar ratio. Note that this is consistent with findings of Meyers and Hunter.[12,13] Table 3 lists the brine chemistry of several field brines.

The ionic strength of these brine ranged from 0.7 to 2.66 M. The Ca concentration ranged from 0.001 to 0.24 M and bottom hole temperature ranged from 99 to 152°C. The solubility products calculated from the inhibitor return concentration, bottom hole pH[15] and calcium concentration of the brine are listed in Table 3. The calculated pKsp values compared favorably to the laboratory measured pKsp values in Table 4.

CONCLUSION

An electrostatic speciation code has been developed for phosphonate water speciation and metal complexation. This speciation code is based on the polymer ionization theory. Both diethylenetriamine N,N,N′,N″,N″-penta(methylenephosphonic acid) (DTPMP) ionization and complexation with Ca have been carefully studied at different ionic strength and temperature. This speciation code has been used to study the precipitation chemistry of a new Ca-DTPMP solid phase. This new solid phase has a Ca/DTPMP stoichiometric ratio of 3/1. Crystals exist in fine needle shape with sharp X-ray diffraction pattern. The solubility product of Ca_3H_4Phn at about 1 M ionic strength and 70°C is $10^{-53.3}$ and the heat of reaction is 33 KJ/mol. Preliminary study indicates that this solid phase probably controls the fate of phosphonates in gas and oil formations following inhibitor squeezes.

ACKNOWLEDGMENTS

This work was supported by the Gas Research Institute under contract No. 5088-212-1717, but in no way does this constitute an endorsement by GRI of any products or views contained herein. In addition, the work was also supported by a consortium of companies including, Champion Technologies, Inc.; Zapata Corporation; Texaco, Inc.; Conoco, Inc.; and FMC Corpation.

REFERENCES

1. Oddo, J. E., M. B. Tomson, 1989, "The Solubility and Stoichiometry of Calcium - Diethylenetriamine penta(Methylene Phosphonate) at 70 Deg C in Brine Solutions at 4.7 and 5.0 pH," App. Geochem. 5, 527-532.
2. Tanford, C., 1961, Physical Chemistry of Macromolecules, John Wiley & Sons, Inc., New York, N.Y.
3. Langmuir, D., 1979, Techniques of Estimating Thermodynamic Properties for Some Aqueous Complexes of Geochemical Interest, American Chemical Society.
4. Smith, R. E., A. E. Martell, 1975, Critical Stability Constants, Plenum Press, New York.
5. Pauling, L., 1970, General Chemistry, W.H. Freeman & Co., Third Edition, San Francisco.
6. Tomson, M. B., G. H. Nancollas, 1982, "Guidelines for the Determination of Stability Constants", 54, 2675-2692.
7. Tikhonova, L. I., J. Russ., 1968, Inorg. Chem., 13, 1384.
8. Greenberg, G. E., R. R. Trusell, L. S. Clesceri, 1985, "Standard Methods for the Examination of Water and Wastewater".
9. Hach, C. C., 1986, Chemical Procedures Explained, Loveland CO.
10. Tomson, M. B., M. L. Johnson, 1991, "How Ferrous Carbonate Kinetics Impacts Oilfield Economics," Int'l Sym. on Oilfield Chem., Anaheim, CA, SPE 21025.
11. Sillen, L. G. a. M., E. Arthur, 1971, Stability Constants of Metal-Ion Complexes, The Chemical Society, Burlington House, London.
12. Meyers, K. O., H. L. Skillman, G. D. Herring, 1985, "Control of Formation Damage at Prudhoe Bay, Alaska, by Inhibitor Squeeze Treatment," J. Petl. Tech., 1019-1034.
13. Hunter, M. A., 1992, "Adsorption/Dissolution of Phosphonates," Rice University, M.S.

26

THE PRACTICAL APPLICATION OF ION ASSOCIATION MODEL SATURATION LEVEL INDICES TO COMMERCIAL WATER TREATMENT PROBLEM SOLVING

R. J. Ferguson,[1] A. J. Freedman,[2] G. Fowler,[3] A. J. Kulik,[4] J. Robson,[5] and D. J. Weintritt [6]

[1] French Creek Software
Kimberton, Pennsylvania 19442
[2] Thomas M. Laronge, Inc
East Stroudsburg, Pennsylvania 18301-9045
[3] Nederlandse Aardolie Maatschappij
Schoonebeek, Netherlands
[4] LEPO Custom Manufacturing
Midland, Texas 79702
[5] Chemco
Vancouver, Washington 98682
[6] Weintritt Consulting Services
Lafayette, Louisiana 70503-5603

ABSTRACT

The availability of fast personal computers has allowed ion association models to be developed for the relatively inexpensive and widely available "PC" platforms. Ion association models predict the equilibrium distribution of species for a cooling water, oil field brine, waste water, or other aqueous solution of commercial interest. Scale potential indices based upon the free ion concentrations estimated by ion association models have been used extensively in the past decade to predict scale problems in industrial cooling water systems. Indices calculated by the models have been found to be transportable between waters of diverse composition and ionic strength. They have overcome many of the problems encountered with simple indices which do not account for ion pairing. This paper discusses the application of ion association model saturation level indices to predicting and resolving scale formation problems in cooling water systems, oil field brines, and for optimizing storage conditions for low level nuclear wastes. Examples are presented in case history format which demonstrate the use of the models in field applications.

Mineral Scale Formation and Inhibition, Edited by Zahid Amjad
Plenum Press, New York, 1995

Table 1. Example ion pairs used to estimate free ion concentrations

CALCIUM
$\sum$[Calcium]=   **[Ca^{+II}]** + [CaSO4] + [CaHCO3^{+I}]+
[CaCO3]+[Ca(OH)$^{+I}$]+[CaHPO4]+
[CaPO4^{-I}]+[CaH2PO4^{+I}]

MAGNESIUM
$\sum$[Magnesium]= **[Mg^{+II}]** + [MgSO4] + [MgHCO3^{+I}]+
[MgCO3]+[Mg(OH)$^{+I}$]+[MgHPO4] +
[MgPO4^{-I}]+[MgH2PO4^{+I}]+[MgF^{+I}]

BARIUM
$\sum$[Barium]=   **[Ba^{+II}]** + [BaSO4] + [BaHCO3^{+I}]+
[BaCO3]+[Ba(OH)$^{+I}$]

STRONTIUM
$\sum$[Strontium]= **[Sr^{+II}]** + [SrSO4] + [SrHCO3^{+I}]+
[SrCO3]+[Sr(OH)$^{+I}$]

SODIUM
$\sum$[Sodium]=   **[Na^{+I}]**+[NaSO4^{-I}]+[Na2SO4]+
[NaHCO3]+[NaCO3^{-I}]+[Na2CO3]+
[NaCl]+[NaHPO4$^{-I]}$

POTASSIUM
$\sum$[Potassium]= **[K^{+I}]**+[KSO4^{-I}]+[KHPO4^{-I}]+[KCl]

IRON
$\sum$[Iron]=   **[Fe^{+II}]+[Fe^{+III}]**+[Fe(OH)$^{+I}$]+
[Fe(OH)$^{+II}$]+[Fe(OH)3^{-I}]+[FeHPO4^{+I}]+
[FeHPO4]+[FeCl^{+II}]+[FeCl2^{+I}]+[FeCl3]+
[FeSO4]+[FeSO4^{+I}]+[FeH2PO4^{+I}]+
[Fe(OH)2^{+I}]+[Fe(OH)3]+[Fe(OH)4^{-I}]+
[Fe(OH)2]+[FeH2PO4^{+II}]

ALUMINUM
$\sum$[Aluminum]= **[Al^{+III}]**+[Al(OH)$^{+II}$]+[Al(OH)2^{+I}]+
[Al(OH)4^{-I}]+[AlF^{+II}]+[AlF2^{+I}]+[AlF3]+
[AlF4^{-I}]+[AlSO4^{+I}]+[Al(SO4)2^{-I}]

INTRODUCTION

Simple indices are widely used by water treatment personnel to predict the formation of scale. In many cases, simple indices are used as the basis for adjusting controllable operating conditions such as pH to prevent scale formation. Simple indices include the Langelier Saturation Index[1], Ryznar Stability Index[2], Stiff-Davis Saturation Index[3], and Oddo-Tomson index[4], for calcium carbonate. Indices have also been developed for other common scales such as calcium sulfate[5] and calcium phosphate. The simple indices provide an indicator of scale potential, but lack accuracy due to their use of total analytical values for reactants. They ignore the reduced availability of ions such as calcium which occurs due to association with sulfate and other ions.[5]

The simple indices assume that all ions in a water analysis are free and available as a reactant for scale forming equilibria. For example, the simple indices assume that all calcium is free. Even in low ionic strength waters, a portion of the analytical value for calcium will be associated with ions such as sulfate, bicarbonate, and carbonate (if present). This leads to over-prediction of the scaling tendency of a water in high ionic strength waters. The impact of these "common ion" effects can be negligible in low ionic strength waters. They can lead to errors an order of magnitude high in high ionic strength brines. Table 1 outlines some of the ion associations which might be encountered in natural waters.

THE CONCEPT OF SATURATION

A majority of the indices used routinely by water treatment chemists are derived from the basic concept of saturation. A water is said to be saturated with a compound (e.g. calcium carbonate) if it will not precipitate the compound and it will not dissolve any of the solid phase of the compound when left undisturbed, under the same conditions, for an infinite period of time. A water which will not precipitate or dissolve a compound is at equilibrium for the particular compound.

By definition, the amount of a chemical compound which can be dissolved in a water and remain in solution for this infinite period of time is described by the solubility product (Ksp). In the case of calcium carbonate, solubility is defined by the relationship:

$$(Ca)(CO_3) = Ksp$$

where:
(Ca) is the activity of calcium activity
(CO_3) is the carbonate activity
Ksp is the solubility product for calcium carbonate at the temperature under study.

In a more generalized sense, the term $(Ca)(CO_3)$ can be called the Ion Activity Product (IAP) and the equilibrium condition described by the relationship:

$$IAP = K_{sp}$$

The degree of saturation of a water is described by the relationship of the ion activity product (IAP) to the solubility product (K_{sp}) for the compound as follows:
If a water is undersaturated with a compound:

$$IAP < K_{sp} \text{ (It will tend to dissolve the compound).}$$

If a water is at equilibrium with a compound:

$$IAP = K_{sp} \text{ (It will not tend to dissolve or precipitate the compound).}$$

If a water is supersaturated with a compound:

$$IAP > K_{sp} \text{ (It will tend to precipitate the compound).}$$

The index called Saturation Level, Degree of Supersaturation, or Saturation Index, describes the relative degree of saturation as a ratio of the ion activity product (IAP) to the solubility product (K_{sp}):

$$\text{Saturation level} = \frac{IAP}{K_{sp}}$$

In actual practice, the saturation levels calculated by the various computer programs available differ in the method they use for estimating the activity coefficients used in the IAP; they differ in the choice of solubility products and their variation with temperature; and they differ in the dissociation constants used to estimate the concentration of reactants (e.g. CO_3 from analytical values for alkalinity, PO_4 from analytical orthophosphate).

Table 2 defines the saturation level for common scale forming species encountered in industrial applications. These formulas provide the basis for discussion of these scales in this paper.

ION PAIRING

The Saturation Index discussed can be calculated based upon total analytical values for the reactants. Ions in water, however, do not tend to exist totally as free ions[6,7,8] Calcium, for example, may be paired with sulfate, bicarbonate, carbonate, phosphate and other species. Bound ions are not readily available for scale formation. The computer program calculates

Table 2. Saturation level definition

Saturation level is the ratio of the Ion Activity Product to the Solubility Product for the scale forming specie.

For calcium carbonate:

$$SL = \frac{(Ca)(CO_3)}{Ksp'}$$

For barium sulfate:

$$SL = \frac{(Ba)(SO_4)}{Ksp'}$$

For calcium sulfate:

$$SL = \frac{(Ca)(SO_4)}{Ksp'}$$

Saturation Levels should be calculated based upon free ion activities using the solubility product for the form typical of the conditions studied (e.g. calcite for low temperature calcium carbonate, aragonite at higher temperatures.)

Saturation levels can be interpreted as follows:

A water will tend to dissolve scale of the compound if the saturation level is less than 1.0

A water is at equilibrium when the Saturation Level is 1.0 . It will not tend to form or dissolve scale.

A water will tend to form scale as the Saturation Level increases above 1.0 .

saturation levels based upon the free concentrations of ions in a water rather than the total analytical value which includes those which are bound.

Early indices such as the Langelier Saturation Index (LSI) for calcium carbonate scale, are based upon total analytical values rather than free species primarily due to the intense calculation requirements for determining the distribution of species in a water. Speciation of a water is time prohibitive without the use of a computer for the number crunching required. The process is iterative and involves:

- Checking the water for a electroneutrality via a cation-anion balance, and balancing with an appropriate ion (e.g sodium or potassium for cation deficient waters, sulfate, chloride, or nitrate for anion deficient waters).
- Estimating ionic strength, calculating and correcting activity coefficients and dissociation constants for temperature, correcting alkalinity for non-carbonate alkalinity.
- Iteratively calculating the distribution of species in the water from dissociation constants (a partial listing is outlined in Table 1).
- Checking the water for balance and adjusting ion concentrations to agree with analytical values.
- Repeating the process until corrections are insignificant.
- Calculating saturation levels based upon the free concentrations of ions estimated using the ion association model (ion pairing).

The use of ion pairing to estimate the free concentrations of reactants overcomes several of the major shortcomings of traditional indices. Indices such as the LSI correct activity coefficients for ionic strength based upon the total dissolved solids. They do not account for "common ion" effects.

Common ion effects increase the apparent solubility of a compound by reducing the concentration of reactants available. A common example is sulfate reducing the available calcium in a water and increasing the apparent solubility of calcium carbonate. The use of indices which do not account for ion pairing can be misleading when comparing waters where the TDS is composed of ions which pair with the reactants versus ions which have less interaction with them.

PITZER COEFFICIENT ESTIMATION OF ION PAIRING IMPACT

The ion association model provides a rigorous calculation of the free ion concentrations based upon the solution of the simultaneous non-linear equations generated by the relevant equilibria.[5,6,7,8] A simplified method for estimating the effect of ion interaction and ion pairing is sometimes used instead of the more rigorous and direct solution of the equilibria.[9] Pitzer coefficients estimate the impact of ion association upon free ion concentrations using an empirical force fit of laboratory data. This method has the advantage of providing a much less calculation intensive direct solution. It has the disadvantage of being based upon typical water compositions and ion ratios, and of unpredictability when extrapolated beyond the range of the original data. The use of Pitzer coefficients is not recommended when a full ion association model is available.

CASE 1: OPTIMIZING STORAGE CONDITIONS FOR LOW LEVEL NUCLEAR WASTE

Storage costs for low level nuclear wastes are based upon volume. Storage is therefore most cost effective when the aqueous based wastes are concentrated to occupy the minimum

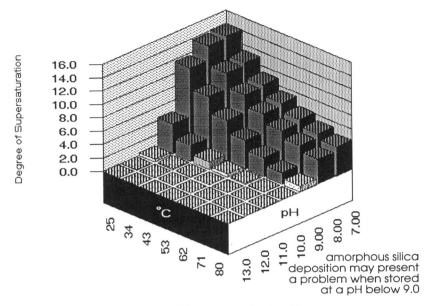

Figure 1. Silica saturation level profile.

volume. Precipitation is not desirable because it can turn a low level waste into a much more costly to store high level waste. Precipitation can also foul heat transfer equipment used in the concentration process. The ion association model approach has been used at the Oak

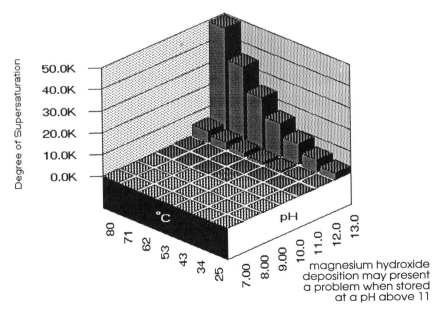

Figure 2. Magnesium hydroxide saturation level profile.

Ridge National Laboratory to predict the optimum conditions for long term storage.[10] Optimum conditions involve the parameters of maximum concentration, pH, and temperature. Figures 1 and 2 depict a profile of the degree of supersaturation for silica and magnesium hydroxide as a function of pH and temperature. It can be seen that amorphous silica deposition may present a problem when the pH falls below approximately 10, and that magnesium hydroxide deposition is predicted when the pH rises above approximately 11. A pH range of 10 to 11 is recommended based upon this preliminary run for storage, and concentration.

Other potential precipitants are screened using the ion association model to provide an overall evaluation of a waste water prior to concentration.

CASE 2: LIMITING HALITE DEPOSITION IN A WET HIGH TEMPERATURE GAS WELL

There are several fields in the Netherlands that produce hydrocarbon gas associated with very high total dissolved solids connate waters. Classical oilfield scale problems are minimal in these fields (e.g. calcium carbonate, barium sulfate, and calcium sulfate). Halite (NaCl), however, is precipitated to such an extent that production can be lost in hours. As a result, a bottom hole fluid sample is retrieved from all new wells. Unstable components are "fixed" immediately after sampling and pH is determined under pressure. A full ionic and physical analysis is carried out in the group central laboratory.

Table 3. Hot gas well water analysis

Analytical values expressed as the ion		Bottom Hole Connate	Boiler Feed Water
Temperature	°C	121	70
Pressure	bars	350	1
pH (site)		4.26	9.10
Density	kg/m^3	1.300	1.000
TDS	mg/l	369,960	< 20
Dissolved CO_2	mg/l	223	< 1
H_2S (gas phase)	mg/Nm^3	50.0	0
H_2S (aqueous phase)	mg/l	< 0.5	0
Bicarbonate	mg/l	16.0	0
Chloride	mg/l	228,485	5.0
Sulfate	mg/l	320	0
Phosphate	mg/l	< 1	0
Borate	mg/l	175	0
Organic acids <C_6	mg/l	12	0
Sodium	mg/l	104,780	< 5
Potassium	mg/l	1,600	< 1
Calcium	mg/l	30,853	< 1
Magnesium	mg/l	2,910	< 1
Barium	mg/l	120	< 1
Strontium	mg/l	1,164	< 1
Total iron	mg/l	38.0	< 1
Lead	mg/l	5.1	< 0.01
Zinc	mg/l	3.6	< 0.01

The analyses are run through an ion association model computer program with the objective of determining susceptibility of the brine to halite (and other scale) precipitation, If a halite precipitation problem is predicted, the ion association model is run in a "mixing" mode to determine if mixing the connate water with boiler feed water will prevent the problem. The computer program used estimates the chemical and physical properties of mixtures over the range of ratios specified, and then calculates the degree of supersaturation for common, and not so common scale forming species.

This approach has been used successfully to control salt deposition in the well with the composition outlined in Table 3. Originally, production on this well was on a test basis. This allowed fluid chemistry and scale data to be studied. Ion association model evaluation of the bottom hole chemistry indicated that the water was slightly supersaturated with sodium chloride under the bottom hole conditions of pressure and temperature. As the fluids cooled in the well bore, the production of copious amounts of halite was predicted.

The ion association model predicted that the connate water would require a minimum dilution of 15% with boiler feed water to prevent halite precipitation (Figure 3). The computer model also predicted that over-injection of dilution water would promote barite (barium sulfate) formation (Figure 4). Although the well produces H_2S at a concentration of 50 mg/l, the program did not predict the formation of iron sulfide due to the combination of low pH and high temperature (Figure 5).

Boiler feed water was injected into the bottom of the well using the downhole injection valve that was normally used for corrosion inhibitor injection. Injection of dilution water at a rate of 25 to 30% has allowed the well to produce successfully since start-up. Barite and iron sulfide precipitation has not been observed, and plugging with salt has not occurred.

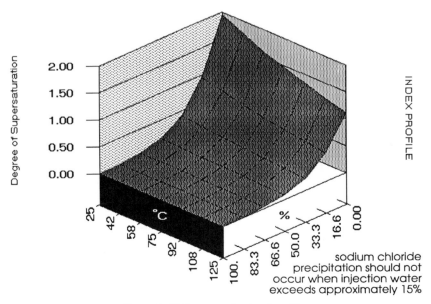

Figure 3. Halite saturation level profile.

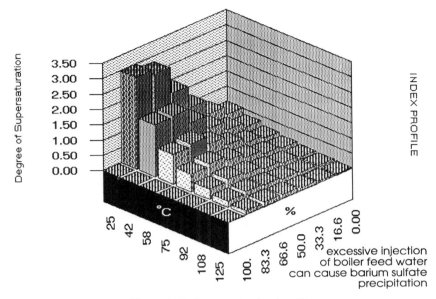

Figure 4. Barite saturation level profile.

CASE 3: IDENTIFYING ACCEPTABLE OPERATING RANGE FOR OZONATED COOLING SYSTEMS

It has been well established that ozone is an effective microbiological control agent in open recirculating cooling water systems (cooling towers). It has also been reported that commonly encountered scales were not observed in ozonated cooling systems under conditions where scale would be expected. The water chemistry of 13 ozonated cooling systems

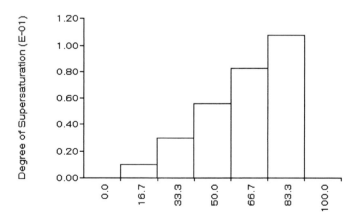

Figure 5. Hydrogen sulfide saturation level profile.

was evaluated using an ion association model.[11] Each system was treated solely with ozone on a continuous basis at the rate of 0.05 to 0.2 mg/l based upon recirculating water flow rates.

The saturation levels for common cooling water scales were calculated, including calcium carbonate, calcium sulfate, amorphous silica, and magnesium hydroxide. Magnesium hydroxide saturation levels were included due to the potential for magnesium silicate formation from the adsorption of silica upon precipitating magnesium hydroxide.

Each system was evaluated by:
1. Estimating the concentration ratio of the systems by comparing recirculating water chemistry to makeup water chemistry;
2. Calculating the theoretical concentration of recirculating water chemistry based upon makeup water analysis and the apparent, calculated concentration ratio from step 1;
3. Comparing the theoretical and observed ion concentrations to determine precipitation of major species;
4. Calculating the saturation level for major species based upon both the theoretical and observed recirculating water chemistry;
5. Comparing differences between the theoretical and actual chemistry to the observed cleanliness of the cooling systems and heat exchangers with respect to heat transfer surface scale buildup, scale formation in valves and on non-heat transfer surfaces, and precipitate buildup in the tower fill and basin.

Three categories of systems were encountered:

Category 1:   Those where the theoretical chemistry of the concentrated water is not scale forming (is undersaturated) for the scale forming species evaluated.

Category 2:   Those systems where the concentrated recirculating water would have a moderate to high calcium carbonate scale forming tendency. Water chemistry observed in these systems is similar to those run successfully using traditional scale inhibitors such as phosphonates.

Category 3:   These systems demonstrated an extraordinarily high scale potential for at least calcium carbonate and magnesium hydroxide. These systems operated with a theoretical recirculating water chemistry more like that of a softener than of a cooling system. The Category 3 water chemistry is above the maximum saturation level for calcium carbonate where traditional inhibitors such as phosphonates are able to inhibit scale formation.

Ozonated Systems Study Results

Table 4 outlines the theoretical versus actual water chemistry for the thirteen (13) systems evaluated. Saturation levels for the theoretical and actual recirculating water chemistries are presented in Table 5.

A comparison of the predicted chemistries to observed system cleanliness revealed the following:

Category 1:   (Recirculating Water Chemistry under-saturated): Category 1 systems showed no scale formation.

Category 2:   (Conventional Alkaline Cooling System Control Range): Scale formation was observed in eight (8) of the nine (9) category 2 systems evaluated.

Category 3:   (The Cooling Tower As A Softener): Deposit formation on heat transfer surface was not observed in most of these systems.

Rationale for Results

Category 2 systems fall into the general operating range for alkaline cooling systems. Typical calcite saturation levels for such systems fall into the range of 10 to 150 ($[Ca][CO_3]$/Ksp). In the absence of chemical treatment, scale would be expected, and was observed, in these systems. In this saturation level range, only a small quantity of the total reactive calcium and carbonate in the system is precipitated.

By comparison, the saturation levels predicted for the concentrated water (before precipitation) range well above 1,000. These levels of supersaturation are typical of softening processes. Under these conditions calcite tends to precipitate as finely divided seed crystals in the bulk solution, rather than by growth on existing active sites. Once such bulk precipitation begins, calcite formation on metal surfaces is greatly reduced because of the overwhelming surface area of suspended calcite crystals. The high flow velocities through heat exchangers in these systems tends to keep the crystals in suspension.

This is the rationale proposed for the unexpectedly low level of scale observed on heat transfer surfaces in the extremely saturated category 3 systems in comparison to the category 2 systems, where scale formation on heat transfer surfaces occurred. It must be noted that precipitation was observed in low flow areas of the Category 3 systems.

Conclusions from the Ozonated Systems Study

Analysis of the water chemistry and system cleanliness data from a number of recirculating cooling systems treated solely with ozone show that calcium carbonate (calcite) scale forms most readily on heat transfer surfaces in systems operating in a calcite saturation level range of 20 to 150 - the typical range for chemically treated cooling water.

At much higher saturation levels, in excess of 1,000, calcite precipitated in the bulk water. Because of the overwhelming high surface area of the precipitating crystals relative to the metal surface in the system, continuing precipitation leads to growth on crystals in the bulk water, rather than on heat transfer surfaces.

The presence of ozone in cooling systems does not appear to influence calcite precipitation and/or scale formation.

Table 4. Theoretical versus actual recirculating water concentrations

System (Category)	Theoretical Recirculating Calcium	Actual Recirculating Calcium	Difference in ppm	Theoretical Recirculating Magnesium	Actual Recirculating Magnesium	Difference in ppm	Theoretical Recirculating Silica	Actual Recirculating Silica	Difference in ppm	System Cleanliness
1(1)	56	43	13	28	36	-8	40	52	-12	No scale observed
2(2)	80	60	20	88	38	50	24	20	4	Basin buildup
3(2)	238	288	-50	483	168	315	38	31	7	Heavy scale
4(2)	288	180	108	216	223	-7	66	48	18	Valve scale
5(3)	392	245	147	238	320	-82	112	101	11	Condenser tube scale
6(3)	803	163	640	495	607	-112	162	143	19	No scale observed
7(3)	1,464	200	1,264	549	135	414	112	101	11	No scale observed
8(3)	800	168	632	480	78	402	280	78	202	No scale observed
9(3)	775	95	680	496	78	418	186	60	126	No scale observed
10(3)	3,904	270	3,634	3,172	508	2,664	3,050	95	2,995	Slight valve scale
11(3)	4,170	188	3,982	308	303	5	126	126	0	No scale observed
12(3)	3,660	800	2,860	2,623	2,972	-349	6,100	138	5,962	No scale observed
13(3)	7,930	68	7,862	610	20	590	1,952	85	1,867	No scale observed

Table 5. Theoretical versus actual recirculating water saturation levels

System (Category)	Theoretical Recirculating Calcite Saturation	Actual Recirculating Calcite Saturation	Theoretical Recirculating Brucite Saturation	Actual Recirculating Brucite Saturation	Theoretical Recirculating Silica Saturation	Actual Recirculating Silica Saturation	System Cleanliness
1(1)	0.03	0.02	<0.001	< 0.001	0.20	0.25	No scale observed
2(2)	49	5.4	0.82	0.02	0.06	0.09	Slight basin buildup
3(2)	89	61	2.4	0.12	0.10	0.12	Heavy scale
4(2)	106	50	1.3	0.55	0.13	0.16	Heavy scale
5(3)	240	72	3.0	0.46	0.21	0.35	Condenser tube scale
6(3)	540	51	5.3	0.73	0.35	0.49	Heavy scale
7(3)	598	28	10	0.17	0.40	0.52	No scale observed
8(3)	794	26	53	0.06	0.10	0.33	No scale observed
9(3)	809	6.5	10	<0.01	0.22	0.27	No scale observed
10(3)	1,198	62	7.4	0.36	0.31	0.35	Slight valve scale
11(3)	1,670	74	4.6	0.36	0.22	0.44	No scale observed
12(3)	3,420	37	254	0.59	1.31	0.55	No scale observed
13(3)	7,634	65	7.6	0.14	1.74	0.10	No scale observed

CASE 4: OPTIMIZING CALCIUM PHOSPHATE SCALE INHIBITOR DOSAGE IN A HIGH TDS COOLING SYSTEM

A major manufacturer of polymers for calcium phosphate scale control in cooling systems has developed laboratory data on the minimum effective scale inhibitor (copolymer) dosage required to prevent calcium phosphate deposition over a broad range of calcium and phosphate concentrations, and a range of pH and temperatures, The data was developed using static tests, but has been observed to correlate well with the dosage requirements for the copolymer in operating cooling systems. The data was developed using relatively low dissolved solids test waters. Recommendations from the data were typically made as a function of calcium concentration, phosphate concentration, and pH. This data base was used to project the treatment requirements for a utility cooling system which used a geothermal brine for make-up water. An extremely high dosage (30 to 35 mg/l) was recommended based upon the laboratory data.

It was believed that much lower dosages would be required in the actual cooling system due to the reduced availability of calcium anticipated in the high TDS recirculating water. As a result, it was believed that a model based upon dosage as a function of the ion association model saturation level for tricalcium phosphate would be more appropriate, and accurate, to use than a simple look-up table of dosage versus pH and analytical values for calcium and phosphate. The ion association model was projected to be more accurate due to its use of free values for calcium and phosphate concentrations rather than the analytical values used by the look-up tables.

Tricalcium phosphate saturation levels were calculated for each of the laboratory data points. Regression analysis was used to develop a model for dosage as a function of saturation level and temperature. Predicted versus observed dosages for the model are depicted in Figure 6.

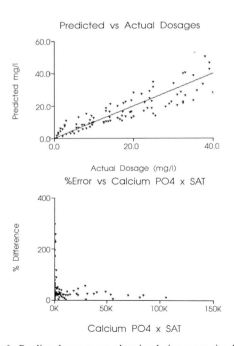

Figure 6. Predicted versus actual recirculating saturation levels.

The model was used to predict the minimum effective dosage for the system with the make-up and recirculating water chemistry found in Table 6. A dosage in the range of 10 to 11 mg/l was predicted rather than the 30 ppm derived from the look-up tables.

A dosage minimization study was conducted to determine the minimum effective dosage. The system was treated with the copolymer initially at a dosage of 30 mg/l in the recirculating water. The dosage was decreased until deposition was observed. Failure was noted when the recirculating water concentration dropped below 10 mg/l, validating the ion association based dosage model.

CASE 5: OPTIMIZING CALCIUM CARBONATE SCALE INHIBITOR DOSAGE IN AN OIL-WATER SEPARATOR

It has been well established that phosphonate scale inhibitors function by extending the induction time prior to crystal formation and growth occurrence.[12,13,14,15] The induction time extension achieved has been reported to be a function of calcite saturation level, temperature as it affects rate, and phopshonate dosage. The formula for the models to which data has been successfully fit is a modification of the classical induction time relationship which adds dosage as a parameter:

$$Time = \frac{k * [Dosage]^m}{[Saturation - 1]^n}$$

Dosage models developed from laboratory and field data have been used successfully to model the minimum effective scale inhibitor dosage in cooling water systems. These models were applied to the problem of calcite scale control in a separator.

Table 6. Calcium phosphate inhibitor dosage opimization example

Water Analysis at 6.2 Cycles		_Deposition Potential Indicators_	
CATIONS		**SATURATION LEVEL**	
Calcium (as $CaCO_3$)	1339	Calcite	38.8
Magnesium (as $CaCO_3$)	496	Aragonite	32.9
Sodium (as Na)	1240	Silica	0.4
		Tricalcium phosphate	1074.
ANIONS		Anhydrite	1.3
Chloride (as Cl)	620	Gypsum	1.7
Sulfate (as SO_4)	3384	Fluorite	0.0
Bicarbonate (as HCO_3)	294	Brucite	<0.1
Carbonate (as CO_3)	36		
Silica (as SiO_2)	62		
Phosphate (as PO_4)	6		
		SIMPLE INDICES	
PARAMETERS		Langelier	1.99
pH	8.40	Ryznar	4.41
Temperature (°C)	36.7	Practical	4.20
1/2 life (hours)	72	Larson-Skold	0.39

TREATMENT RECOMMENDATION
100% Active Copolymer 10.53 mg/l

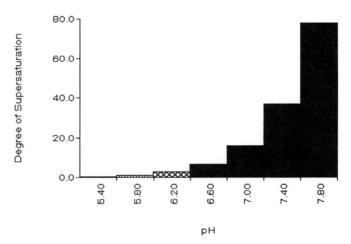

Figure 7. The pH of the brine rises as CO_2 flashes across the separator.

Carbon dioxide flashes as oil field brines pass through a separator, and go from a high partial pressure of CO_2 to atmospheric. Figure 7 depicts the impact of CO_2 flashing upon pH, and calcite saturation levels. Dosage were predicted from the models for the phosphonate HEDP (1,1-hydroxyethylidene diphosphonic acid).

Figure 8 depicts the dosage requirement increase as CO_2 flashes, and pH rises, across the separator. The dosages predicted by the model for this application are within 30% of those determined through field evaluations.

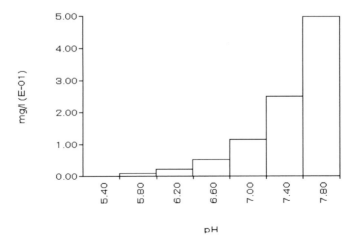

Figure 8. Scale inhibitor requirements increase as pH and calcite saturation level increases as the brine passes through the separator.

CONCLUSIONS

Ion association model saturation levels provide an effective tool for predicting and resolving scale problems in a wide variety of commercial applications. The use of personal computer versions of the models make their use in the field as an engineering tool practical, and removes the restriction of their use solely as a research tool in a laboratory environment.

REFERENCES

1 Langelier, W.F., 1936, The Analytical Control Of Anti-Corrosion Water Treatment, *JAWWA*, Vol. 28, No. 10, 1500-1521.

2 Ryznar, J.W., 1944, A New Index For Determining The Amount Of Calcium Carbonate Scale Formed By Water, *JAWWA*, Vol. 36, 472.

3 Stiff Jr., H.A., and L.E. Davis, 1952, A Method For Predicting The Tendency of Oil Field Water to Deposit Calcium Carbonate, Pet. Trans. *AIME* 195;213.

4 Oddo, J.E. and M.B. Tomson, Scale Control, Prediction and Treatment Or How Companies Evaluate A Scaling Problem and What They Do Wrong, CORROSION/92, Paper No. 34, (Houston, TX:NACE INTERNATIONAL 1992).

5 Ferguson, R.J., Computerized Ion Association Model Profiles Complete Range of Cooling System Parameters, International Water Conference, 52nd Annual Meeting, Pittsburgh, PA, IWC-91-47.

6 Chow W., J.T. Aronson, and W.C. Micheletti, Calculations Of Cooling Water Systems: Computer Modeling Of Recirculating Cooling Water Chemistry, International Water Conference, 41st Annual Meeting, Pittsburgh, PA, IWC-80-41.

7 Johnson, D.A. and K.E. Fulks, Computerized Water Modeling In The Design And Operation of Industrial Cooling Systems, International Water Conference, 41st Annual Meeting, Pittsburgh, PA, IWC-80-42.

8 Truesdell, A.H., and B.F. Jones, 1974, WATEQ - A Computer Program For Calculating Chemical Equilibria Of Natural Waters, *J. Research, U.S. Geological Survey*, Volume 2, No. 2, p. 233-248.

9 Musil, R.R., and H.J. Nielsen, Computer Modeling Of Cooling Water Chemistry, International Water Conference, 45th Annual Meeting, Pittsburgh, PA, IWC-84-104.

10 Fowler, V.L. and J.J. Perona, Evaporation Studies On Oak Ridge National Laboratory Liquid Low Level Waste, ORNL/TM-12243.

11 Ferguson, R.J. and A.J. Freedman, A Comparison Of Scale Potential Indices With Treatment Program Results In Ozonated Systems, CORROSION/93,Paper No. 279,(Houston, TX:NACE INTERNATIONAL 1993).

12 Gill, J.S., C.D. Anderson, and R.G. Varsanik, Mechanism Of Scale Inhibition By Phosphonates, International Water Conference, 44th Annual Meeting, Pittsburgh, PA, IWC-83-4.

13 Amjad, Z. and W.F. Masler III, The Inhibition Of Calcium Sulfate Dihydrate Crystal Growth By Polyacrylates And The Influence Of Molecular Weight, CORROSION/85, Paper No. 357, (Houston, TX: NACE INTERNATIONAL, 1985).

14 Ferguson, R.J., Developing Scale Inhibitor Models, WATERTECH, Houston, TX, 1992.

15 Ferguson, R.J., O. Codina, W. Rule, and R. Baebel, Real Time Control Of Scale Inhibitor Feed Rate, International Water Conference, 49th Annual Meeting, Pittsburgh, PA, IWC-88-57.

16 Ferguson, R.J. and D.J. Weintritt, Modeling Scale Inhibitor Dosages For Oil Field Operations, CORROSION/94, Paper No. 46, (Houston, TX:NACE INTERNATIONAL, 1994).

27

CALCIUM OXALATE NUCLEATION AND GROWTH ON OXIDE SURFACES

Lin Song, Bruce C. Bunker, Gordon L. Graff, Michael J. Pattillo, and
Allison A. Campbell

Materials and Chemical Sciences Center[*]
Pacific Northwest Laboratory
Richland, Washington 99352

ABSTRACT

The heterogeneous nucleation of calcium oxalate onto colloidal oxides from aqueous solution has been studied as a model system for understanding the role of surface chemistry in biomimetic processes. The Constant Composition technique was used for measuring nucleation induction times. Results show that the dependence of nucleation on supersaturation fit well with classical nucleation theories. Surfaces which are anionic appear to promote calcium oxalate nucleation better that neutral or cationic surfaces. Modifications to positive surfaces via the adsorption of anionic surfactants, lower the effective energy barrier for nucleation, and stimulates the heterogeneous nucleation of calcium oxalate

INTRODUCTION

Bone, teeth, and mollusk shells are examples of highly ordered ceramic materials produced by organisms from aqueous solutions at ambient conditions. Much scientific effort has been aimed at the elucidation of the mechanisms involved in the formation of hard tissues in nature.[1,2,3] Results of these studies show that while the biomineralization process is quite complex, a general scheme of events necessary of mineralization has been determined.[4] For instance, it is known that organisms create an organic matrix which acts as a template for nucleation and growth. In addition, they control the solution concentrations and supersaturation levels of the forming phase as well as the mineral growth rate.[5] Biomimetic research uses the basic concepts of the biomineralization process and applies those concepts to the development and synthesis of ceramic materials.[6]

---

[*] Operated for the U.S. Department of Energy by Battelle Memorial Institute under Contract DE-AC06-76RLO 1830.

Mineral Scale Formation and Inhibition, Edited by Zahid Amjad
Plenum Press, New York, 1995

The heterogeneous nucleation of mineral phases is a very important aspect of biomimetic research. Understanding the effects that surface energy has on lowering the barrier to heterogeneous nucleation may help in developing schemes for the synthetic alteration of surfaces in order to either promote or retard mineral formation. Classical homogeneous nucleation theory[7] predicts that the solid/liquid interfacial energy (σ), temperature (T), and supersaturation (S) control the number of nuclei that form (N), the nucleation induction time (t), and the particle growth rate (dr/dt) via the following equations:

$$\ln (1/t) = \ln B - (16p\sigma^3 u^2)/3(kT)^3(\ln S)^2 \tag{1}$$

$$\ln N = \ln P - Qbs^3u^2/((kT)^3(\ln S)^2 \tag{2}$$

$$dr/dt = K(S-2)^n \tag{3}$$

where K, B, P, Q, b, and n are constants, k is Boltzmann's constant, and u is the molecular volume of the forming phase. For heterogeneous nucleation, similar expression apply, except that the surface energy term is described by three terms: σ_{cl}, σ_{cs}, and σ_{sc} where c, s, and l refer to the crystalline particle, the substrate, and the liquid phases, respectively. Thus lowering the net interfacial energy between the substrate surface and the growth particle will lower the energy barrier for heterogeneous nucleation.

This aim of this work is to investigate the effects of surface chemistry on the heterogeneous nucleation of calcium oxalate (CaOx). Calcium oxalate was chosen as a model mineral system because; (i) CaOx solubility isotherms are accurately know as a function of solution pH, ionic strength, and temperature, (ii) the solubility is relatively constant other a pH range of 4 to 11, and (iii) homogeneous nucleation and growth kinetics are well known for a variety of solution conditions.

Three oxide materials, Al_2O_3, TiO_2, and SiO_2 where used as substrates for CaOx heterogeneous nucleation because they possess very different surface charge characteristics (Figure 1). At pH 6.5 Al_2O_3 has a positive surface charge, SiO_2 has a negative surface charge, and TiO_2 surfaces are near neutral.

Changes in solution pH alter the surface charges of these oxide surface without altering CaOx solubility, thus providing a systematic method for investigating surface chemistry effects on CaOx nucleation.

METHODS

Solutions, prepared using reagent grade chemicals (Fisher Scientific) and deionized, reverse osmosis (Millipore) CO_2-free water, were filtered (0.22 μm Millipore) prior to use. The filters were prewashed to remove any residual wetting agents or surfactants.

α-Al_2O_3 (AKP-30, Sumitomo), TiO_2 (P25, Degussa), and SiO_2 (PST-3, Nissan) were used in the nucleation experiments. The specific surface area, measured by BET methods were 6.8 m^2 gm^{-1}, 56.9 m^2 gm^{-1}, and 15.0 m^2 gm^{-1}, respectively. Suspensions were prepared by ultrasonic dispersion in 0.01 M NaCl, pH 6.5 for at least three days prior to use in nucleation experiments. Suspension pH was adjusted by the addition of either sodium hydroxide or hydrochloric acid solutions. Surface charge characteristics were measured using a Matec ESA-8000 instrument and measurements were conducted in a 0.01 M NaCl aqueous solution at 25°C.

The adsorption of dextran sulfate (DS) (Sigma) on alumina was determined at various pH values. Suspensions were prepared by the addition of various concentrations of DS to a known amount of alumina particles. The slurry pH was adjusted to the desired value by the

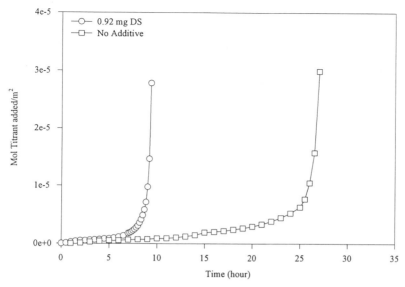

Figure 1. Zeta potential of Al_2O_3, TiO_2, and SiO_2 (0.5 v%) versus solution pH. T=25°C.

addition of HCl or NaOH. Solutions were then placed onto a wrist action shaker for 15-18 hours in order for adsorption equilibrium to be reached. Following the equilibrium time, the suspensions were centrifuged and the supernatant analyzed for polysaccharide content using colorimetric and UV-Vis adsorption.[8]

Nucleation and growth experiments were performed using the Constant Composition (CC) technique developed by Nancollas.[9] A calcium ion selective electrode (Orion) was calibrated by the addition of aliquots of calcium chloride to sodium chloride medium electrolyte in a reaction vessel. Supersaturated CaOx solutions were then prepared by the slow addition of potassium oxalate solution. When equilibrium was attained, a known volume of the oxide seed slurry (total surface area = 0.2670 m²) was added to the mixture. Any decrease in calcium ion activity due to nucleation and/or growth, was sensed by the calcium electrode which triggered an automatic titrator potentiostat (Metrohm) to add titrant solutions of calcium chloride and potassium oxalate from mechanically coupled burets. The titrant solutions also contained sodium chloride in order to maintain the ionic strength constant (0.01M NaCl). The time between seed addition and the addition of titrants as well as the volume of titrant consumption as a function of time was recorded.

RESULTS AND DISCUSSION

Crystal growth occurs only when the solution is supersaturated with respect to the mineralizing phase. In the case of calcium oxalate monohydrate (CaOx), the supersaturation may be written in terms of the lattice ion activities as:

$$S = (IP/K_{sp})^{1/2} \qquad (4)$$

where $IP = (Ca^{2+})(C_2O_4^{2-})$ and K_{sp} is the solubility product. To determine the activities of ionic species, calculations were performed as described by Nancollas[10] using expressions for mass balance, electronegativity, and the equilibrium and ion pair association constants.

Table 1. Interfacial Energies of the Oxide Particles (0.01 M NaCl, 25°C). In order to check this observation, a highly negatively charged surfactant, dextran sulfate (DS), was adsorbed onto Al$_2$O$_3$ (Figure 2). It can be seen that the amount of DS that is adsorbed increases with decreasing pH

Surface	Surface charge	pH	s (mJ m^{-2})
Al$_2$O$_3$	positive	6.5	37
TiO$_2$	neutral	6.5	25
TiO$_2$	negative	8.0	25
TiO$_2$	positive	5.0	37
SiO$_2$	negative	6.5	24
Homogeneous			69 (6)
Spontaneous		6.5	31

Constant Composition nucleation and growth experiments were performed on high surface area colloidal oxide particles. In unseeded supersaturated solutions, the induction time for the nucleation of calcium oxalate is long and is associated with the heterogeneous nucleation of calcium oxalate onto small foreign particles which are always present in solution ("spontaneous" nucleation). When seeded with CaOx crystals, nucleation induction times are zero due to the lack of an energy barrier for further CaOx formation. Induction times for the three oxide powders lie in between the spontaneous and CaOx seeded results.[6] Results indicate that the presence of the oxide particles lowers the net interfacial energy, σ, for heterogeneous CaOx nucleation. A plot of 1/t vs. (log S)$^{-2}$ gives a straight lines whose slope can be used to calculate σ (see equation 1). Table 1 lists the interfacial energies of the oxide powders calculated from the CC data and equation 1. The interfacial energy for homogeneous CaOx nucleation has been reported to be 69 mJ m^{-2}.[6] The value for spontaneous nucleation obtained in this study is in close agreement with the literature.[11] The oxide particles appear to fall into two groups, those that stimulate CaOx nucleation (σ< 31 mJ m^{-2}) and those that are less effective than the materials responsible for spontaneous nucleation (s>31mJ m^{-2}).

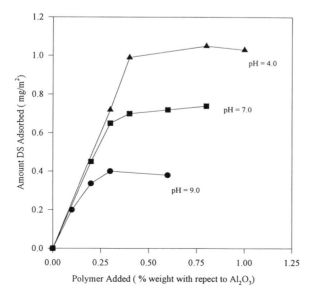

Figure 2. Adsorption of dextran sulfate onto Al2O3 as a function of solution pH.

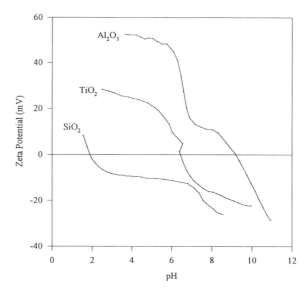

Figure 3. CaOx crystallization on alumina in the presence of dextran sulfate. (S = 2.8 at 25°C).

Initial results indicate that surfaces which have a positive surface charge are not effective at inducing the heterogeneous nucleation of CaOx, but that neutral and negatively charged surfaces (Table 1) act as nucleation stimulates.

This behavior can be explained by examination of the surface charge of alumina as a function of pH (Figure 1). Under acidic conditions (pH=4), alumina has a highly positive relative surface charge, while at basic conditions (pH=9), the relative surface charge near neutral. Because DS is fully dissociated (pK<1) over the entire pH range, there is an increased electrostatic attraction between the more positively charged alumina surfaces and the highly negative DS molecules at lower pH values.

Zeta potential measurements show that the adsorption of dextran sulfate onto alumina surfaces causes a charge reversal from positive to negative values at pH 6.5. In fact, at approximately 1% weight % DS/Al_2O_3, the zeta potential decreases from approximately 20 mV (no DS) to approximately -15 mV.

The adsorption of DS onto the alumina also had a dramatic affect on the time required for CaOx nucleation to occur on the surface. Data shows that the induction time for nucleation of CaOx on alumina with preadsorbed DS is identical to those measured for silica at pH 6.5 (250 min) (Figure 3).

Since the alumina with preadsorbed DS has a similar zeta potential as SiO2, this suggests that surface charge is an important parameter for CaOx nucleation.

These results show that the surface chemistry of the oxide can have a dramatic effect on the heterogeneous nucleation of calcium oxalate. However, the surface charge effect does not correspond to the change in interfacial energies evidenced by the fact that TiO_2 (pH 8) and SiO_2 (pH 6.5) have nearly identical s values (24-25 mJ m^{-2}) for CaOx growth. Hence the surface complexation of anions and cations may be more important. Future experiments are planned that will investigate the role of surface complexation verses lowering of the net interfacial energy as effective means to stimulate calcium oxalate nucleation from aqueous solutions. Understanding these relationships may help researchers modify surfaces in such a manner as to promote or retard mineral formation.

REFERENCES

1. B. R. Heywood and E.D. Eanes, 1987, *Calcif. Tissue Int.*, **41**, 192,.
2. L. Addadi and S. Weiner, 1985, Proc. Natl. Acad. Sci. U.S.A., **82**, 4110.
3. A. Miller, D. Phillips, and R. J. P. Williams, 1984, Philos. Trans. R. Soc. London Ser. B, **304**, 409.
4. S. Weiner, 1986,CRC Crit. Rev. *Biochem.,* **20**, 365.
5. H. A. Lowenstam,1981, Science, **211**, 1126.
6. B. C. Bunker, P. C. Rieke, B. J. Tarasevich, A. A. Campbell, G. E. Fryxell, G. L. Graff, L. Song, J. Liu, J. W. Virden, and G. L. McVay, 1994, *Science*, 264, 48.
7. R. Mohanty, S. Bhandarkar, and J. Estrin, 1990, AIChE Journal, **36**(10), 1536.
8. M. DuBois, K. A. Gilles, J. K. Hamilton, P. A. Rebers, and F. Smith, 1956, *Anal. Chem.*, **28**(3), 350.
9. P. G. Koutsoukos, M. E. Sheehan, and G. H. Nancollas, 1981, Invest. Urol., **18**, 358.
10. G. H. Nancollas, 1966, *Interactions in Electrolyte Solutions*, Elsevier, Amsterdam.
11. C. M. Brown, D. K. Ackermann, D. L. Purich, and B. Finlayson, 1991, *J. Crystal Growth*, **108**, 455.

INDEX